AF527728

A Petrographic Atlas of Ophiolite

Naresh Chandra Ghose
Nilanjan Chatterjee • Fareeduddin

A Petrographic Atlas of Ophiolite

An Example from the Eastern India–Asia Collision Zone

Springer

Naresh Chandra Ghose
Geological Society of India
Bangalore
India

Nilanjan Chatterjee
Department of Earth, Atmospheric
and Planetary Science
Massachusetts Institute of Technology
Cambridge, MA
USA

Fareeduddin
PPOD Division
Geological Survey of India
Bangalore
India

ISBN 978-81-322-3492-0 ISBN 978-81-322-1569-1 (eBook)
DOI 10.1007/978-81-322-1569-1
Springer New Delhi Heidelberg New York Dordrecht London

© Springer India 2014
Softcover reprint of the hardcover 1st edition 2014
This work is subject to copyright. All rights are reserved by the Publisher, whether the whole or part of the material is concerned, specifically the rights of translation, reprinting, reuse of illustrations, recitation, broadcasting, reproduction on microfilms or in any other physical way, and transmission or information storage and retrieval, electronic adaptation, computer software, or by similar or dissimilar methodology now known or hereafter developed. Exempted from this legal reservation are brief excerpts in connection with reviews or scholarly analysis or material supplied specifically for the purpose of being entered and executed on a computer system, for exclusive use by the purchaser of the work. Duplication of this publication or parts thereof is permitted only under the provisions of the Copyright Law of the Publisher's location, in its current version, and permission for use must always be obtained from Springer. Permissions for use may be obtained through RightsLink at the Copyright Clearance Center. Violations are liable to prosecution under the respective Copyright Law.
The use of general descriptive names, registered names, trademarks, service marks, etc. in this publication does not imply, even in the absence of a specific statement, that such names are exempt from the relevant protective laws and regulations and therefore free for general use.
While the advice and information in this book are believed to be true and accurate at the date of publication, neither the authors nor the editors nor the publisher can accept any legal responsibility for any errors or omissions that may be made. The publisher makes no warranty, express or implied, with respect to the material contained herein.

Printed on acid-free paper

Springer is part of Springer Science+Business Media (www.springer.com)

One who illumines the intellect

Foreword

The basic tenet of the plate tectonic paradigm is that a newly generated oceanic plate is subducted at a trench. However, it is widely accepted today that some material is not subducted, but is accreted to form an accretionary wedge or prism, within which there may be a preserved section of the oceanic crust/mantle, known as an ophiolite. And many detailed studies now reveal that there are many types of ophiolites that may have formed, in e.g. a mid-oceanic ridge, close to a continental margin, and (the majority) in a suprasubduction zone. Ophiolites commonly retain evidence of the low-temperature conditions of the ocean floor, and a few are closely associated with blueschists and eclogites, the metamorphic petrology of which provides information on the depths to which the material was subducted and exhumed. An associated accretionary wedge may contain relicts of ocean plate stratigraphy that records the travel history of the ocean floor as it moved laterally from ridge to trench. Most important are structures (e.g. extensional faults, thrusts, folds, lineations, kinematic indicators and S-C fabrics), which can only be obtained from detailed field work that will tell us if the ophiolite was rooted, for example in a ridge or an oceanic transform fault, and if it went through transpressional or orthogonal shear during accretion.

Following that, and most importantly, microstructures related to textural relations and mineralogy of structurally oriented samples give us information about the physico-chemical conditions during the magmatic and metamorphic history.

However, it seems to be accepted today that most papers on ophiolites provide just a brief list of the rock types, and then concentrate on the trace element rock chemistry, the mineral chemistry, the isotopic systematics and the zircon age, but with little petrographic information. But if we are to understand the origin and evolution of this key component of the Wilson Cycle, we need more information on the petrography and mineralogy, because it is that which relates the magmatic development to the metamorphic and structural history. And above all, the non-ophiolite specialist, who is keen to understand the general nature of an ophiolite and how the petrology and mineralogy provide the key knowledge to work out the magmatic-metamorphic-structural history, increasingly requires innovative syntheses of such relations in different ophiolites, and such reviews are rare.

This book on the Naga Hills ophiolite by Ghose, Chatterjee and Fareeduddin is an excellent example of a comprehensive review emphasising petrographic relations that are necessary for the ophiolite specialist and the general reader to obtain an up-to-date understanding of the processes experienced by a slice of ocean floor preserved on a convergent margin. It should be widely read, and I hope it will stimulate more such books on ophiolites from other tectonic settings.

Department of Geology, University of Leicester, U.K. Brian Windley

Preface

Microstructure, texture and mineralogy provide crucial clues regarding the evolutionary history of a rock. An atlas documenting the petrographic and mineralogical details of common rock types is integral to the understanding of physico-chemical conditions of magmatic, metamorphic and sedimentary processes of rock formation for both beginners and advanced researchers alike. Complex processes involved in the development of a variety of rock types at convergent plate boundaries and continental collision zones make a fascinating subject for discussion among earth scientists. But a comprehensive account of texture and mineralogy of rocks from such a tectonic setting is currently lacking. This book attempts to fill that gap.

The idea of creating a petrographic atlas occurred to the senior author Naresh C. Ghose in the late 1970s while supervising the research projects of three Ph.D. candidates—R. N. Singh, O. P. Agrawal, and M. P. Shrivastava—in the newly discovered Naga Hills Ophiolite (NHO) belt at the eastern continental margin of India. Early investigations carried out over a decade in this virgin and challenging terrain were aimed at establishing the geology, stratigraphy and classification of rocks. Despite limited funds and lack of instrumental facilities, two books—"Ophiolite and Indian Plate Margin" (1986) edited jointly by Dr. Ghose and S. Varadarajan, and "Phanerorozoic Ophiolites of India" (1989) edited by Dr. Ghose—were published. These books are considered to be among the first of their kind on the application of plate tectonics in the study of ophiolite sequences from the Indian continental margin.

This book emerges from an intensive study of the eastern part of the India-Eurasia collision zone by the authors. It deals with a segment of the oceanic lithosphere that originated under the Neotethys Ocean in Late Mesozoic. The lithospheric segment was emplaced within the accretionary wedge of the subduction zone where the Indian plate subducted below the Myanmar (Burmese) microplate. Remnants of the lithosphere are now exposed as ophiolite along the suture zone where India and Eurasia collided to form the Himalayan Mountain Range at the northern margin and the Indo-Myanmar Range (IMR) at the eastern margin of India in Early Eocene. Detached outcrops of ophiolite are exposed all along the IMR. The ophiolites in the northern part of IMR in the Indian states of Nagaland and Manipur are known as the NHO. The fascinating tectonic history of the Naga Hills, wide variety of lithological assemblages, seismic activity and mineral resources associated with the mantle-derived ophiolitic rocks have received global attention for intensive research in recent years.

As the surface manifestations of the oceanic lithosphere, ophiolites serve as natural museums for studying the three-dimensional characteristics of the earth's oceanic crust and upper mantle. However, because of incomplete preservation and alteration, the crust and upper mantle in ophiolite formed at convergent plate margins present a major challenge in characterizing the oceanic lithosphere. About 1,200 thin sections of rocks from the ophiolite suite collected across the NHO by the three aforementioned Ph.D. scholars and Dr. Ghose were re-examined between 2008 and 2012 for the preparation of this atlas. The black-and-white field photographs of the inaccessible Tizu gorge section are supplied by O. P. Agrawal, who led an expedition in the region in the late 1970s.

This book is an outcome of the integration of collective ideas of the authors. Dr. Ghose and Fareeduddin collaborated to acquire and interpret the optical photomicrographs, Nilanjan Chatterjee acquired the Backscattered Electron (BSE) images, and Nilanjan Chatterjee and Dr. Ghose interpreted the chemical data obtained with the Electron Probe Microanalyzer (EPMA). Each chapter has been reviewed by experts in different disciplines to streamline the text. The book focuses on rocks typically associated with ophiolite including metamorphic peridotite (tectonite), cumulate mafic-ultramafics rocks of the layered sequence (peridotite-pyroxenite-gabbro-plagiogranite-anorthosite), volcanic and volcaniclastic rocks of mid-oceanic ridge origin, zeolite-prehnite and greenschist facies metamorphic rocks of the ocean floor, blueschists and C-type eclogites associated with convergent plate boundaries, pelagic sediments, and syngenetic (podiform chromitite, platinum and gold) and epigenetic (nickeliferous magnetite and Cu–Mo sulfides) economic minerals. Photomicrographs of rocks and field photographs showing mutual relationships are representative of the structure, mineralogy and petrology of the NHO rocks. Optical photomicrographs and BSE images of magmatic, metamorphic and sedimentary rocks, and associated economic mineral deposits are systematically presented. We hope this book will generate interest among beginners and professionals in the field of ophiolite research and facilitate better understanding of the chain of events regarding the origin and emplacement of ophiolite.

This book adopts a simple approach by creating a 'pictorial representation of ophiolite' to study the complex processes of its formation. Descriptions of myriad events affecting the oceanic lithosphere from the time of its birth to its emplacement on the continental margin are never complete. Every effort has been made by the authors to capture accurately the different facets of the problems related to the ophiolite. As in any major endeavour and given the vast magnitude of the subject matter discussed in this book, expectations may fall short of target. The lead author (Naresh Chandra Ghose) takes full responsibility for any shortcomings in the presentation and interpretation.

May 5, 2013

Naresh Chandra Ghose
Nilanjan Chatterjee
Fareeduddin

Acknowledgments

Preparation of this atlas was made feasible by the Geological Survey of India (GSI) that allowed the use of the petrological microscope and the Electron Microprobe (Cameca SX-100) facilities at the PPOD Laboratory, Bangalore, for quantitative analysis of critical samples. We are especially grateful for the cooperation of the authorities of the GSI for bringing out the concept of this atlas to reality. We are also indebted to the Electron Microprobe Laboratory at the Department of Earth, Atmospheric and Planetary Sciences, Massachusetts Institute of Technology (MIT), Cambridge, Massachusetts, USA for allowing us to use their JEOL JXA-733 Superprobe Electron Probe Microanalyzer (EPMA) to verify mineral identifications and identify minerals under high magnifications in several samples. Rock sections were studied with the EPMA through Backscattered Electron (BSE) imaging, and mineral phases were qualitatively identified through Energy Dispersive Spectrometry (EDS) and confirmed through a limited number of quantitative analysis with Wavelength Dispersive Spectrometry (WDS). One of the authors (NCG) expresses his gratitude to the Geological Society of India for constant motivation and support towards the endeavour of writing this book.

We express our sincere thanks to a host of reviewers in different fields who at various levels made useful suggestions in improving accuracy and presentation of the manuscript or contributed by examining the samples. The following is a list of notable reviewers:

Geology: S. V. Srikantia, Om Prakash Agrawal, Radha Nand Singh, K. T. Vidyadharan and Jadaba Nanda Das.

Metamorphic Petrology: Ram S. Sharma.

Sedimentary Petrology: Subhasish Das.

Igneous Petrology: Gautam Sen, Mei-Fu Zhou, Asish Basu, M.S. Rao, Jyotishankar Ray, Rajesh K. Srivastava, Nittala V. Chalapathi Rao, B. V. Rao, Biswajit Ghosh and Athokpam K. Singh.

Structure and Tectonics: H. M. Ramachandran, Abhinaba Roy and Manoranjan Mohanty.

Geophysics: Bimalendu B. Bhattacharya and Sikhendra K. Dey.

Ore Geology: Sisir Mondal and Ramkrishna Sawkar.

EPMA (Geological Survey of India): Mahesh M. Korakoppa.

Several important suggestions were made by Subhrangsu K. Acharyya and Roger H. Mitchell.

We also gratefully appreciate the invaluable assistance of Dipankar Saha and Lokendra Kumar, who helped in the digitization of maps and sketches.

Financial support from the Department of Science and Technology, New Delhi for book writing to retired scientists is gratefully acknowledged by Dr. Ghose.

Naresh Chandra Ghose
Nilanjan Chatterjee
Fareeduddin

Contents

Part I

Description of Ophiolite Suites and the Naga Hills Ophiolite

1 Introduction

An ophiolite is a section of the oceanic lithosphere emplaced on continental crust or within the accretionary prism sediments of a subduction zone. Much of our knowledge of the vertical structure of the ocean floor comes from the study of ophiolite sequences. From bottom to top, a typical and complete ophiolite sequence comprises depleted mantle peridotite with tectonite at the base, layered ultramafic–mafic cumulates, massive (isotropic) gabbro, sheeted dikes and extrusive volcanic rocks represented by pillow basalts (Anonymous 1972; Coleman 1977). The sequence is typically overlain by deep-sea pelagic sediments and chert. The Semail massif section in the Oman Mountains is an example of a complete ophiolite sequence (Fig. 1.1). Direct drilling on the ocean floor has shown that its structure is variable and dependent on the tectonic environment. Although drilling has been achieved only down to the top of the oceanic lower crust ($\sim$1.5 km), a correlation between the Semail section and drill core at the IODP Hole 1256D drill site is noticeable (Fig. 1.1). At this drill site, hyaloclastite breccia, diabase dike and gabbro are encountered with increasing depth (Wilson et al. 2006).

Ophiolite sequences have been traditionally described as "Tethyan-type" or "Cordilleran-type" on the basis of their tectonic settings (Moores 1982). A Tethyan-type ophiolite overlies continental crust and shallow-water continental margin/platform sediments, and is overlain by pelagic sediments with little or no volcaniclastic material. On the other hand, a Cordilleran-type ophiolite has no clear relationship with the continental basement or platform sedimentary rocks. They commonly include arc-type volcaniclastic deposits within their extrusive units, and are spatially and temporally associated with tectonic mélange containing high-grade metamorphic rocks. However, we will see that the ophiolites related to the closure of the Neotethys Ocean exhibit characteristics of both of the above-mentioned end-member types.

Ophiolitic lava exhibits a wide range of chemical compositions similar to normal and enriched mid-oceanic ridge basalt (N-MORB and E-MORB), and importantly, highly depleted arc-type basalts. The presence of arc-type lavas in some ophiolites, first recognised by Miyashiro (1973) in the Troodos Ophiolite of Cyprus, indicates that they form in supra-subduction zones, i.e. on the overriding oceanic plate of a subduction zone. However, Moores et al. (2000) argued that the composition of lava erupted at spreading centres depends on the history of melting and contamination of the mantle source through previous spreading and subduction events, and therefore, although some ophiolitic lavas display arc-type characteristics, they do not necessarily originate in supra-subduction zones. Metcalf and Shervais (2008) refute this historical contingency model by pointing out that compositions of slab-influenced basalt erupting from modern ridges show only a subtle difference from MORB, and they are unlike basalts from both subduction zones and supra-subduction zone ophiolites. Rather, supra-subduction zone ophiolitic lava strongly resembles primitive island-arc lava with high magmatic water contents. Metcalf and Shervais (2008) also cite flaws in the mantle convection model used by Moores et al. (2000) with arguments based on incompatible element and isotope compositions.

Petrographic textures and mineral assemblages of ophiolitic rocks provide important clues to their environment of formation. They are complementary to field and geochemical data used in tectonic and petrological interpretations. For example, it has been shown from a combination of mineral chemical, textural and bulk compositional data that dunite in many ophiolites originated through metasomatic alteration of upper mantle peridotite by infiltrating basaltic melt that dissolves pyroxene and precipitates olivine (Dick 1977; Boudier and Nicolas 1977; Kelemen etal. 1992; Edwards and Malpas 1995; Kelemen and Dick 1995). Therefore, in this book, we present a comprehensive account of the textural relations

N. C. Ghose et al., *A Petrographic Atlas of Ophiolite*,
DOI: 10.1007/978-81-322-1569-1_1, © Springer India 2014

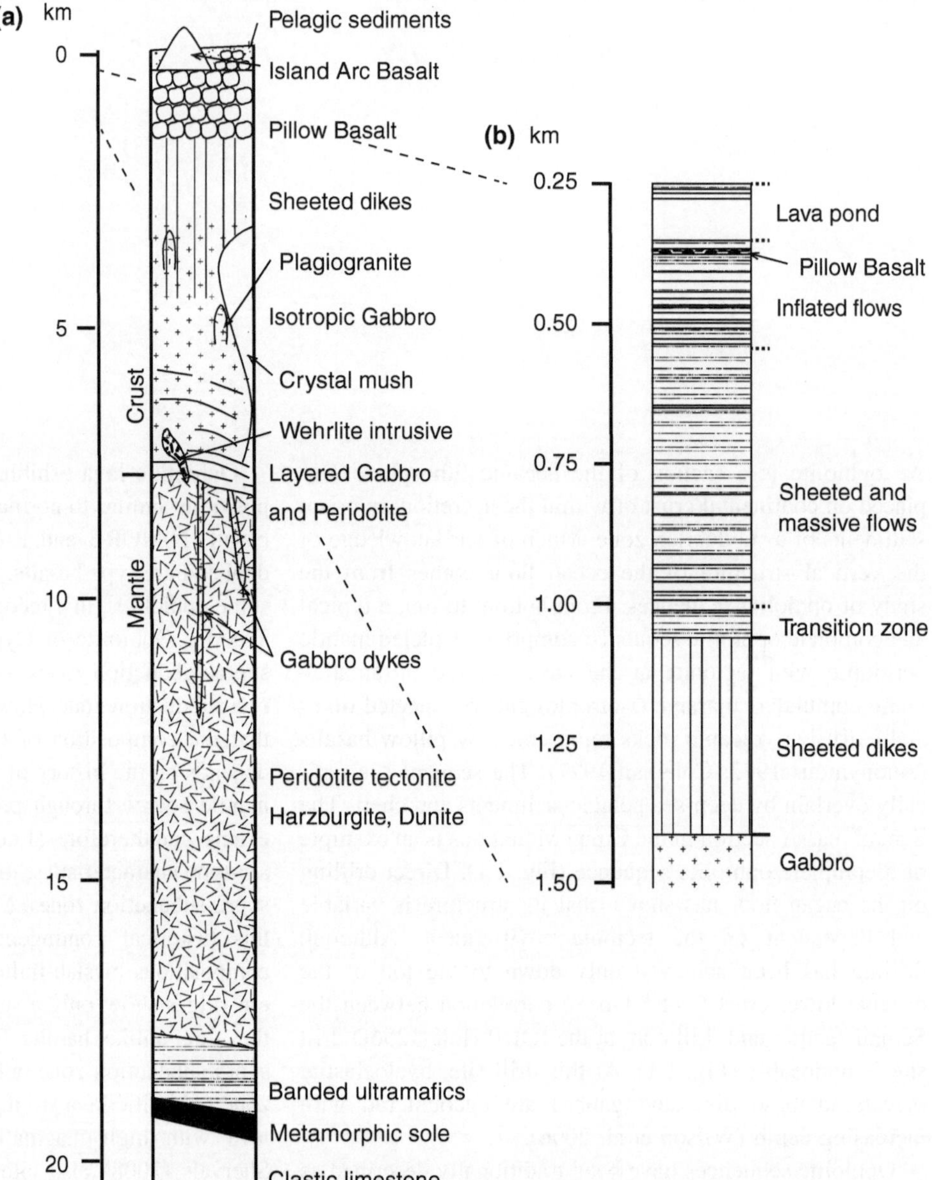

Fig. 1.1 **a** Simplified columnar section of the Semail massif of the Oman Ophiolite (after Hacker et al. 1996; Searle and Cox 2002). **b** Ocean floor section at drill site IODP Hole 1256 D drilled by the Integrated Ocean Drilling Program to a depth of 1257 m (1507 m below sea floor) in the oceanic sub-basement of a fast-moving plate in equatorial Pacific Ocean off the coast of Central America (after Wilson et al. 2006)

found in ophiolitic rocks that may be used as a reference for petrological and tectonic interpretations. Below we provide an introduction to the Oman and Troodos ophiolites, the two best known ophiolite sequences in the world (Fig. 1.2). In Chap. 2, we summarise the tectonic settings, geochemical signatures, chronology and where available, petrographic details of ophiolites at the northern and eastern margins of the Indian plate. In chap. 3, we discuss in detail the geological and lithological relationships in the Naga Hills Ophiolite (NHO) at the eastern margin of the Indian plate. The fourth chapter deals with the structure and tectonics of the NHO. The geophysical characteristics of the Indo-Myanmar (Burma) collision zone, one of the most active seismic zones in the world, are briefly discussed in this chapter. The fifth chapter presents salient petrographic features and metamorphic history of the NHO based on observations from optical photomicrographs and EPMA data. In the sixth chapter, a brief account of the petrogenesis of basalts, evolutionary path of eclogite and emplacement history of the NHO are presented. In the second part of this book, we present photomicrographs and back-scattered electron images of textures of ophiolitic rocks from the NHO.

1.1 Oman Ophiolite

The ophiolite complex comprising several massifs along the approximately 500 km long arcuate coast of Oman bordering the Gulf of Oman (Fig. 1.2) is the best studied and one of the largest known examples of a Tethyan-type

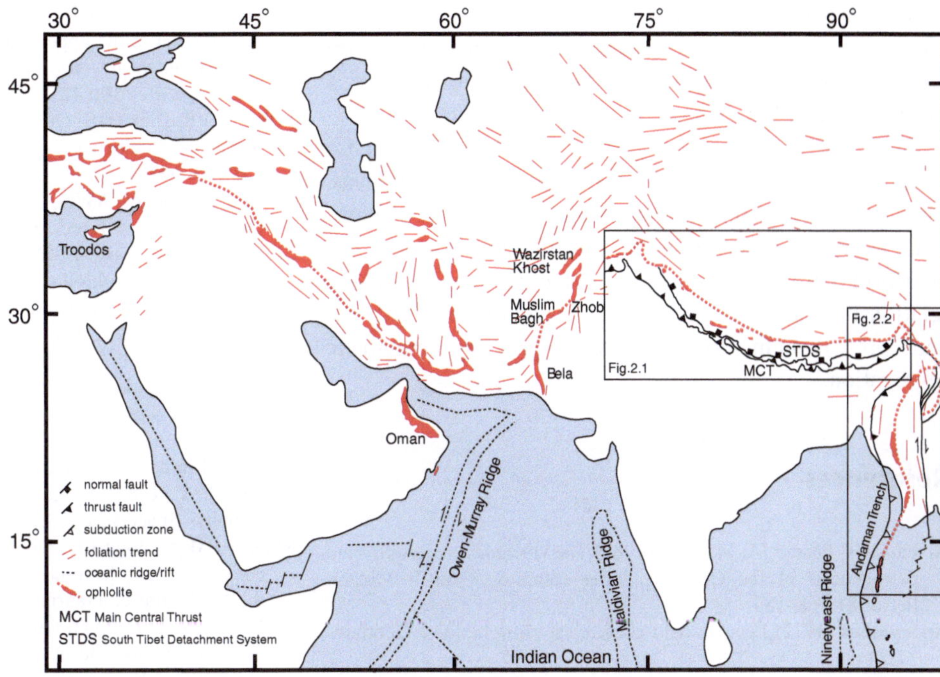

Fig. 1.2 Ophiolite complexes along the northern continental margins of India, Arabia and eastern Africa. Compiled from Dilek et al. (2008); Gnos et al. (1997); Hébart et al. (2012); Lippard et al. (1986); Mitchell et al. (2007); Nicolas and Boudier (2001); Pedersen et al. (2001); Yin and Harrison (2000)

ophiolite in the world. It is located within the Phanerozoic Alpine-Himalayan orogenic belt and contains complete assemblages of ophiolitic rocks. Geological observations indicate that the sea-floor rocks now emplaced as ophiolite probably originated in a medium- to fast-spreading ridge such as the East Pacific Rise or the Juan de Fuca Ridge (Coleman 1971, 1981; Tilton et al. 1981; Nicolas 1989). The northern massifs consist of MORB-like lavas near the base, but the upper lavas contain dacite and andesite, and exhibit geochemical similarities with some subduction zone lavas (Pearce et al. 1981; Alabasteret al. 1982). Also, the Cr-content of spinel in mantle harzburgite and dunite in the northern massifs is higher than that of abyssal peridotites (Lippard et al. 1986; Dick and Bullen 1984). Although the southern massifs such as Semail and Wadi Tayin do not show significant subduction zone signatures, it is likely that the Oman ophiolite originated at a spreading centre in the fore-arc of an NE-dipping intra-oceanic subduction zone (Lippard et al. 1986; Robertson 2006) in early Late Cretaceous (93.5–97.9 Ma, Tilton et al. 1981; Warren et al. 2005). As the Neotethys closed, these rocks were obducted on the Arabian continental passive margin to the south. A well-developed high-temperature granulitic sole occurs beneath the mantle tectonites of the ophiolite (Fig. 1.1) (Searle and Malpas 1980). Mesozoic deep-sea volcano-sedimentary rocks occur as an accretionary prism under the metamorphic sole. The Arabian margin rocks are mostly unmetamorphosed. However, the occurrence of eclogite and blueschists in a promontory of the Arabian margin indicates that some of these continental rocks were subducted to pressures of 17–20 kb and subsequently exhumed (El-Shazly et al. 1990; Warren and Waters 2006).

1.2 Troodos Ophiolite

The early Late Cretaceous (91.6 ± 1.4 Ma, Mukasa and Ludden 1987) Troodos ophiolite of Cyprus (Fig. 1.2) comprises a relatively undeformed and complete ophiolite sequence emplaced on continental crust. It was initially interpreted as oceanic lithosphere formed by sea-floor spreading at a mid-oceanic ridge (Gass 1968). However, the presence of calc-alkalic series basaltic lava prompted Miyashiro (1973) to suggest that the Troodos Ophiolite formed at a spreading centre above a northward-dipping intra-oceanic subduction zone. Other evidence for a subduction-related origin includes a characteristic crystallization sequence, glass with high water content, selective enrichment in LILE and depletion in HFSE, high oxygen fugacity, high Cr# of mantle spinel, isotope composition and presence of boninite (reviewed in Pearce and Robinson 2010). The high-SiO_2, moderate-FeO and low K_2O/H_2O (at 8 wt% MgO) tholeiites overlain by boninites are consistent with a near-trench origin, and a 1400 °C potential temperature for the lower lavas is consistent with rapid slab roll-back and/or sideways influx of hot mantle (Pearce and Robinson 2010). The ophiolite was thrust over the Mesozoic deep-water volcano-sedimentary Mamonia Formation of Africa-Arabia continental passive margin (Robertson and Woodcock 1979).The process of juxtaposition of the Mamonia

Formation with the Troodos Ophiolite has been variously interpreted as subduction/accretion, oceanic transform faulting, strike-slip, collision and/or paleo-rotation of a Troodos microplate (Robertson and Xenophontos 1993).

Both the Oman and Troodos ophiolites are known for massive sulphide deposits associated with the pillow lavas. They are also the second most significant source of chromite after the Precambrian stratiform ore deposits. The podiform chromite deposits in ophiolite show a spatial relationship with cumulate ultramafics and layered gabbros and are thought to have developed by magmatic segregation.

References

Alabaster T, Pearce JA, Malpas J (1982) The volcanic stratigraphy and petrogenesis of the Oman ophiolite complex. Contrib Mineral Petrol 81:168–183

Anonymous (1972) Penrose field conference on ophiolites. Geotimes 17:24–25

Boudier F, Nicolas A (1977) Structural controls on partial melting in the Lanzo peridotites. In: Dick HJB (ed) Magma genesis. Oregon Dept Geol Miner Ind Bull 96:39–56

Coleman RG (1971) Plate tectonic emplacement of upper mantle peridotites along continental edges. J Geophys Res 76:1212–1222

Coleman RG (1977) Ophiolites: ancient oceanic lithosphere. Springer, Berlin 229 p

Coleman RG (1981) Tectonic setting for ophiolite obduction in Oman. J Geophys Res 86:2497–2508

Dick HJB (1977) Evidence of partial melting in the Josephine peridotite. In: Dick HJB (ed) Magma genesis. Oregon Dept Geol Miner Ind Bull 96:59–62

Dick HJB, Bullen T (1984) Chromian spinel as a petro genetic indicator in abyssal and alpine-type peridotites and spatially associated lavas. Contrib Mineral Petrol 86:54–76

Dilek Y, Furnes H, Shallo M (2008) Geochemistry of the Jurassic Mirdita Ophiolite (Albania) and theMORB to SSZ evolution of a marginal basin oceanic crust. Lithos 100:174–209

Edwards SJ, Malpas J (1995) Multiple origins for mantle harzburgites: examples from the Lewis Hills massif, Bay of Islands ophiolite, Newfoundland. Can J Earth Sci 32:1046–1057

El-Shazly AK, Coleman RG, Liou JG (1990) Eclogites and blueschists from the northeastern Oman: petrology and P-T evolution. J Petrol 31:629–666

Gass IG (1968) Is the Troodos massif of Cyprus a fragment of Mesozoic ocean floor? Nature 220:39–42

Gnos E, Immenhauser A, Peter T (1997) Late Cretaceous/early Tertiary convergence between the Indian and Arabian plates recorded in ophiolites and related sediments. Tectonophysics 271:1–19

Hacker BR, Mosenfelder JL, Gnos E (1996) Rapid emplacement of the Oman ophiolite: Thermal and geochronological constraints. Tectonics 15:1230–1247

Hébert R, Bezard R, Guilmette C, Dostal J, Wang CS, Liu ZF (2012) The Indus–Yarlung Zangbo ophiolites from Nanga Parbat to Namche Barwa syntaxes, southern Tibet: First synthesis of petrology, geochemistry, and geochronology with incidences on geodynamic reconstructions of Neo-Tethys. Gondwana Res 22(2):377–397

Kelemen PB, Dick HJB (1995) Focused melt flow and localized deformation in the upper mantle: juxtaposition of replacive dunite and ductile shear zones in the Josephine peridotites, SW Oregon. J Geophys Res 100:423–438

Kelemen PB, Dick HJB, Quick JE (1992) Formation of harzburgite by pervasive melt/rock reaction in the uppermantle. Nature 358:635–641

Lippard SJ, Shelton AWG, Gass IG (1986) The Ophiolite of Northern Oman. Geol Soc Lond Mem 11

Metcalf RV, Shervais JW (2008) Suprasubduction-zone ophiolites: is there really an ophiolite conundrum? In: Wright JE, Shervais JW (eds) Ophiolites, arcs, and batholiths: a tribute to Cliff Hopson. Geol Soc Am Spec Pap 438:191–222

Mitchell AHG, Htay MT, Htun KM, Win MN, Oo T, Hlaing T (2007) Rock relationships in the Mogok metamorphic belt, Tatkon to Mandalay, central Myanmar. J Asian Earth Sci 29:891–910

Miyashiro A (1973) The Troodos ophiolitic complex was probably formed in an island arc. Earth Planet Sci Lett 19/2:218–224

Moores EM (1982) Origin and emplacement of ophiolites. Rev Geophys Space Phys 20:735–760

Moores EM, Kellogg LH, Dilek Y (2000) Tethyan ophiolites, mantle convection, and tectonic "historical contingency"; a resolution of the "ophiolite conundrum." In: Dilek Y, Moores EM, Elthon D, Nicolas A (eds) Ophiolites and oceanic crust: new insights from field studies and the ocean drilling program. Geol Soc Am Spec Pap 349:3–12

Mukasa SB, Ludden JN (1987) Uranium–lead isotopic ages of plagiogranites from the Troodos ophiolite, Cyprus, and their tectonic significance. Geology 15:825–828

Nicolas A (1989) Structure of ophiolites and dynamics of oceanic lithosphere. Kluwer Academic, Boston, 367 p

Nicolas A, Boudier F (2001) Oman and UAE ophiolite map. Marine Geophys Res 21

Pearce JA, Robinson PT (2010) The Troodos ophiolitic complex probably formed in a subduction initiation, slab edge setting. Gondwana Res 18/1:60–81

Pearce JA, Alabaster T, Shelton AW, Searle MP (1981) The Oman ophiolite as a Cretaceous arc-basin complex: evidence and implications. Philos Trans R Soc Lond 300:299–317

Pedersen RB, Searle MP, Corfield RI (2001) U-Pb zircon ages from the Spontang Ophiolite, Ladakh Himalaya. J Geol Soc Lond 158:513–520

Robertson AHF (2006) Contrasting modes of ophiolite emplacement in the Eastern Mediterranean region. Geol Soc Lond Mem 32:235–261

Robertson AHF, Woodcock NH (1979) The Mamonia Complex, southwest Cyprus: the evolution and emplacement of a Mesozoic continental margin. Geol Soc Am Bull 90:651–665

Robertson AHF, Xenophontos C (1993) Development of concepts concerning the Troodos ophiolite and adjacent units in Cyprus. Geol Soc Lond Spec Pub 76:85–119

Searle MP, Cox J (2002) Subduction zone metamorphism during formation and emplacement of the Semail ophiolite in the Oman Mountains. Geol Mag 139(3):241–255

Searle MP, Malpas J (1980) Structure and metamorphism of rocks beneath the Semail ophiolite of Oman and their significance in ophiolite obduction. Trans R Soc Edinb Earth Sci 71:247–262

Tilton GR, Hopson CA, Wright JE (1981) Uranium–lead isotopic ages ofthe Semail Ophiolite, Oman, with applications to Tethyan ridge tectonics. J Geophys Res 86:2763–2775

Warren CJ, Parrish RR, Waters DJ, Searle MP (2005) Dating the geologic history of Oman's Semail Ophiolite: insights from U–Pb geochronology. Contrib Mineral Petrol 150:403–422

Warren CJ, Waters DJ (2006) Oxidized eclogites and garnet-blueschists from Oman: P–T path modelling in the NCFMASHO system. J Metamorph Geol 24:783–802

Wilson DS, Teagle DAH, Alt JA, Banerjee NR, Umino S, Miyashita S, Acton GD, Anma R, Barr SR, Belghoul A, Carlut J, Christie DM, Coggon RM, Cooper KM, Cordier C, Crispini L, Durand SR, Einaudi F, Galli L, Gao Y, Geldmacher J, Gilbert LA, Hayman NW, Herrero-Bervera H, Hirano N, Holter S, Ingle S, Jiang S, Kalberkamp U, Kerneklian M, Koepke J, Laverne C, Lledo Vasquez HL, Maclennan J, Morgan S, Neo N, Nichols HJ, Park S-H, Reichow MK, Sakuyama T, Sano T, Sandwell R, Scheibner B, Smith-Duque CE, Swift SA, Tartarotti P, Tikku AA, Tominaga M, Veloso EA, Yamasaki T, Yamazaki S, Ziegler C (2006) Drilling to gabbro in intact ocean crust. Science 312:1016–1020

Yin A, Harrison TM (2000) Geologic evolution of the Himalayan-Tibetan orogen. Ann Rev Earth Planet Sci 28:211–280

2 Ophiolite Around the Indian Plate Margin

Ophiolite and ophiolitic rocks were emplaced along the western, northern and eastern margins of the Indian plate during the northward movement of the Indian plate and the collision of the Indian continental block first with intra-oceanic island arcs within the Neotethys, and then with the Helmand and Kabul continental blocks to the west (Tapponnier et al. 1981), the Asian continental block to the north (Gansser 1964) and the Myanmar continental block to the east (Mitchell 1993) (Fig. 1.2).

2.1 The Western Margin of the Indian Plate

The ophiolite belt along the western margin comprises the Bela, Muslim Bagh, Zhob and Waziristan-Khost ophiolites in Pakistan and bordering Afghanistan (Alleman 1979; Gnos et al. 1997) (Fig. 1.2). These and the related ophiolite belt along the southeastern coast of Oman were emplaced around the Cretaceous/Tertiary boundary. The western ophiolite belt of Pakistan consists of pyroxenite, gabbro and sheeted dikes deformed and metamorphosed at granulite to greenschist facies conditions. They are underlain by an accretionary wedge mélange containing slabs of pillow basalt intruded by microgabbro and associated with Early to Middle Cretaceous pelagic sediments. The mélange grades into calcareous turbidites of the Indian continental shelf.

2.1.1 Bela

The Bela Ophiolite is a dismembered sequence that presumably once formed a single thrust sheet. It consists of serpentinised mantle harzburgite overlain by layered peridotite and gabbro, foliated and isotropic gabbro, sheeted dikes and extrusive rocks (Gnos et al. 1998). The mantle rocks are underlain by a poorly preserved metamorphic sole of amphibolite to greenschist facies rocks. Foliated and undeformed granitoid plutons are common within the crustal section that yielded $^{40}Ar/^{39}Ar$ plateau ages of 69.5 ± 0.7 Ma and 68.7 ± 0.7 Ma from amphibole separates (Gnos et al. 1998). The granitoid plutons comprise gabbro, diorite, trondhjemite, plagiogranite and granite. Chemical composition of the plagiogranite is consistent with its origin from fractional crystallisation of basaltic magma, and U–Pb zircon dating of trondhjemite and granite yielded an age of 68 ± 3 Ma (Ahmed 1993). These ages constrain the age of origin and emplacement of the ophiolite. The accretionary wedge unit below the ophiolitic unit contains imbricate sheet of pillow basalt with E-MORB-type chemistry covered by radiolarian chert, shale and calci-turbidites and capped by Reunion hotspot-related lavas.

2.1.2 Muslim Bagh and Waziristan

The Muslim Bagh ophiolite is related to a WSW-ENE trending thrust, and occurs in the uppermost part of a pile of nappes accreted onto the Indian continental margin (Asrafullah et al. 1979; Ahmad and Abbas 1979). The ^{40}Ar-^{39}Ar plateau age of the Muslim Bagh ophiolite is 70.7 ± 5.0 Ma for the metamorphics, and 68.7 ± 1.8 Ma for the amphibolites at the base of the sheeted dykes (Mahmood et al. 1995). It seems that the floor of the Neotethys Sea was subducted 65–70 many years ago in the Pakistan sector (Valdiya 2010).

The Paleocene-Early Eocene Waziristan ophiolite (~ 500 km^2) is the third largest ophiolite complex in Pakistan after Bela and Zhob (Jan et al. 1985). It occurs as dismembered thrust slices tectonically overriding Jurassic-Cretaceous calcareous sediments of the Indian plate. It consists of ultramafics rocks, gabbro, sheeted dikes, pillow lava, plagiogranite, anorthosite and pelagic sediments. The sedimentary rocks associated with the ophiolite are Mesozoic and Early Tertiary in age. The ultramafics include harzburgite, dunite, pyroxenite and secondary serpentinite, and display deformational features and cataclastic fabrics similar to those in the Zhob and Jijal complexes. The sheeted dikes show chilled margins. Some trondhjemite

N. C. Ghose et al., *A Petrographic Atlas of Ophiolite*,
DOI: 10.1007/978-81-322-1569-1_2, © Springer India 2014

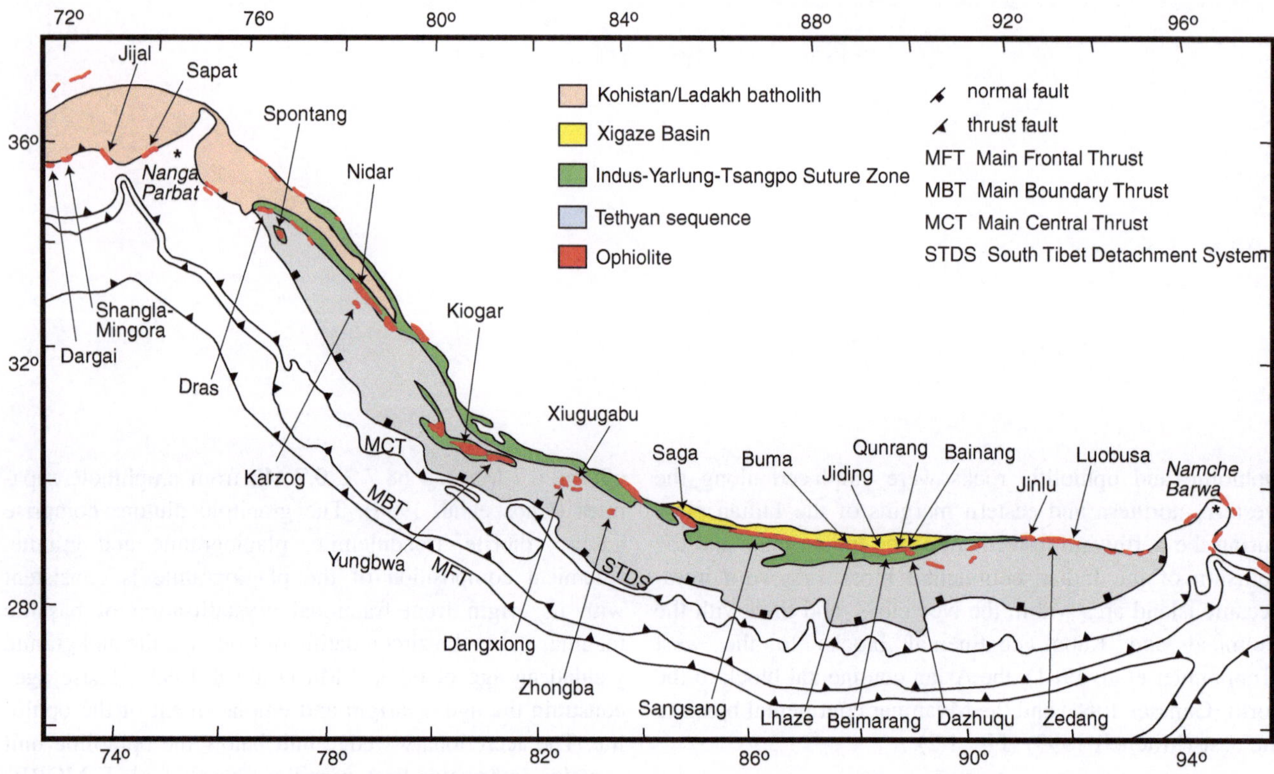

Fig. 2.1 Himalayan ophiolite complexes (after Hébart et al. 2012)

dikes containing sodic plagioclase show a calc-alkaline affinity. The volcanic rocks commonly showing pillow structures comprise about half of the total area of the ophiolite complex. Subordinate occurrence of agglomerate and volcanic tuffs are noteworthy features.

2.2 The Northern Margin of the Indian Plate

According to our present understanding, ophiolitic rocks at the northern margin of the Indian plate originated between Jurassic and Early Cretaceous in mid-oceanic ridges and supra-subduction zones of intra-oceanic arc systems within the Neotethys Ocean (Aitchison et al. 2003; Aitchison and Davis 2004). Geochemical data indicate that the ophiolitic rocks formed in fore-arc, arc and back-arc setting from mixed magmas including calc-alkaline, mid-oceanic ridge basaltic and oceanic-island basaltic magmas (Hébart et al. 2012). These rocks were obducted southwards onto the Indian plate margin or accreted within the Indus-Yarlung-Tsangpo Suture Zone (Thakur 1981) as the Indian continental block collided with the intra-oceanic arc in Late Cretaceous and with the Asian continent in Paleocene-Eocene as the Neotethys closed (Tapponnier et al. 1981; Allègre et al. 1984; White and Lister 2012). The Indus-Yarlung-Tsangpo Suture continues into the Central Myanmar Basin in the east and to Baluchistan in the west (Valdiya 1988, 2010). A variety of assemblages of sediments that were deposited in the northwestern part of the suture zone are known as the Indus Flysch. Some of the ophiolites along the suture zone were emplaced as thrust sheets on the Indian continental margin and presently occur as klippen on deep-water marine sedimentary sequences of the Indian passive margin, whereas others occur as well-preserved or dismembered sections and mélanges within the Indus-Yarlung-Tsangpo Suture Zone (Gansser 1964, 1980).

2.2.1 The Indus Suture Zone

The ophiolite occurrences along the Indus Suture Zone include the Dargai, Shangla-Mingora, Jijal and Sapat complexes on the western side of the Nanga Parbat syntaxis in northern Pakistan (reviewed in Arif and Jan 2006) and the Dras, Spontang and Nidar-Karzog complexes on the eastern side of the syntaxis in eastern Jammu and Kashmir, India (reviewed in Mahéo et al. 2004) (Fig. 2.1).

2.2.1.1 Sapat

The Sapat and Jijal complexes are located on the hanging wall of the north-dipping Indus Suture Zone and constitute the southernmost part of the Kohistan Arc (Fig. 2.1).The Sapat Complex consists of two lithological units. The lower unit comprises ultramafic rocks such as meta-harzburgite, dunite and intrusive clinopyroxenite (Bouilhol et al. 2009). Meta-harzburgite displays a massive, fine-grained texture with relict orthopyroxene and disseminated spinel grains. The olivine porpyroblasts are lobate in shape with sub-grain boundaries. They show undulose extinction and are surrounded by fine-grained, undeformed olivine neoblasts. Small euhedral spinel grains are present within the olivine pophyroblasts, and amoeboid spinel grains are associated with the olivine neoblasts. The matrix also contains orthopyroxene decomposition products such as olivine, talc and tremolite that indicate fluid-assisted lower amphibolite facies re-equilibration. Secondary chlorite around spinel, serpentine and rare diopside around olivine and calcite vein in dunite have been observed. The dunites are fresh and coarser and contain trails/pods of chromite spinel and clinopyroxene that may have originated through melt percolation. Olivine-clinopyroxene micro-veins and thin dikes are interpreted as melt crystallisation products. Mineral and bulk rock compositions and U-shaped REE patterns suggest that the meta-harzburgites represent a refractory, metasomatised mantle.

The ultramafic unit is bounded by a fault zone at the top and is overlain by meta-gabbro and tonalite–trondhjemite with clinopyroxenite intrusives. The olivine-bearing clinopyroxenites occur as sub-horizontal bands below meta-plutonic rocks, vertical intrusive bodies with gabbro and tonalite-trondhjemite cores, and dikes. Clinopyroxene overgrows olivine in the clinopyroxenites. Amoeboid clinopyroxene occurs at the grain boundaries of olivine porphyroblasts. Trails of lobate clinopyroxene porphyroblasts are also present. Secondary chlorite and hornblende are associated with clinopyroxene. Geochemical modelling indicates that the parental melt of the pyroxenites and meta-gabbros was a highly depleted primitive arc melt of supra-subduction affinity (Bouilhol et al. 2009).

2.2.1.2 Dras

In the Ladakh-Zanskar area of the northwestern Himalayas, the NW–SE Indus Suture Zone separates the Ladakh Batholith of the Asian margin to the NE from the Zanskar and Tethys Himalayas of the Indian margin to the SW (Fig. 2.1). The NW part of the Indus Suture Zone comprises a northern unit known as the Indus Group with continental molasse-type sediments, and an ophiolite-bearing southern unit containing the Dras Arc. The Dras Complex is composed of island arc calc-alkaline and tholeiitic basalts intruded by granodioritic plutons (Dietrich et al. 1983; Radhakrishna et al. 1984). The arc-related units are stratigraphically underlain by blocks of ophiolitic rocks such as serpentinised harzburgite, lherzolite tectonite, cumulate dunite, pyroxenite and intrusive gabbro and diabase (Srikantia and Razdan 1980 , 1985; Srikantia 1986; Radhakrishna et al. 1987; Reuber et al. 1989). The Dras Ophiolite is marked by a basal fault and is underlain by the Sapi-Shergol ophiolitic mélange containing 90 Ma blueshists (Honegger et al. 1989). The mélange, the ophiolite and the Dras Arc units are tectonically overlain by the Lamayuru Formation and the Zanskar shelf succession of the Indian margin along a south-dipping back-thrust (Sinha and Upadhyay 1993; Robertson and Degnan 1993).

U–Pb zircon ages of 103 ± 3 Ma (Honegger et al. 1982) and 101 ± 2 Ma (Schärer et al. 1984) of granodiorite intrusives in the Dras volcanic and volcaniclastic rocks constrain the minimum age of the Dras Ophiolite. K–Ar ages of 77.5 ± 1 Ma (Sharma et al. 1979) and 78.5 ± 2.9 Ma (amphibole separate, Reuber et al. 1989) indicate Late Cretaceous metamorphism of the Dras Complex.

2.2.1.3 Spontang

About 75 km SE of Dras and south of the Indus Suture, the Spontang Ophiolite occurs in a tectonic thrust slice above the Paleocene-Eocene carbonate sequence of the Zanskar Himalaya (Fig. 2.1). The ophiolite sequence consists of weakly foliated mantle harzburgite and spinel-lherzolite intruded by diorite and rare plagiogranite, cumulates of gabbro and ultramafic rocks, isotropic gabbro, highly tectonised sheeted dikes with intrusive plagiogranite feeding pillow lava of N-MORB-type geochemical signature (Corfield et al. 2001). The mantle units may be divided into a lower and an upper section separated by a south-dipping thrust. The lower section consists of serpentinised peridotites (up to 98 % serpentine near the top), whereas the upper section consists of fresh peridotites (Mahéo et al. 2004). Both sections are cross-cut by dunitic and dioritic dikes and sills. The mantle rocks are composed of porphyroclastic olivine and pyroxenes. Orthopyroxene commonly shows lobate boundaries interpreted as evidence of mantle melting. Temperatures of 1000–1200 °C at low deviatoric stress have been inferred from kinked olivine in peridotite that originated through high-temperature plastic flow during ascent of the upper mantle and partial melting below a spreading centre (Reuber 1986). Plagioclase laths in the upper crustal gabbro are well preserved but commonly altered, indicating that they crystallised before clinopyroxene in a normal mid-oceanic ridge environment (Corfield et al. 2001). The diorites dikes and sills cross-cutting the mantle peridotites have ophitic to cumulate textures. They are dominated by metamorphic amphibole and albitic plagioclase with minor amounts of oxides, apatite, epidote and titanite (Mahéo et al. 2004). Magmatic relics occur in the plagioclase cores and as diopsidic clinopyroxene inclusions in hornbende and edenitic amphibole. Hornblende is locally replaced by actinolite or ferro-actinolite. The ophiolite sequence is unconformably overlain by basalt, andesite and volcano-sedimentary rocks of the Spong arc that developed over the fore-arc of the supra-subduction zone. The

Spontang ophiolitic mélange containing Permo-Triassic and Albian alkaline magmatic rocks tectonically underlies the ophiolite sequence. Structural, tectonic and palaeomagnetic constraints indicate that the Spong Arc and the Dras Arc were separate but coexisting intra-oceanic island arcs (Corfield et al. 2001).

U–Pb zircon dating of N-MORB-type plagiogranite intrusive in sheeted basaltic dike indicates that the oceanic crust of the Spontang Ophiolite formed at 177 ± 1 Ma in a mid-oceanic ridge environment, making it one of the oldest of the Tethyan ophiolites discovered so far (Pedersen et al. 2001). Moreover, incompatible trace element data including REE and $^{39}Ar/^{40}Ar$ dating of amphiboles in intrusive diorite in peridotite indicate that the mantle source of the N-MORB-type basalts was metasomatised in a supra-subduction zone at c. 130 Ma during the initiation of subduction (Mahéo et al. 2004). Recent fossil evidence suggests that intra-oceanic subduction may have continued until c. 55 Ma (Baxter et al. 2010). An 88 ± 5 Ma date from andesite in the Spong arc indicates that the arc was still active at c. 88 Ma and obduction of the ophiolite on the Indian continental margin occurred after c. 88 Ma (Pedersen et al. 2001). Based on field evidence, Corfield et al. (2005) suggest that thrusting of the ophiolite over the Cenozoic Zanskar sediments was a younger event that happened during the India-Asia collision in late Paleocene-early Eocene. However, Garzanti et al. (2005) disagree with the field evidence of Corfield et al. (2005) and contend that the ophiolite was obducted on the Indian margin in Early Eocene.

2.2.1.4 Nidar and Karzog

Further east, the Nidar Ophiolite delineates the southern margin of the Indus Suture Zone in eastern Ladakh (Fig. 2.1). It has a south-dipping thrust contact with the Indus Formation in the north, and a north-dipping thrust contact with the blueschist-bearing Zildat ophioliticmélange (Virdi et al. 1977; Virdi 1987, 1989; Thakur and Misra 1984). Both the ophiolite and the Zildat mélange are thrust over the Tso-Morari crystalline complex in the south. Petrological and geochemical data including isotopes indicate that the ophiolite originated in an intra-oceanic subduction environment (Ahmad et al. 2008). The ophiolite is composed of ultramafic rocks at the base, gabbroic rocks in the middle and volcano-sedimentary rocks at the top. The ultramafics include spinel–harzburgite near the base, spinel–dunite with chromite veins at higher levels and minor pyroxenite intrusives that also occur in the overlying gabbros. Both microgabbro with ophitic texture and cumulate gabbros are present. The gabbros consist of olivine, clinopyroxene, plagioclase and amphibole with minor interstitial magnetite and spinel (Ahmad et al. 2008). Olivine occurs as large lobate grains, smaller euhedral grains at margins of porphyroclasts and as inclusion within plagioclase and clinopyroxene. Clinopyroxene occurs as oikocrysts with olivine and plagioclase inclusions, and smaller interstitial grains. Some large plagioclase grains show kink bands and bent cleavage overgrown by undeformed rims. The cores of the large clinopyroxene have lower calcium contents than the undeformed rims. Amphibole occurs as rims around clinopyroxene or in late-stage veins. Many gabbro samples consist of metamorphic assemblage of hornblende + albitic plagioclase (Mahéo et al. 2004). The pargasitic amphiboles range in composition from magnesio-hornblende to actinolite characteristic of oceanic metamorphism. Anorthite-rich cores of plagioclase low-Ti diopside inclusions in hornblende are interpreted as magmatic relics. Plagiogranite intrusives within the gabbro showing nearly flat REE patterns originated through fractional crystallisation of sub-alkalinetholeiitic magmas (Rao et al. 2004).

The volcano-sedimentary unit contains porphyritic basalt and andesite with poorly preserved pillow structures near its base, and gradually passes into volcaniclastics comprising chert, jasper, shale, siltstone, volcanic sandstones and conglomerates. The upper part of the volcano-sedimentary unit is dominated by andesite and rhyolite intercalated with ash, tuff, lapilli and volcanic breccia. The phenocryst in basalts show evidence of hydrothermal alteration and is dominated by albitic plagioclase laths, diopside-augite and serpentinised olivine, and minor Fe–Ti oxides. The groundmass is composed of fine-grained augite, plagioclase and rare glass. Epidote and radiating crystals of chlorite are also present in the groundmass. The rhyolites show banding and are dominated by volcanic glass and feldspar spherules with minor mafic phases such as pyroxene and fibrous chlorite. Palagonite is commonly associated with the glass. Calcite and quartz veins and vesicles with quartz filling are observed in some samples.

The Tso-Morari crystalline complex, an exhumed block of the subducted Indian continental crust containing eclogite, is located to the immediate south of the Nidar Ophiolite (Berthelsen 1953). A small section of ophiolite consisting of highly deformed and altered chromitite, serpentinite and minor gabbro occurs at the southern margin of the Tso-Morari dome near Karzog. The Karzog ophiolite is presumably a detached klippe of the Nidar Ophiolite (de Sigoyer et al. 2004). The mafic rocks comprise large poikilitic metamorphic edenitic amphibole with relics of low-Ti diopside, albite, oxides and greenschist facies minerals such as actinolite, titanite and epidote (Mahéo et al. 2004).

Gabbro and basalt from the Nidar Ophiolite yielded an Sm–Nd isochron age of 140 ± 32 Ma that indicates the time of formation of the ophiolitic crust (Ahmad et al. 2008). Amphiboles in the Nidar gabbro yielded ages between 100 and 120 Ma by $^{39}Ar/^{40}Ar$ step-heating technique that probably corresponds with metamorphism related to the initiation of subduction (Mahéo et al. 2004). Radiolarian taxonomy and biostratigraphy reveal an upper Barremian to Upper Aptian age for the Nidar volcano-sedimentary section (Zyabrev et al. 2008), consistent with the $^{39}Ar/^{40}Ar$ amphibole age of Nidar gabbro.

2.2.2 The Yarlung-Tsangpo (Yarlung-Zangbo) Suture Zone

2.2.2.1 The Western Part

The ophiolites in the western part of the Yarlung-Tsangpo Suture Zone occur along the suture zone itself as well as to the south of suture zone as klippen overlying an ophiolitic tectonic mélange (Hodges 2000; Searle et al. 1988). The klippe-type ophiolites include the Kiogar, Yungbwa, Dangxiong, Xiugugabu and Zhongba massifs (Fig. 2.1). Recent U–Pb zircon dating on diabase and tholeiite yielded ages of 123.4 ± 0.9 Ma and 123.9 ± 0.9 Ma for the Yungbwa and 126.7 ± 0.4 Ma and 123.4 ± 0.8 Ma for the Dangxiong (Chan et al. 2007), 120.2 ± 2.3 Ma for the Yungbwa (Li et al. 2008) and 122.3 ± 2.4 Ma for the Xiugugabu (Wei et al. 2006) and 125.7 ± 0.9 Ma for the Zhongba (Dai et al. 2012) massifs. These dates are younger than the 159.7 ± 0.5 Ma U–Pb zircon age of the Kiogar gabbro (Chan et al. 2007), and the earlier determined 152 ± 33 Maage of hornblende in the Yungbwa massif by the $^{40}Ar/^{39}Ar$ isotope method (Miller et al. 2003). The tectonic mélange underlying the ophiolites contains clasts of Late Jurassic to Early Cretaceous radiolarite (Dai et al. 2011) and older rocks including reef and micritic limestone and mafic rocks. The mélange overlies rocks of the Indian continental passive margin.

The Yungbwa Ophiolite is a large peridotitic complex tectonically overlying a mélange that contains blocks of Permian reef limestone, Late Cretaceous micritic limestone and volcanic rocks. It is located in SW Tibet, about 20 km south of the Indus-Yarlung-Tsangpo Suture Zone, south of Mount Kailas, and to the immediate NW of the Gurla Mandhata crystalline complex. The ophiolite and the mélange are thrust over sediments of the Tethyan series to the south. The main constituent of the ophiolite is clinopyroxene-bearing, serpentinised spinel-harzburgite that is cross-cut by dikes of basalt, gabbro and gabbronorite (Miller et al. 2003; Liu et al. 2010). Similar rocks occur in the Kiogar Ophiolite located to the immediate NW of the Yungbwa massif (Gansser 1964). Fresh harzburgite from the Yungbwa massif shows porphyroclastic texture and contains elongated olivine with kink-bands and orthopyroxene with exsolution lamellae and deformed cleavage showing undulatory extinction, and arranged sub-parallel to the foliation. Large spinel grains have "holly-leaf" shape. Orthopyroxene compositions are within the range of orthopyroxenes in abyssal peridotites, and temperatures based on two-pyroxene equilibria are 900–960 °C and on olivine-spinel Mg–Fe exchange are 760 °C (Miller et al. 2003). The lower temperature is similar to temperatures calculated in peridotite tectonite from slow spreading ridges (Constantin 1999). Pyroxene inclusions have been observed in plagioclase. The interstitial grains comprise neoblasts of olivine, orthopyroxene and clinopyroxene with well-defined crystal shapes, rounded spinel and disseminated sulphides. Mineral chemistry and Os-isotopic ratio of the peridotites are similar to those of abyssal peridotites, and Nd model age of clinopyroxene indicates that the peridotites underwent melt extraction in Jurassic (Miller et al. 2003). Minor plagioclase and amphibole are unevenly distributed in the harzburgites (Liu et al. 2010). Amphibole also occurs as blebs within orthopyroxene, and as interstitial grains aligned with the clinopyroxene exolution lamellae. Anhedral plagioclase commonly replaces clinopyroxene and orthopyroxene. The pargasitic amphibole and anorthitic plagioclase originated through metasomatic alteration of the mantle with a high-Ca, high-LREE and low-Ti arc-type hydrous melt in a subduction zone setting (Liu et al. 2010).The bulk REE and Nd-isotopic compositions of peridotite also indicate secondary syn- or post-emplacement modification of the ophiolite (Miller et al. 2003). The basaltic dikes show subophitic orhypidiomorphic-granular texture and consist of diopsidic augite, plagioclase, minor hornblende and accessory ilmenite. They are tholeiitic and show N-MORB-type REE patterns. Sericitisation of plagioclase and replacement of clinopyroxene and hornblende byactinolite indicate low-grade metamorphism. The gabbronorite dikes show sharp contacts with the host peridotites and consist of clinopyroxene, orthopyroxene and plagioclase with 120° triple junctions. Their primary textures are modified by recrystallisation and granulation. Plagioclase shows undulatory extinction and is partially transformed into prehnite-bearing saussurite. Clinopyroxene is partially replaced by metamorphic hornblende, tremolite or actinolite with a typical flaser texture.

The Xiugugabu massif is dominated by variably serpentinised harzburgite intruded by amphibole-bearing microgabbro and microgabbronoritesills sub-parallel to foliation, and isolated outcrops of red and green chert. Geochemical characteristics of the harzburgites indicate that they are residues of 5–25 % of partial melting of a depleted mantle that has been enriched by percolating metasomatic melts in a supra-subduction environment (Bezard et al. 2011). Dunite is rare and occurs in patches within harzburgite. The harzburgites display porphyroclastic to porphyromylonitic texture and consist of kinked and highly stretched olivine and orthopyroxene porphyroclasts showing undulatory extinction. Orthopyroxene contains clinopyroxene exsolution lamellae and are surrounded by neoblasts. Minor spinel occurs as large euhedral grains and small vermicular intergrowths. Accessory magnetite, chlorite and amphibole are present. The microgabbro and microgabbronorite sills comprise plagioclase and pyroxene microphenocrysts oriented along the foliation, and a groundmass of amphibole, plagioclase, clinopyroxene and orthopyroxene, ilmenite and magnetite with secondary epidote, chlorite and sericite. The textural relation indicates that plagioclase crystallised before pyroxene. The amphiboles likely originated in a high-T ($\sim$800 °C) hydrothermal environment. Metamorphic textures such as magnetite + clinopyroxene symplectite, amphibole + clinopyroxene symplectite and amphibole overgrowth around clinopyroxene have been observed. The intrusive rocks have tholeiitic compositions and show N-MORB-type REE pattern, Th enrichment and Nb, Ta and Ti negative anomalies, indicating a back-arc, supra-subductionzone setting.

The Zhongba massif consists of mantle ultramafics, and a crust composed of diabase dikes and pillow basalts (Dai et al. 2012). It is surrounded by a mélange containing blocks of massive basalt associated with layers of deformed limestone, black shale and purple-red and green-grey chert. The mantle ultramafics are dominated by fresh foliated harzburgite with minor dunite. A diabase dike shows granular to medium-grained intersertal texture with idiomorphic plagioclase, chloritised pyroxene and olivine and minor Fe–Ti oxides. Porphyritic pillow and massive basalts consist of phenocrysts of plagioclase, clinopyroxene and olivine in a fine-grained, intersertal to intergranular groundmass of plagioclase, clinopyroxene and Fe–Ti oxides, and secondary veins and amygdules of carbonate, quartz and chlorite. Some of the phenocrysts are partially or completely replaced by albite or chlorite. Texture showing clinopyroxene cores with radiating plagioclase laths is present in some samples. The mafic rocks exhibit an ocean-island and Hawaiian alkali basalt-type REE pattern that suggests an origin from a series of seamounts within the Neo-Tethys during the Early Cretaceous (Dai et al. 2012).

2.2.2.2 The Central Part

The central part of the Yarlung-Tsangpo Suture Zone contains a series of ophiolite massifs including the Saga and Sangsang ophiolites in the west and the Xigaze ophiolites (from west to east: Buma, Lhaze, Jiding, Beimarang, Qunrang, Bainang and Dazhuqu massifs) in the east (Fig. 2.1). U–Pb zircon dating in diorite, gabbro and diabase yielded ages of 125.2 ± 3.4 Ma for the Sangsang massif (Xia et al. 2008b), 128 ± 2 Ma for the Jiding massif Wang et al. 2006), 125.6 ± 0.8 Ma for the Bailang massif (Li et al. 2009) and 126 ± 1.5 Ma for the Dazhuka massif (Malpas et al. 2003). Plagiogranite from the Xigaze ophiolitic crust yielded an age of 120 ± 10 Ma (Göpel et al.1984), and apegmatitic gabbro yielded an age of 132.0 ± 2.9 Ma (Chan et al. 2007). Hornblende in amphibolite in the Buma and Bailang massifs yielded similar dates of 123.6 ± 2.9 Ma, 127.7 ± 2.2 Ma and 127.4 ± 2.3 Ma with the $^{40}Ar/^{39}Ar$ step-heating technique (Guilmetteet al. 2008, 2009). Radiolarites from above the volcanic rocks in the Bailang massif confirm these ages (Zyabrev et al. 1999, 2008).

The Saga and Sangsang ophiolites are located about 200–350 km west of Xigaze. They consist of mantle peridotites overlain by leuco- and melano-gabbros, and a mafic crust consisting of brecciated basalt, lava, diabase and amphibolite overlain by sandstone with chert. Field and geochemical data suggest that they formed in the arc or back-arc of an intra-oceanic supra-subduction zone(Bedard et al. 2009). In the south, the massifs are underlain by an ophioliticmélange containing blocks of strongly foliated garnet- and clinopyroxene-bearing amphibolite, serpentinite, diabase, gabbro and sedimentary rocks. The high-grade metamorphic rocks in the mélange, showing peak P–T conditions of 12 kb/850 °C, ages of 123–128 Ma and bulk composition similar to MORB, E-MORB and back-arc basalt indicate the presence of a metamorphic sole under the ophiolite massifs that formed contemporaneously with the ophiolitic crust (Guilmette et al. 2009, 2012). The blocks of gabbro show LREE depletion and are composed of magmatic plagioclase, diopside-augite clinopyroxene and brown amphiboles such as tschermakite and magnesiohorn blende and minor ilmenite and titanite (Dupuis et al. 2005). Further south, the ophiolitic mélange is underlain by the Yamdrock tectonic mélange comprising imbricate thrust sheets representing slices of the ocean floor including red siliceous shale, radiolarian chert and minor alkaline basalt. Mafic rocks from the Yamdrock mélange and from the underlying Triassic flysch of the Indian passive margin show LREE enrichment, and are similar to the volcanic rocks of the Reunion hotspot and the Deccan basalts of India (Dupuis et al. 2005).

The mantle lherzolites and harzburgites in the Saga massif are relatively fresh with secondary serpentine only in small veins or localised domains, whereas harzburgite in the Sangsang ophiolite is more affected by serpentinisation especially near faults or shear zones. The large olivine and pyroxene grains show minor folds or kinks and undulatory extinction. Olivine inclusions are present in pyroxene. Minor spinel occurs as euhedral grains. Recrystallised, interstitial fine-grained olivine around clinopyroxene, olivine + clinopyroxene symplectite around orthopyroxene, orthopyroxene + spinel symplectite and vermicular spinel have been interpreted as evidence of melt-rock interactions (Bedard et al. 2009 and references therein). Minor chlorite, actinolite, carbonate and prehnite occur as products of secondary alteration and metasomatism. The gabbros are composed of sub-idiomorphic plagioclase altered to albite + prehnite, sub-idiomorphic to allotriomorphic clinopyroxene replaced by metamorphic green amphibole or minor chlorite, rare brown igneous amphibole and rare orthopyroxene. Minor titanite and prehnite veinlets are also present. The uppermost part of the mafic crust contains brecciated lava with clasts of pyroxene altered to actinolite and chlorite, plagioclase and quartz, set in a greenish-brown matrix of recrystallised plagioclase, minor chlorite, quartz, carbonate and prehnite in veins. Mafic to intermediate lavas contain altered clinopyroxene, plagioclase microlites and glomerocrysts (with pyroxene inclusions) altered to albite, prehnite and/or sericite and minor oxides. Amygdules filled with chlorite, quartz and carbonate and carbonate and chlorite veinlets are also observed in the lavas. Diabase contains fresh plagioclase, altered clinopyroxene, orthopyroxene and minor secondary actinolite needles and epidote. Amphibolite containing cloudy plagioclase, metamorphic green hornblende and minor oxides and titanite is highly altered to albite-prehnite-epidote-chlorite assemblages. Sandstone contains clasts of deformed quartz, feldspar, pseudomorphs of pyroxene and hornblende with actinolite and chlorite, spinel, felsic plutonics, mafic to intermediate volcanics, rare radiolarian chert and carbonate oolite and minor chlorite and carbonate.

In the Xigaze area, several dismembered ophiolite complexes occur along the Yarlung-Tsangpo Suture between the Xigaze fore-arc flysh in the north and the Mesozoic Yamdrock mélange and Triassic flysch of the Tethyan series in the south (Nicolas et al. 1981; Girardeau et al. 1985a, b; Girardeau and Mercier 1988). The mantle units consist of well foliated, variably serpentinised, spinel-bearing dunite and harzburgite at the top grading downward to Cr-diopside harzburgite and lherzolite that equilibrated at 1155–1255 °C and 5–11 kb indicating a slow-spreading ridge environment (Girardeau and Mercier 1988). Harzburgites in the western massifs contain porphyroclastic orthopyroxene with clinopyroxene exsolution lamellae that are surrounded by granoblastic olivine showing slip planes and kink bands. Their geochemical and isotopic characteristics indicate that they are residues of 7–12 % melting of an N-MORB-type mantle source in a back-arc setting (Dubois-Côté et al. 2005). On the other hand, peridotites in the eastern massifs consisting of granular orthopyroxene and olivine with interstitial Cr-diopside represent metasomatically altered residues of 30–40 % melting of a depleted mantle source in an intra-oceanic arc environment. Concordant to discordant orthopyroxene bands, websterite veins, and irregular and tabular dunite lenses are common in the peridotites. Rodingitised diabase and gabbro dikes commonly occur in the upper part of the mantle unit. The crustal rocks are primarily made of diabase sheeted dikes and sills, and extrusives. Lower crustal ultramafic–mafic rocks are minor, and volcaniclastic rocks are very rare. Pillow lava is altered and contains a few vesicles filled with quartz, calcite, chlorite and/orepidote. Microphenocrysts of plagioclase and clinopyroxene in basalts display intersertal texture. Rare hornblende phenocrysts are present in a diabase dike. Subordinate glass and opaques are present in the groundmass. Secondary actinolite and clay minerals are commonly observed.

2.2.2.3 The Eastern Part

The eastern part of the Yarlung-Tsangpo Suture Zone includes the Jinlu, Zedang and the Luobusa massifs (Fig. 2.1, 2.2a). The intrusives and volcanics of the Zedang massif have been dated at 161 ± 2.3 Ma by the U–Pb zircon method in dacite, and 152.2 ± 3.3 Ma by $^{40}Ar/^{39}Ar$ step heating of hornblende in andesitic dike (McDermid et al. 2002). U–Pb dating of zircon in diabase yielded ages of 149.7 ± 3.4 Ma, 150.0 ± 5.0 Ma (Chan et al. 2007) and 162.9 ± 2.8 Ma (Zhong et al. 2006) for the Luobusa massif.

The Luobusa Ophiolite is overthrust northward onto the Gangdese Batholith of the Asian margin (Lhasa Terrane) and onto the Tertiary molasse-type deposits of the Luobusa Formation (Zhou et al. 1996). A south-dipping thrust separates the ophiolite complex from a thick Triassic flysch sequence in the south. The mantle sequence consists of coarse-grained harzburgite and diopsidic harzburgite well-developed porphyroclastic texture. Chromite compositions indicate that the harzburgites are residua left after extraction of MORB-type magmas. Rare olivine porphyroclasts contain orthopyroxene inclusions and show deformation lamellae and kink bands. Olivine replaces orthopyroxene as indicated by fine-grained olivine neoblasts in the granoblastic matrix surrounding subhedral to anhedral orthopyroxene porphyroclasts with clinopyroxene lamellae. Accessory chromite is anhedral and ubiquitous. Dunites occur as lenses or as envelopes around podiform chromitites in the peridotites. The upper part of the peridotite sequence, described as a transition zone (Zhou et al. 1996), is sheared and primarily consists of dunite grading into wehrlite. Chromite occurs as euhedral inclusions in olivine in dunite as well as fine-disseminated grains in thin layers. A mélange zone tectonically overlies the transition zone and contains deformed cumulate lenses, pillow lava and mid-Cretaceous marine sedimentary rocks of the Zedong Formation in a serpentinised ultramaficmatrix.

2.2.2.4 The Namche Barwa Syntaxis

Sporadic lenses of metamorphosed mafic–ultramafic rocks of a dismembered mélange suite occur in the Yarlung-Tsangpo Suture Zone of this area (Quanru et al. 2006) (Figs. 2.1, 2.2a). These rocks include diabase and amphibolite of boninite, arc tholeiite and back-arcbasalt affinity indicating origin in a supra-subduction zone environment.

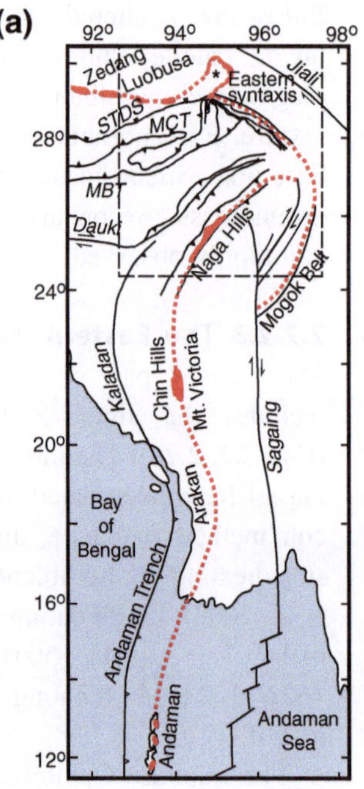

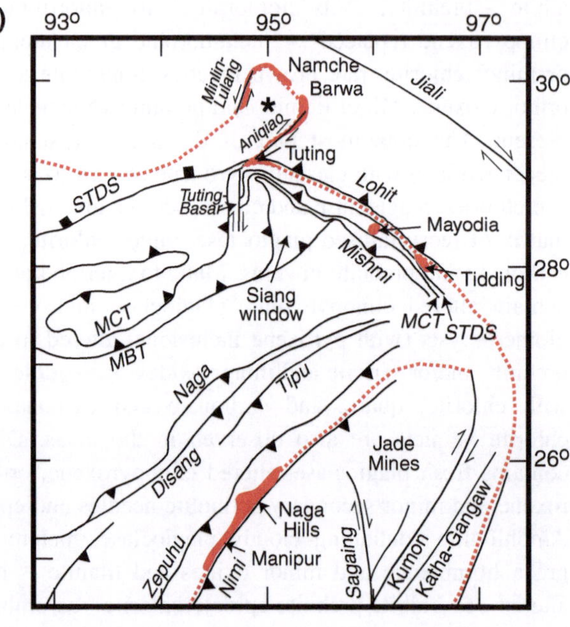

Fig. 2.2 Ophiolite complexes along the eastern margin of the Indian plate: **a** regional setting, **b** detailed view of the area demarcated by *dashed lines* in (**a**). The symbols are as in Figs. 2.1 and 2.2. The names of the faults are italicised. Compiled from Bhattacharjee 1991; Ding et al. 2001; Ghosh et al. 2007; Misra 2009; Mitchell 1993; Mitchell et al. 2007; Saha et al. 2012; Searle et al. 2007; Singh and Singh 2011; Quanru et al. 2006; Yin et al. 2010

2.3 The Eastern Margin of the Indian Plate

At the eastern Himalayan syntaxis, the collision of the Indian block with Asia is accommodated by a pair of roughly NE–SW strike-slip faults along the northwestern (Minlin-Lulang sinistral fault) and southeastern (Aniqiao dextral fault) boundaries of the Namche Barwa massif (Burg et al. 1998; Ding et al. 2001; Quanru et al. 2006). The Yarlung-Tsangpo suture follows these faults, swerves around Namche Barwa and its trend becomes southwestward (Fig. 2.2). Southwest of Namche Barwa, the Aniqiao fault is continuous with the N–S trending Tuting-Basar dextral strike-slip fault (Acharyya and Saha 2008; Saha et al. 2012) that offsets the Main Central Thrust (MCT) and the South Tibet Detachment System (STDS) (Fig. 2.2b). On the eastern side of the Tuting-Basar fault, the MCT and STDS swerve around the Siang window and follow a NW–SE trend. The Siang window is a domal structure defined by a closed trace of the Main Boundary Thrust (MBT) that exposes rocks structurally below the Lesser Himalayan sequence (Acharyya and Saha 2008). Southeast of the Siang window, the MBT is known as the Mishmi thrust that also trends NW–SE. The Trans-Himalayan Lohit batholith, equivalent to the Gangdese batholith of Tibet, is thrust southwestward over a narrow strip of carbonate and argillaceous Tethyan sediments along the NW–SE trending Lohit thrust. Thus all the major faults, e.g. MBT, MCT, STDS and Lohit thrust, are parallel and trend NW–SE in this region. At the intersection of the Aniqiao fault and Lohit thrust, the Yarlung-Tsangpo suture sharply changes direction and becomes roughly parallel to the Lohit thrust (Fig. 2.2b).

The continuation of the Yarlung-Tsangpo suture follows an arcuate path convex to the east until it encounters the Sagaing fault, a major N–S dextral strike slip fault in Myanmar that separates the Burma microplate to the west from the Shan plateau of Asia (Sundaland) to the east (Fig. 2.2a). The northern end of the Sagaing fault splays into several branches, one of which may be continuous with the Mishmi thrust (Vigny et al. 2003). High-grade kyanite and garnet-bearing schists occur in the Katha-Gangaw and Kumon belts parallel to and on the concave side of the suture near the intersection of the Sagaing fault (Fig. 2.2b). An eclogite boulder probably originating from the Kumon range yielded P–T conditions of 12–13 kb and 530–615 °C (Enami et al. 2012). Similar P–T conditions of >14 kb and 550–600 °C were also determined from the high pressure rocks of the Jade Mines belt at the northwestern branch of the Sagiang fault (Goffé et al. 2000). U–Pb zircon dates of 146.5 ± 3.4 Ma (Shi et al. 2008) and 158 ± 2 Ma (Qiu et al. 2009) in jadeitites from the Jade Mines belt indicate subduction in Late Jurassic perhaps beneath the Sundaland margin (Searle and Morley 2011).

2.3.1 Tuting Metavolcanics

The Tuting metavolcanics are structurally continuous with the Jiaresa-Pangxin and Yigongbai ophiolitic sheets at the eastern margin of the Namche Barwa massif (Quanru et al. 2006; Saha et al. 2012). They occur at the intersection of the Lohit thrust and the Anaqiao fault. They are folded, sheared and banded, and are composed of fine-grained greenschist facies rocks with an albite + epidote + chlorite +

sericite ± titanite ± magnetite assemblage (Saha et al. 2012). The primary mineralogy and texture are obliterated, and the porphyroblasts are composed of albite enclosing chlorite, epidote and sericite. Slender actinolite is embedded in chlorite. Late stage quartz-calcite veins are common. Plagiogranite intrusives occur locally within the metavolcanics. The Lohit granodiorite intrudes the Tuting metavolcanics and contains xenoliths of mafic rocks. Geochemistry of the Tuting metavolcanics is consistent with an origin in a supra-subduction zone environment (Saha et al. 2012).

2.3.2 Tidding Serpentinite

In the Dibang and Lohit valleys of northeastern Arunachal Pradesh and southeast of Tuting, the Tethyan sequence is replaced by the Tidding formation comprising imbricate thrust sheets of ophiolitic affinity (Acharyya 1986). The Tuting-Tidding shear zone shows strike-slip movement along the NW–SE direction and represents the Indus Suture Zone in this region (Valdiya, 2010). Extending further southeast into western Myanmar, the Tuting-Tidding shear zone is offset by the transcurrent, right-lateral, strike-slip Sagaing fault. The Myanmar block has moved 300–500 km northward with respect to Southeast Asia along the Sagaing fault (Swe 1972; Hla Maung 1987).

The Tidding formation consists of folded garnetiferous metavolcanics, actinolite-tremolite schist, chlorite schist and marble with slivers of graphite-garnet schist, epidote and leucogranite intrusives (Misra 2009; Singh and Singh 2011). A tabular body of serpentinised peridotite with relict olivine and Cr-spinel occurs in the upper part of the formation. Compositions of Cr-spinels suggest that the serpentinised peridotite is possibly a residue of a depleted mantle source that underwent a high degree of partial melting in a marginal ocean basin (Singh and Singh 2011).

2.3.3 Mayodia Ophiolite

A probable klippen of ophiolitic rocks occurs above the High Himalayan Crystallines at Mayodia, Dibang Valley, northwest of Tidding (Ghosh and Ray 2003; Ghosh et al. 2007). These rocks may be correlated with the Tidding ophiolitic rocks. The Mayodia ophiolitic sequence comprises two units. The lower unit consists of foliated and deformed serpentinised peridotite (wehrlite with Fo_{89-92} olivine) tectonite intruded by hornblendite dikes and overlain by banded amphibolite. The upper unit consists of metabasalt interlayered with metapelite. The metabasalt is folded and comprises actinolite-chlorite-albite-epidote schist indicating greenschist facies metamorphism. The chemical composition of the metabasalt shows similarities with MORB (Ghosh et al. 2007).

2.3.4 Indo-Myanmar Range

The Sagaing fault offsets the continuation of the Yarlung-Tsangpo suture such that the latter is displaced northward on the western side of the fault (Fig. 2.2). Further south, the suture enters the Indo-Myanmar Range (IMR) that consists of predominantly Paleogene flyschoid sediments with Triassic (Carnian) feldspathic turbidites, ophiolite and Triassic-Cretaceous metamorphic rocks including a 30-km-wide mica schist belt (Brunnschweiler 1966; Bender 1983; Mitchell 1993). The suture here represents collision between the Indian plate and the Burma microplate, a westward convex, 1,300 km long plate including continental crust and is demarcated by an ophiolite belt traceable through Nagaland and Manipur including the Naga Hills, the Chin Hills (Mt. Victoria), the Arakan Hills and the Arakan coast of western Myanmar. The Naga/Disang thrust is equivalent to the Main Boundary Thrust (MBT) of the Himalayas and marks the western border of the Paleogene Indo-Myanmar Range. A closely spaced thrust fault to the Disang thrust known as the Tipu thrust is *en echelon* with the Kaladan fault that continues southward and connects with the arcuate Andaman–Sumatra–Java subduction zone (Yin et al. 2010). The India-Burma plate boundary has jumped westward several times during the Cenozoic, and the Kaladan fault marks the present-day plate boundary (Maurin and Rangin 2009). At present, the Indian plate is obliquely subducting beneath the Burma-microplate at the Andaman subduction zone with a velocity of ~ 4 cm.a^{-1} toward N20°E (Paul et al. 2001; Vigny et al. 2005). The geology of IMR is discussed in detail in Chap. 5.

2.3.5 Andaman Ophiolite

The 850-km-long chain of islands between the Bay of Bengal and the Andaman Sea constitute the Andaman Island Arc. It comprises two nearly parallel arcuate belts. The Andaman Ophiolite is exposed intermittently on the Andaman-Nicobar Islands, a forearc ridge that constitutes a part of the accretionary complex in the outer part of the Andaman Arc referred to as the western arc (Karunakaran et al. 1964; Bandyopadhyay et al. 1973; Halder 1985; Prasad 1985; Vohra et al. 1989; Bandopadhyay 2005). This belt is the southern extension of Naga Hills and Arakan orogenic belt, which swerves southeastwards and continues through the subduction zone offshore of Sumatra. Currently, active arc volcanoes on Barren Island (Ball 1973; 1888;

Haldar and Luhr 2003; Luhr and Haldar 2006) and Narcondam Island (Pal et al. 2007) are located east of the Andaman Islands along the eastern arc that continues southeastward through the Java–Sumatra volcanic chain, and northward into Myanmar where several extinct volcanoes such as Mt Popa and Mt Taungthonlon (Miocene-Quaternary) are found. A back-arc basin exists further east of the Andaman Arc, where the sea floor is spreading from several ENE–WSW ridge segments offset by several N–S right-lateral strike-slipfaults since c. 4 Ma (Raju et al. 2004; Curray 2005). The N–S strike-slip faults are part of an extensive horsetail structure at the southern end of the Sagaing fault of Myanmar (Maurin and Rangin 2009). A 95 ± 2 Ma age of trondhjemite (Pedersen et al. 2010) and 93.6 ± 1.3 Ma age of plagiogranite (Sarma et al. 2010), late-stage derivatives of the Andaman ophiolitic gabbro, constrain the age of formation of the oceanic lithosphere emplaced in the Andaman Ophiolite. The age of obduction of the ophiolite is constrained by the 83.5–70.6 Ma age of radiolarian chert above the ophiolite (Ling et al.1996). On the basis of similarity in age of the 95 ± 2 Ma Andaman Ophiolite, 93.5–97.9 Ma Samail Ophiolite in Oman (Tilton et al. 1981; Warren et al. 2005) and 91.6 ± 1.4 Ma Troodos Ophiolite in Cyprus (Mukasa and Ludden 1987), Pedersen et al. (2010) concluded that these ophiolites formed simultaneously in the fore-arcs of several coexisting trans-Tethyan subduction zones.

The Andaman Ophiolite Group comprises subhorizontal thrust slices that form the basement on which Tertiary accretionary prism sediments were deposited (Sengupta et al. 1990). The ophiolite group is overlain by the Mithakhari Group containing polymictic conglomerate, sandstone, arkose, shale, foraminiferal limestone and pyroclasts in the upper part (Bandopadhyay 2011).

The volcanic rocks of the Andaman Ophiolite belong to the tholeiitic series. Trace element and REE content of the hyaloclastites and pillow basalts are similar to MORB and indicate a back-arc environment (Jafri et al. 2010). Chromite compositions, oxygen fugacity values and trace element compositions of the mantle peridotite units indicate formation of the ophiolite at a mid-oceanic ridge or a supra-subduction zone spreading centre (Pal 2011). However, the upper crust in the southern part also contains andesite and dacite, which suggests that arc-related crust is built on top of the MORB-type oceanic crust (Pedersen et al. 2010). The mantle tectonites are serpentinised and crudely foliated. The orthopyroxene porphyroclasts (En_{87-88}) with olivine (Fo_{89-94}) and chromite inclusions are resorbed and surrounded by olivine neoblasts (Pal 2011). Large diopsidic clinopyroxene grains are embayed and surrounded by orthopyroxene. The Cr# (100 Cr/(Cr + Al)) of chromite ranges from 20–39 in peridotite to 73–80 in chromitite pods (Ghosh et al. 2009). The composition of the chromites is consistent with their origin in peridotite in a supra-subduction zone. The chromitite pods also contain chromian andradite garnet with up to 50 mol % uvarovite (Ghosh and Morishita 2011). The layered ultramafic–mafic cumulates contain sieve-textured olivine and resorbed chromite. Unaltered olivine (Fo_{93-90}) and euhedral chromite are also present in some samples (Pal 2011). Lherzolite cumulates contain resorbed orthopyroxene. Less altered diopside-augite clinopyroxene contains olivine, orthopyroxene and chromite (Cr# 71–73) inclusions. Cumulate gabbro contains olivine (Fo_{87-78}) and orthopyroxene (En_{78}) in some samples. P–T estimates based on clinopyroxene thermobarometry yield a wide range of temperatures (500–1000 °C) at 7–8.6 kb pressure for the cumulate pyroxenite and gabbros (Saha et al. 2010).

Alteration is common in the cumulate gabbros with plagioclase showing saussuritisation and pyroxene showing uralitisation and chloritisation. Magnetite-ulvospinel occurs as an accessory phase. Clinopyroxene is altered to amphibole and plagioclase is saussuritised also in the intrusive gabbro and dolerite. The intrusive gabbro grades into quartz-bearing plagiogranite with An_{93} plagioclase, minor hornblende altered to chlorite and accessory magnetite andilmenite. The basaltic volcanics contain phenocrysts of augite and olivine near the base, and augite and plagioclase near the top. The groundmass is subophitic, intersertal or vitrophyric and consists of albite, epidote, chlorite, altered glass and magnetite. The groundmass of the upper lavas contains glass droplets and intricate veins of carbonate and silica and zeolite. The red charts consist of microcrystalline quartz, clay and iron oxides. The pelagic sedimentary units consist of micritic foraminiferal limestone and carbonate-bearing red clay.

2.4 Summary of Petrographic Features

2.4.1 Waziristan

Ultramafic rocks:

- Dunite layers and lenses in harzburgite contain strained olivine.
- Harzburgite contains olivine (Fo_{91}),orthopyroxene (En_{91}, $Al_2O_3$1.6 wt %) and minor plagioclase.
- Pyroxenite dikes are composed of orthopyroxene ± diopside ± olivine.
- Anorthosite contains zoned plagioclase (An_{77-73}) with higher K_2O in the core, and minor clinopyroxene (Mg# 56).
- Podiform chromitite commonly contains zoned alumino-chromite (Cr_2O_3 49–61 wt %) with ferri-chromite rim, and minor amounts of chlorite, serpentine and magnetite (Jan et al. 1985).

2.4.2 Sapat

Meta-harzburgite and dunite with fine-grained, massive texture:

- Relict orthopyroxene and lobate olivine with euhedral spinel inclusions, sub-grain boundaries and undulose extinction; matrix has fine-grained, undeformed olivine neoblasts, disseminated amoeboid spinel and orthopyroxene break down products like olivine, talc and tremolite: fluid-assisted lower amphibolite facies re-equilibration (Bouilhol et al. 2009).
- Trails/pods of chromite spinel and clinopyroxene in dunite, micro-veins and thin dikes of olivine-clinopyroxene: originated through melt percolation.
- Secondary chlorite around spinel, serpentine and rare diopside around olivine and calcite vein in dunite: late stage alteration.

Meta-gabbro and tonalite–trondhjemite with clinopyroxenite intrusives:

- Clinopyroxene overgrows olivine in clinopyroxenite, amoeboid clinopyroxene at grain boundaries of olivine porphyroblasts; trails of lobate clinopyroxene porphyroblasts; secondary chlorite and hornblende: melt-rock interactions.

2.4.3 Spontang

Peridotite with porphyroclastic texture:

- Olivine and orthopyroxene have lobate margins: evidence of mantle melting.
- Exsolution lamellae of diopside-augite in orthopyroxene, and of orthopyroxene in clinopyroxene: sub-solidus equilibration.
- Undulatory extinction and kinks in olivine: deformation during ascent of mantle below spreading center at 1000–1200 °C and low deviatoric stress (Reuber 1986).

Diorite dikes and sills with ophitic to cumulate texture cross-cutting mantle peridotite:

- Hornblende/edenite and albite, minor epidote, oxides, titanite and apatite; hornblende is locally replaced by actinolite: amphibolite to greenschist facies metamorphism.
- Anorthite-rich plagioclase core and diopside inclusions in hornbende/edenite: magmatic relics.

Upper crustal gabbro:

- Plagioclase laths, though commonly altered, are well-preserved: probably crystallised before clinopyroxene in a normal mid-oceanic ridge environment (Corfield et al. 2001).

2.4.4 Nidar

Crustal micro-gabbro with ophitic texture and cumulate gabbro:

- Plagioclase contains olivine inclusions, clinopyroxene contains olivine and plagioclase inclusions and pargasitic hornblende rims clinopyroxene: olivine—plagioclase—clinopyroxene MORB-type crystallisation sequence.
- Large lobate olivine grains and interstitial neoblasts of euhedral olivine and some clinopyroxene: evidence of magmatic reactions.
- An-rich plagioclase core with kink and bent cleavage is rimmed by undeformed plagioclase; lower-Ca clinopyroxene is rimmed by higher-Ca undeformed clinopyroxene, pargasite rim around diopside: deformation followed by amphibolite facies metamorphism.
- Actinolitic amphibole in veins: late stage alteration.

Crustal basalts:

- Phenocrysts of albitic plagioclase laths, diopside-augite, serpentinised olivine, minor Fe–Ti oxides: hydrothermal alteration.
- Groundmass of augite, plagioclase, rare glass, epidote with radiating crystals of chlorite, banded rhyolite with glass and palagonite, feldspar spherules, fibrous chlorite, veins/vesicles with quartz and calcite: hydrothermal alteration.

2.4.5 Yungbwa

Harzburgite with porphyroclastic texture:

- Elongated olivine with kink-bands, orthopyroxene with exsolution lamellae and deformed cleavage showing undulatory extinction and sub-parallel to foliation: mantle deformation.
- Large spinel grains have "holly-leaf" shape.
- Pyroxene inclusions in plagioclase.
- Euhedral neoblasts of olivine, orthopyroxene and clinopyroxene, rounded spinel and disseminated sulphides.
- Minor anorthitic plagioclase and pargasitic amphibole: metasomatic alteration mantle.

Basalt and gabbronorite dikes with sub-ophitic orhypidiomorphic-granular texture:

- Sericitisation of plagioclase and replacement of clinopyroxene and hornblende byactinolite: low-grade metamorphism.
- Recrystallisation and granulation, plagioclase with undulatoryextinction and partial transformation to prehnite-bearing saussurite, clinopyroxene partially replaced by metamorphic hornblende, tremolite or actinolite with a typical flaser texture: low-grade metamorphism.

2.4.6 Xiugugabu

Harzburgite with porphyroclastic to porphyromylonitic texture:

- Kinked and highly stretched olivine and orthopyroxene porphyroclasts with undulatory extinction: mantle deformation.
- Orthopyroxene containing clinopyroxene lamellae surrounded by neoblasts: recrystallisation.
- Minor large euhedral spinel and small vermicular intergrowths, and accessory magnetite, chlorite and amphibole: probably alteration.

Microgabbro and microgabbronoritesills intruding harzburgite:

- Plagioclase crystallised before pyroxene: MORB-type crystallisation sequence.
- Foliation defined by parallel orientation of plagioclase and pyroxene microphenocrysts; magnetite + clinopyroxene symplectite, amphibole + clinopyroxene symplectite and amphibole overgrowth around clinopyroxene: amphibolite facies hydrothermal metamorphism.
- Epidote, chlorite and sericite: secondary alteration.

2.4.7 Zhongba

Harzburgite with minor dunite:

- Fresh and foliated.

Diabase dike with granular- to medium-grained intersertal texture:

- Idiomorphic plagioclase, chloritised pyroxene and olivine and minor Fe–Ti oxides. Texture showing clinopyroxene cores with radiating plagioclase laths is present in some samples.

Porphyritic pillow and massive olivine basalts with intersertal to intergranular texture:

- Phenocrysts are partially replaced by albite or chlorite.
- Secondary veins and amygdules of carbonate, quartz and chlorite.

2.4.8 Saga and Sangsang

Lherzolite and harzburgite:

- Serpentinised near faults or shear zones.
- Large olivine and pyroxene grains show minor folds or kinks and undulatory extinction: mantle deformation.
- Olivine inclusions in pyroxene, minor euhedral spinel.
- Recrystallised, interstitial olivine around clinopyroxene, olivine + clinopyroxene symplectite around orthopyroxene, orthopyroxene + spinel symplectite and vermicular spinel: melt-rock interactions.
- Minor chlorite, actinolite, carbonate and prehnite: secondary alteration and metasomatism.

Gabbro:

- Plagioclase altered to albite + prehnite, clinopyroxene replaced by metamorphic green amphibole or minor chlorite: low-temperature metamorphism.

Mafic to intermediate lava:

- Clinopyroxene altered to actinolite and chlorite, plagioclase microlites and glomerocrysts with pyroxene inclusions altered to albite, prehnite and/or sericite: low-temperature alteration.
- Amygdules filled with chlorite, quartz and carbonate, veins of carbonate and chlorite: secondary alteration.
- Uppermost part brecciated.

Diabase:

- Minor secondary actinolite needles and epidote.

Amphibolite:

- Cloudy plagioclase, metamorphic green hornblende and minor oxides and titanite are highly altered to albite-prehnite-epidote-chlorite assemblages.

Sandstone:

- Clasts of deformed quartz, feldspar, pseudomorphs of pyroxene and hornblende with actinolite and chlorite, spinel, felsic plutonics, mafic to intermediate volcanics, rare radiolarian chert and carbonate oolite and minor chlorite and carbonate: low temperature alteration.

2.4.9 Xigaze

Foliated and serpentinised, spinel and clinopyroxene bearing dunite, harzburgite and lherzolite:

- Porphyroclastic orthopyroxene with clinopyroxene lamellae surrounded by granoblastic olivine showing slip planes and kink bands, or granular orthopyroxene and olivine with interstitial Cr-diopside: mantle deformation.

Pillow basalt with intersertal texture:

- Altered and contains vesicles with quartz, calcite, chlorite and/orepidote.

Diabase:

- Rare hornblende phenocrysts: hydrous magma.
- Subordinate glass and opaques in the groundmass, secondary actinolite and clay minerals.

2.4.10 Luobusa

Diopsidic harzburgite with porphyroclastic texture:

- Rare olivine porphyroclasts with orthopyroxene inclusions show deformation lamellae and kink bands: mantle deformation.

- Fine-grained olivine neoblasts in the granoblastic matrix surrounding subhedral to anhedral orthopyroxene porphyroclasts with clinopyroxene lamellae: olivine replaces orthopyroxene.
- Accessory chromite occurs as euhedral inclusions in olivine and anhedral, ubiquitous grains in the matrix.

2.4.11 Andaman

Foliated, serpentinised mantle tectonite:

- Resorbed orthopyroxene porphyroclasts with olivine and chromite inclusions surrounded by olivine neoblasts; embayed diopside grains surrounded by orthopyroxene: mantle melting reactions.

Layered ultramafic–mafic cumulates:

- Sieve-textured olivine and resorbed chromite; some unaltered olivine and euhedral chromite.
- Lherzolite cumulates contain resorbed orthopyroxene. less altered diopside-augite with olivine, orthopyroxene and chromite inclusions.
- Cumulate gabbro contains olivine and orthopyroxene in some samples, saussuritised plagioclase and uralitised and chloritised pyroxene and accessory magnetite-ulvospinel.

Intrusive gabbro and dolerite:

- Clinopyroxene altered to amphibole and plagioclase is saussuritised, minor hornblende altered to chlorite and accessory magnetite andilmenite.

Basaltic volcanics with subophitic, intersertal or vitrophyric groundmass:

- Phenocrysts of augite and olivine near the base, and augite and plagioclase near the top.
- Groundmass consists of albite, epidote, chlorite, altered glass and magnetite, with glass droplets and intricate veins of carbonate and silica and zeolite in the upper lavas.

Sediments:

- Chert contains microcrystalline quartz, clay and iron oxides.
- The pelagic sediments consist of micritic foraminiferal limestone and carbonate-bearing red clay.

References

Acharyya SK, (1986) Cenozoic plate motions creating the EasternHimalaya and Indo-Burmese range around the northeast corner ofIndia. In:Ghose, N.C.and Varadarajan, S., (eds.), “Ophiolites and Indian Plate Margins”, Patna University, Patna, 143–160

Acharyya SK, Saha P (2008) Geological setting of the Siang Dome located at the Eastern Himalayan Syntaxis. Extended abstracts: 23rd Himalayan-Karakoram-Tibet workshop, India. Himal J Sci 5:16–17

Ahmad T, Tanaka T, Sachan HK, Asahara Y, Islam R, Khanna PP (2008) Geochemical and isotopic constraints on the age and origin of the Nidar Ophiolitic Complex, Ladakh, India: implications for the Neo-Tethyan subduction along the Indus suture zone. Tectonophysics 451:206–224

Ahmad Z, Abbas SG (1979) The Muslim Bagh ophiolites. In: Farah A, Delong KA (eds) Geodynamics of Pakistan. Geological Survey of Pakistan, Quetta, 243–249

Ahmed Z (1993) Leucocratic rocks from the Bela ophiolite, Khuzadrdistrict, Pakistan. In: Searle MP, Treloar PJ (eds.) Himalayan Tectonics. Geological Society of London Special Publication 74, pp 89–100

Aitchison JC, Davis AM (2004) Evidence for the multiphase nature of the India–Asia collision from the Yarlung Tsangpo suture zone, Tibet. In: Malpas JG, Fletcher CJN, Ali JR, Aitchison JC (eds) Aspects of the tectonic evolution of China. Geological Society of London Special Publication 226: 217–233

Aitchison JC, Davis AM, Abrajevitch AV, Ali JR, Badengzhu Liu J, Luo H, McDermid IRC, Ziabrev SV (2003) Stratigraphic and sedimentological constraints on the age and tectonic evolution of the Neotethyan ophiolites along the Yarlung Tsangpo suture zone, Tibet. In: Dilek Y, Robinson PT (eds) Ophiolites in earth history. Geological Society, London, Special Publications, 218: 147–164

Allègre CJ, Courtillot V, Tapponnier P, Hirn A, Mattauer M, Coulon C et al (1984) Structure and evolution of the Himalaya–Tibet orogenic belt. Nature 307:17–22

Alleman F (1979) Time of emplacement of the Zhob valley ophiolites and Bela ophiolites, Balochistan (preliminary report). In: Farah A, DeJong KA (eds) Geodynamics of Pakistan. Geological Survey of Pakistan, Quetta, 215–242

Arif M, Jan Q (2006) Petrogenetic significance of the chemistry of chromite in the ultramafic–mafic complexes. J Asian Earth Sci 27:628–646

Asrarullah AZ, Abbas SG (1979) Ophiolites in Pakistan: introduction. In: Farah A, Dejong KA (eds) Geodynamics of Pakistan. Geological Survey of Pakistan, Quetta, 181–192

Ball V (1888) The volcanoes of Barren Island and Narcondam in the Bay of Bengal. Geol Mag 9:404–408

Bandopadhyay PC (2005) Discovery of abundant pyroclasts in the Namunagarh grit, South Andaman: evidence for arc volcanism and active subduction during the Palaeogene in the Andaman area. J Asian Earth Sci 25:95–107

Bandopadhyay PC (2011) Re-interpretation of the age and environment of deposition of paleogene turbidites in the Andaman and Nicobar Islands, Western Sunda Arc. J Asian Earth Sci. doi: 10.1016/j.jseaes.2011.08.018

Bandyopadhyay S, Subramanyam MR, Sharma PN (1973) The geology and mineral resources of the Andaman and Nicobar Islands. Rec Geol Surv India 105:25–68

Baxter AT, Aitchison JC, Ali JR, Zyabrev SV (2010) Early Cretaceous radiolarians from the Spongtang massif, Ladakh, NW India: implications for Neo-Tethyan evolution. J Geol Soc Lond 167:511–517

Bédard É, Hébert R, Guilmette C, Lesage G, Wang CS, Dostal J (2009) Petrology and geochemistry of the Saga and Sangsang ophiolitic massifs, Yarlung-Zangbo Suture Zone, Southern Tibet: evidence for an arc–back-arc origin. Lithos 113:48–67

Bender F (1983) Geology of Burma. Gebruder Borntraeger, Berlin 293p

Berthelsen A (1953) On the geology of the Rupshu District, N.W. Himalaya: Meddelelser fra Dansk Geologisk Forening 12:350–415

Bezard R, Hébert R, Wang C, Dostal J, Dai J, Zhong H (2011) Petrology and geochemistry of the Xiugugabu ophiolitic massif, western Yarlung Zangbo suture zone, Tibet. Lithos 125:347–367

Bhattacharjee CG (1991) The ophiolites of northeast India—a subduction zone ophiolite complex of the Indo-Burman orogenic belt. Tectonophysics 191:213–222

Bouilhol P, Burg JP, Bodinier JL, Scmidt MW, Dawood H, Hussain S (2009) Magma and fluid percolation in arc to forearc mantle: evidence from Sapat (Kohistan, Northern Pakistan). Lithos 107:17–37

Brunnschweiler RO (1966) On the geology of the Indoburman ranges. J Geol Soc Aust 11:117–194

Burg JP, Nievergelt P, Oberli F, Seward D, Davy P, Maurin JC, Diao ZZ, Meier M (1998) The Namche Barwa syntaxis: evidence for exhumation related to compressional crustal folding. J Asian Earth Sci 16:239–252

Chan GH-N, Crowley Q, Searle M, Aitchison JC, Horstwood M (2007) U–Pb zircon ages of the Yarlung Zangbo suture zone ophiolites, south Tibet. In: 22th Himalaya–Karakoram–Tibet workshop, Hong Kong, China, workshop abstract, vol 12

Constantin M (1999) Gabbroic intrusions and magmatic metasomatismin harzburgites from the Garrett transform fault: implications for the nature of the mantle—crust transition at fast spreading ridges. Contrib Miner Petrol 136:111–130

Corfield RI, Searle MP, Pedersen RB (2001) Tectonic setting and obduction history of the Spontang ophiolite, Ladakh Himalaya, NW India. J Geol 109:715–736

Corfield RI, Watts AB, Searle MP (2005) Subsidence of the North Indian Continental Margin, Zanskar Himalaya, NW India. J Geol Soc Lond 162:135–146

Curray JR (2005) Tectonics and history of the Andaman Sea region. J Asian Earth Sci 25:187–232

Dai J-G, Wang C-S, Hébert R, Santosh M, Li Y-L, Xu J-Y (2011) Petrology and geochemistry of peridotites in the Zhongba ophiolite, Yarlung Zangbo Suture Zone: Implications for the Early Cretaceous intra-oceanic subduction zone within the Neo-Tethys. Chem Geol 288:133–148

Dai J, Wang C, Li Y (2012) Relicts of the Early Cretaceous seamounts in the central-western Yarlung Zangbo Suture Zone, southern Tibet. J Asian Earth Sci 53:25–37

de Sigoyer J, Guillot S, Dick P (2004) Exhumation of the ultrahigh-pressure Tso Morari unit in eastern Ladakh (NW Himalaya): a case study: Tectonics 23:TC3003

Dietrich VJ, Frank W, Honegger K (1983) A Jurassic–Cretaceousisland arc in the Ladakh–Himalaya. J Volcanol Geoth Res 18:405–433

Ding L, Zhong DL, Yin A, Kapp P, Harrison TM (2001) Cenozoic structural and metamorphic evolutionof the eastern Himalayan syntaxis (Namche Barwa). Earth Planet Sci Lett 192:423–438

Dubois-Côté V, Hébert R, Dupuis C, Wang CS, Li YL, Dostal J (2005) Petrological and geochemical evidence for the origin of the Yarlung Zangbo ophiolites, southern Tibet. Chem Geol 214:265–286

Dupuis C, Hébert R, Dubois-Côté V, Wang CS, Li YL, Li ZJ (2005) Petrology and geochemistry of mafic rocks from mélange and flysch units adjacent to the Yarlung Zangbo Suture Zone, southern Tibet. Chem Geol 214:287–308

Enami M, Ko ZW, Win A, Tsuboi M (2012) Eclogite from the Kumon range, Myanmar: petrology and tectonic implications. Gondwana Res 21:548–558

Gansser A (1964) Geology of the Himalayas. Wiley Interscience, London 289 p

Gansser A (1980) The significance of the Himalayan suture zone. Tectonophysics 62:37–52

Garzanti E, Sciunnach D, Gaetani M, Corfield RI, Watts AB, Searle MP (2005) Discussion on subsidence history of the north Indian continental margin, Zanskar–Ladakh Himalaya, NW India. J Geol Soc Lond 162:889–892

Ghosh B, Morishita T (2011) Andradite-uvarovite solid solution from hydrothermally altered podiform chromitite, Rutland ophiolite, Andaman, India. Can Mineral 49:573–580

Ghosh B, Ray J (2003) Mineral chemistry of ophiolitic rocks of Mayodia–Hunli area of Dibang valley district, Arunachal Pradesh, Northeastern India. Mem Geol Soc India 52:447–471

Ghosh B, Mahoney J, Ray J (2007) Mayodia Ophiolites of Arunachal Pradesh, Northeastern Himalaya. J Geol Soc India 70:595–604

Ghosh B, Pal T, Bhattacharya A, Das D (2009) Petrogenetic implications of ophiolitic chromite from Rutland Island, Andaman—a boninitic parentage in suprasubduction setting. Mineral Petrol 96:59–70

Girardeau J, Mercier J-CC (1988) Petrology and texture of the ultramafic rocks of the Xigaze ophiolite (Tibet): constraints for mantle structure beneath slow-spreading ridges. Tectonophysics 147:33–58

Girardeau J, Mercier J-CC, Xibin W (1985a) Petrology of the mafic rocks of the Xigaze ophiolite, Tibet: implications for the genesis of the oceanic lithosphere. Contribrib Mineral Petrol 90:309–321

Girardeau J, Mercier J-CC, Yougong Z (1985b) Origin of the Xigaze ophiolite, Yarlung Zangbo suture zone, southern Tibet. Tectonophysics 119:407–433

Gnos E, Immenhauser A, Peter T (1997) Late Cretaceous/early Tertiary convergence between the Indian and Arabian plates recorded in ophiolites and related sediments. Tectonophysics 271:1–19

Gnos E, Khan M, Mahmood K, Villa IM, Khan AS (1998) Bela oceanic lithosphere assemblage and its relation to the Réunion hotspot. Terra Nova 10:90–95

Göpel C, Allègre CJ, Xu R-H (1984) Lead isotopic study of the Xigaze ophiolite (Tibet): the problem of the relationship between magmatites (gabbros, dolerites, lavas) and tectonites (harzburgites). Earth Planet Sci Lett 69:301–310

Guilmette C, Hébert R, Dupuis C, Wang CS, Li ZJ (2008) Metamorphic history and geodynamic significance of high-grade metabasites from the ophiolitic melange beneath the Yarlung Zangbo ophiolites, Xigaze area, Tibet. J Asian Earth Sci 32:423–437

Guilmette C, Hébert R, Wang C, Villeneuve M (2009) Geochemistry and geochronology of the metamorphic sole underlying the Xigaze Ophiolite, Yarlung Zangbo Suture Zone, South Tibet. Lithos 112:149–162

Guilmette C, Hébert R, Dostal J, Indares A, Ullrich T, Bédard E, Wang C (2012) Discovery of a dismembered metamorphic sole in the Saga ophiolitic mélange, South Tibet: assessing an Early Cretaceous disruption of the Neo-Tethyan supra-subduction zone and consequences on basin closing. Gondwana Res 22(2):398–414

Halder D (1985) Some aspect of the Andaman ophiolite complex. Geol Surv India 115Pt(2):1–11

Halder D, Luhr JF (2003) The Barren Island volcanism during 1991 and 1994–95: Eruption style and lava petrology. Mem Geol Surv India 52:313–338

Hébert R, Bezard R, Guilmette C, Dostal J, Wang CS, Liu ZF (2012) The Indus–Yarlung Zangbo ophiolites from Nanga Parbat to Namche Barwa syntaxes, southern Tibet: First synthesis of petrology, geochemistry, and geochronology with incidences on geodynamic reconstructions of Neo-Tethys. Gondwana Res 22(2):377–397

Hodges KV (2000) Tectonics of the Himalaya and southern Tibet from two perspectives. Geol Soc Am Bull 112:324–350

Honegger K, Dietrich V, Frank W, Gansser A, Thoni M, Trommsdorf V (1982) Magmatic and metamorphism in the Ladakh Himalayas (the Indus–Tsangpo suturezone). Earth Planet Sci Lett 60:253–292

Honegger K, Le Fort P, Mascle G, Zimmermann JL (1989) The blueschists along the Indus Suture Zone in Ladakh, NW Himalaya. J Metamorph Geol 7:57–72

Jafri SH, Sarma DS, Sheikh JM (2010) Hyaloclastites in pillow basalts, South Andaman Island, Bay of Bengal, India. Curr Sci 99(12):1825–1829

Jan MQ, Windley BF, Khan A (1985) The Waziristan ophiolite, Pakistan: general geology and chemistry of chromite and associated phases. Econ Geol 80:294–306

Karunakaran C, Pawde MB, Raina VK, Ray KK (1964) Proceeding 22nd International Geological Congress, New Delhi, vol 11, 79–100

Li JF, Xia B, Liu LW, Xu LF, He GS, Wang H, Zhang YQ, Yang ZQ (2009) SHRIMP U–Pb dating for the gabbro in Qunrang Ophiolite, Tibet: the geochronology constraint for the development of eastern Tethys basin. Geotectonica et Metallogenia 33:294–298

Ling HY, Chandra R, Karkare SG (1996) Tectonic significance of Eocene and Cretaceous radiolaria from South Andaman Island, northeast Indian Ocean. Island Arc 5:166–179

Liu C-Z, Wu F-Y, Wilde SA, Yu L-J, Li JL (2010) Anorthitic plagioclase and pargasitic amphibole in mantle peridotites from the Yungbwa ophiolite (southwestern Tibetan Plateau) formed by hydrous melt metasomatism. Lithos 114:413–422

Luhr JF, Haldar D (2006) Barren Island Volcano (NE Indian Ocean): Island arc high-alumina basalts produced by troctolite contamination. J Volcanol Geotherm Res 149:177–212

Mahéo G, Bertrand H, Guillot S, Villa IM, Keller F, Capiez P (2004) The South Ladakh ophiolites (NW Himalaya, India): an intra-oceanic tholeiitic arc origin with implication for the closure of the Neo-Tethys. Chem Geol 203:273–303

Mahmood K, Boudier F, Gnos E, Monie P, Nicolas A (1995) 40Ar/39Ar dating of the emplacement of the Muslim Bagh ophiolite, Pakistan. Tectonophysics 250:169–181

Malpas J, Zhou MF, Robinson PT, Reynolds PH (2003) Geochemical and geochronological constraints on the origin and emplacement of the Yarlung Zangbo ophiolites, Southern Tibet. In: Dilek Y, Robinson PT (eds) Ophiolites in earth history. Geological Society of London Special Publication 218, pp 191–206

Maung Hla (1987) Transcurrent movements in the Burma–Andaman Sea region. Geology 15:911–912

Maurin T, Rangin C (2009) Structure and kinematics of the Indo-Burmese Wedge: recent and fast growth of the outer wedge. Tectonics 28:TC2010

McDermid IRC, Aitchison JC, Davis AM, Harrison TM, Grove M (2002) The Zedong Terrane: a Late Jurassic intra-oceanic magmatic arc within the Yarlung Zangbo suture zone, southeastern Tibet. Chem Geol 187:267–277

Miller C, Thöni M, Frank W, Schuster R, Melcher F, Meisel T, Zanetti A (2003) Geochemistry and tectonomagmatic affinity of the Yungbwa ophiolite, SW Tibet. Lithos 66:155–172

Misra DK (2009) Litho-tectonic sequence and their regional correlation along the Lohit and Dibang Valleys, Eastern Arunachal Pradesh. J Geol Soc India 73:213–219

Mitchell AHG (1993) Cretaceous-Cenozoic tectonicevents in the western Myanmar (Burma)-Assam region. J Geol Soc Lond 150:1012–1089

Mitchell AHG, Htay MT, Htun KM, Win MN, Oo T, Hlaing T (2007) Rock relationships in the Mogok metamorphic belt, Tatkon to Mandalay, central Myanmar. J Asian Earth Sci 29:891–910

Mukasa SB, Ludden JN (1987) Uranium–lead isotopic ages of plagiogranites from the Troodos ophiolite, Cyprus, and their tectonic significance. Geology 15:825–828

Nicolas A, Girardeau J, Marcoux J, Dupré B, Wang X, Cao Y, Zheng H, Xiao X (1981) The Xigaze ophiolite (Tibet): a peculiar oceanic lithosphere. Nature 294:414–417

Pal T (2011) Petrology and geochemistry of the Andaman ophiolite: melt–rock interaction in a suprasubduction-zone setting. J Geol Soc Lond 168:1031–1045

Pal T, Mitra SK, Sengupta S, Katari A, Bandopadhyay PC, Bhattacharya AK (2007) Dacite–andesites of Narcondam volcano in the Andaman Sea—an imprint of magma mixing in the inner arc of the Andaman–Java subduction system. J Volcanol Geotherm Res 168:93–113

Paul J, Burgmann R, Gaur VK, Bilham R, Larson KM, Ananda MB, Jade S, Mukal M, Anupama TS, Satyal G, Kumar D (2001) The motion and and active deformation of India. Geophys Res Lett 28:647–650

Pedersen RB, Searle MP, Corfield RI (2001) U–Pb zircon ages from the Spontang Ophiolite, Ladakh Himalaya. J Geol Soc Lond 158:513–520

Pedersen RB, Searle MP, Carter A, Bandopadhyay PC (2010) U–Pb zircon age of the Andaman ophiolite: implications for the beginning of subduction beneath the Andaman–Sumatra arc. J Geol Soc Lond 167:1105–1112

Prasad U (1985) Ophiolites of India. Rec Geol Surv India 115Pt(2):13–24

Qiu Z, Wu F, Yang S, Zhu M, Sun J, Yang P (2009) Age and genesis of the Myanmar jadeite: constraints from U–Pb ages and Hf isotopes of zircon inclusions. Chin Sci Bull 54(4):658–668

Quanru G, Pan GT, Zheng LL, Chen ZL, Fisher RD, Sun ZM, Ou CS, Dong H, Wang XW, Li S, Lou XY, Fu H (2006) The eastern Himalayan syntaxis: major tectonic domains, ophiolitic mélanges and geologic evolution. J Asian Earth Sci 27:265–285

Radhakrishna T, Divakara Rao V, Murali AV (1984) Geochemistry of Dras volcanics and the evolution of the Indus suture ophiolites. Tectonophysics 108:135–153

Radhakrishna T, Divakara Rao V, Murali AV (1987) Geochemistry and petrogenesis of ultramafic and mafic plutonic rocks of the Dras ophiolitic melange, Indus suture (northwest Himalaya). Earth Planet Sci Lett 82:136–144

Raju KA, Ramprasad T, Rao PS, Rao R, Varghese J (2004) New insights into the tectonic evolution of the Andaman basin, northeast Indian Ocean. Earth Planet Sci Lett 221:145–162

Rao DR, Rai H, Senthil Kumar J (2004) Origin of oceanic plagiogranite in the Nidar ophiolitic sequence of eastern Ladakh, India. Curr Sci 87(7):999–1005

Reuber I (1986) Geometry of accretion and oceanic thrusting of Spontang ophiolite, Ladakh-Himalaya. Nature 321:592–596

Reuber I, Montigny R, Thuizat R, Heitz A (1989) K-Ar ages of ophiolites and arc volcanics of the Indus suture zone: clues on the early evolution of the Neo-Tethys. Eclogae Geologische Helvetiae 82/2:699–715

Robertson AHF, Degnan PJ (1993) Sedimentology and tectonic implications of the Lamayuru Complex: deep-water facies of the Indian passive margin, Indus Suture Zone, Ladakh Himalaya. Geol Soc Lond Spec Publ 74:299–321

Saha A, Dhang A, Ray J, Chakraborty S, Moecher D (2010) Complete preservation of ophiolite suite from south Andaman, India: a mineral-chemical perspective. J Earth Syst Sci 119/3:1–16

Saha P, Acharyya SK, Balaram V, Roy P (2012) Geochemistry and tectonic setting of Tuting metavolcanic rocks of possible ophiolitic affinity from Eastern Himalayan Syntaxis. J Geol Soc India 80:167–176

Sarma DS, Jafri SH, Fletcher IR, McNaughton NJ (2010) Constraints on the Tectonic Setting of the Andaman Ophiolites, Bay of Bengal, India, from SHRIMP U–Pb Zircon Geochronology of Plagiogranite. J Geol 118:691–697

Schärer U, Hamet J, Allegre J (1984) The trans-Himalaya (Gangdese) plutonism in the Ladakh region:a U–Pb and Rb-Sr study. Earth Planet Sci Lett 67:327–339

Searle MP, Morley CK (2011) Tectonic and thermal evolution of Thailand in the regional context of SE Asia. In: Ridd MF, Barber AJ, Crow MJ (eds) The geology of Thailand. Geological Society, London, 539–571

Searle MP, Cooper DJW, Rex AJ (1988) Collision tectonics of the Ladakh-Zanskar Himalaya. In: Shackleton RM, Dewey JF, Windley BF (eds) Tectonic evolution of the Himalayas and Tibet. The Royal Society, London, 117–149

Sengupta S, Ray KK, Acharyya SK, de Smeth JB (1990) Nature of ophiolite occurrences along the eastern margin of the Indian plate and their tectonic significance. Geology 18:439–442

Sharma KK, Sinha AK, Bagdasarian GP, Gukasian P (1979) Potassium-argon dating of Dras volcanics. Shyok volcanics and Ladakh granite, Ladakh, Northwest Himalaya. In: Nautiyal SP (ed) Himalayan geology. Wadia Institute of Himalayan Geology, vol 8. Hindustan Publishers, New Delhi, pp 288–295

Shi G, Cui W, Cao S, Jiang N, Jian P, Liu D, Miao L, Chu B (2008) Ion microprobe zircon U–Pb age and geochemistry of the Myanmar jadeitite. J Geol Soc Lond 165:221–234

Singh AK, Singh RKB (2011) Zn- and Mn-rich chrome-spinels in serpentinite of Tidding Suture Zone, Eastern Himalaya and their metamorphism and genetic significance. Curr Sci 100(5):743–749

Sinha AK, Upadhyay R (1993) Mesozoic neo-tethyan pre-orogenic deep marine sediments along the Indus–Yarlung Suture, Himalaya. Terra Res 271–281

Srikantia SV (1986) Tectonic design of the Ladakh region, India. In: Ghose NC, Varadarajan S (eds) Ophiolites and Indian plate margin. Sumna Publications, Patna, pp 29–47

Srikantia SV, Razdan ML (1980) Geology of part of central Ladakh Himalaya with particular reference to the Indus Tectonic Zone. J Geol Soc India 21:523–545

Srikantia SV, Razdan ML (1985) The Indus tectonic zone of the Ladakh Himalaya: its geology, tectonics and ophiolite occurrence. Geol Surv India Rec 115:61–92

Swe W (1972) Strike-slip faulting in Central Burma. In: Haile NH (ed) In: Regional conference on geology of Southeast Asia, Geological Society of Malayasia, Kuala Lampur, pp 59–61

Tapponnier P, Mercier JL, Proust F, Andrieux J, Armijo R, Bassoullet JP et al (1981) The Tibetan side of the India–Eurasia collision. Nature 294:405–410

Thakur VC (1981) Regional framework and geodynamic evolution of the Indus–Tsangpo suture zone in the Ladakh Himalayas. Trans R Soc Edinb Earth Sci 72:89–97

Thakur VC, Misra DK (1984) Tectonic framework of Indus and Shyok Suture Zones in eastern Ladakh. Northwest Himalaya. Tectonophysics 101:207–220

Tilton GR, Hopson CA, Wright JE (1981) Uranium–lead isotopic ages of the Semail Ophiolite, Oman, with applications to Tethyan ridge tectonics. J Geophys Res 86:2763–2775

Valdiya KS (1988) Tectonic evolution of the central sector of Himalaya. Philos Trans Royal Soc Lond A326:151–175

Valdiya KS (2010) The making of India : Geodynamic evolution. Macmillan Publishers India Ltd., New Delhi 816 p

Vigny C, Socquet A, Rangin C, Chamot-Rooke N, Pubellier M, Bouin M-N, Bertrand G, Becker M (2003) Present-day crustal deformation around Sagaing fault, Myanmar. J Geophys Res 108/B11:2533

Vigny C, Simons WJF, Abu S et al (2005) Insight into the 2004 Sumatra–Andaman earthquake from GPS measurements in Southeast Asia. Nature 436:201–206

Virdi NS (1987) Northern margin of the Indian plate—some lithotectonic constraints. Tectonophysics 134:29–38

Virdi NS (1989) Glaucophane metamorphism in the ophiolite belt of the Indus–Tsangpo zone in the Himalaya. In: Ghose NC (ed) Phanerozoic ophiolites of India. Sumna Publishers, Patna, pp 73–91

Virdi NS, Thakur VC, Kumar S (1977) Blueschist facies metamorphism from the Indus suture zone of Ladakh and its significance. Himal Geol 7:479–482

Vohra CP, Haldar D, Ghosh Roy AK (1989) The Andaman-Nicobar ophiolite complex and associated mineral resources—current appraisal. In: Ghose NC (ed) Phanerozoic ophiolites of India. Sumna Publishers, Patna, pp 381–315

Wang R, Xia B, Zhou G, Zhang Y, Yang Z, Li W, Wei D, Zhong L, Xu L (2006) SHRIMP zircon U-Pb dating for gabbro from the Tiding ophiolite in Tibet. Chinese Science Bulletin 51:1776–1779

Warren CJ, Parrish RR, Waters DJ, Searle MP (2005) Dating the geologic history of Oman's Semail Ophiolite: insights from U–Pb geochronology. Contrib Miner Petrol 150:403–422

Wei ZQ, Xia B, Zhang YQ, Wang R (2006) SHRIMP zircon dating of diabase in the Xiugugabu ophiolite in Tibet and its geological implications. Geotectonica et Metallogenia 30:93–97

White LT, Lister GS (2012) The collision of India with Asia. J Geodyn 56–57:7–17

Xia B, Li JF, Liu LW, Xu LF, He GS, Wang H, Zhang YQ, Yang ZQ (2008) SHRIMP U–Pb dating for diabase in Sangsang ophiolite, Xizang, China: Geochronological constraint for development of eastern Tethys basin. Geochimica 37:399–403

Yin A, Dubey CS, Kelty TK, Webb AAG, Harrison TM, Chou CY, Célérier J (2010) Geologic correlation of the Himalayan orogen and Indian craton: part 2. structural geology, geochronology, and tectonic evolution of the Eastern Himalaya. Geol Soc Am Bull 122:360–395

Zhong LF, XIa B, Zhang YQ, Wang R, Wei DL, Yang ZQ (2006) SHRIMP age determination of Diabase in Luobusa ophiolite, southern Xizang (Tibet).Geol Rev 52:224–229

Zhou M-F, Robinson PT, Malpas J, Li Z (1996) Podiform chromitites in the Luobusa Ophiolite (Southern Tibet): implications for melt-rock interaction and chromite segregation in the upper mantle. J Petrol 37(1):3–21

Zyabrev SV, Aitchison JC, Badengzhu, Davis AM, Luo H, Malpas J (1999) Radiolarian biostratigraphy of supra-ophiolite sequences in the Xigaze area, Yarlung Tsangpo suture, southern Tibet (preliminary report). Radiolaria 17:13–19

Zyabrev SV, Kojima S, Ahmad T (2008) Radiolarian biostratigraphic constraints on the generation of the Nidar ophiolite and the onset of Dras arc volcanism: Tracing the evolution of the closing Tethys along the Indus–Yarlung-Tsangpo suture. Stratigraphy 5/1:99–112

3 Geology of the Naga Hills Ophiolite

3.1 Introduction

The NE–SW trending Naga Hills constitute the northern part of Indo-Myanmar Ranges (IMR) in the northeastern Indian states of Manipur, Nagaland, parts of Arunachal Pradesh and the adjoining areas of western Myanmar. The central part of this hilly range is known for the occurrence of ophiolite as a narrow arcuate body. The Phanerozoic IMR fold belt defines a convergent plate boundary, a seismically active zone where the eastern margin of the Indian continental plate is overridden by the Myanmar continental microplate from the east. The name Naga Hills Ophiolite (NHO) has been adopted in this book because of the large exposures, better defined geological sections and presence of complete lithological assemblage in Nagaland compared to its southern counterpart in Manipur where the ophiolite belt splits into thin slices of limited lithological variation (cumulate and spilite). The ophiolitic rocks of mantle and oceanic crust parentage at the continental plate margin have vast potential for intensive research and economic growth.

The NHO consists of a variety of Mesozoic and Cenozoic magmatic, metamorphic and sedimentary rocks that originated at the India-Myanmar convergent plate boundary (Figs. 3.1, 3.2, 3.3). From east to west, the ophiolite and associated rocks are broadly classified into three distinct tectono-stratigraphic units, viz., (1) the Nimi Formation consisting of low- to medium-grade accretionary wedge metasediments of possible Mesozoic age, (2) the NHO and (3) the Disang Formation consisting of a thick pile of folded Late Cretaceous-Eocene flysch-type sediments. A fourth tectono-stratigraphic unit known as the Jopi Formation is a post-orogenic molasse that consists of ophiolite-derived conglomerate-grit-greywacke-sandstone-shale/polymictic tuff breccia belonging to a paralic sedimentary facies and occurring as a cover sequence over the NHO. In addition, some Neogene high-level terrace deposits composed of boulder-conglomerate-sand are exposed along Tizu River over the Disang Formation.

In the southern part of the ophiolite belt in Ukhrul district of Manipur two distinct types of mélanges are delineated in the Phungyar-Kamjong sector (Das et al. 2008):

(i) The eastern serpentinite mélange dominantly carries sheared blocks of serpentinised dunite and harzburgite with pockets and pods of deformed chromitite and minor blocks of gabbro, plagiogranite, volcanic and pelagic sediments. They are exposed as blocky and barren outcrop with red soil profile covering highest elevation.

(ii) The western argillaceous mélange is interspersed within an arenaceous-argillaceous turbidite. This mélange is characterised by a chaotic mixture of blocks of conglomerate, greywacke, rhythmite, siltstone, lensoid limestone, red and black shale, chert, highly altered mafic volcanic of variable dimensions in a pervasive sheared argillite matrix. Large bodies of cement grade limestone add significance to this occurrence. The faunal assemblages of limestone indicate a Late Cretaceous-Paleocene age.

Two distinct metallogenic episodes have been established in the ophiolite belt, viz., (i) pre-orogenic (Late Cretaceous - Eocene) resulted in syngenetic deposit of podiform chromite containing traces of nickel, cobalt, gold and platinum group of elements (PGE) in cumulate perodotite and serpentinite; and native gold in olivine gabbro of layered cumulates (ii) syn-to late-orogenic (Late Paleogene) involving epigenetic remobilization of disseminated and vein-type copper-molybdenum sulfides associated with mafic volcanics (Agrawal and Ghose 1989; Singh et al. 2012a; Ghose 2013). Secondary processes such as serpentinisation, weathering and metasomatism have resulted in the development of magnetite, asbestos, talc, rodingite and minor laterite. Occurrence of important economic minerals and their relationship with the host rock in NHO are dealt with at the end of this chapter and in Chap. 5.

3.2 Age

On the basis of faunal assemblages in sediments associated with ophiolite at the western margin of the IMR, the NHO has been assigned ages ranging from Cretaceous to Paleocene

N. C. Ghose et al., *A Petrographic Atlas of Ophiolite*,
DOI: 10.1007/978-81-322-1569-1_3, © Springer India 2014

Fig. 3.1 Bird's eye view of ophiolite, Shiloi Lake, Nagaland

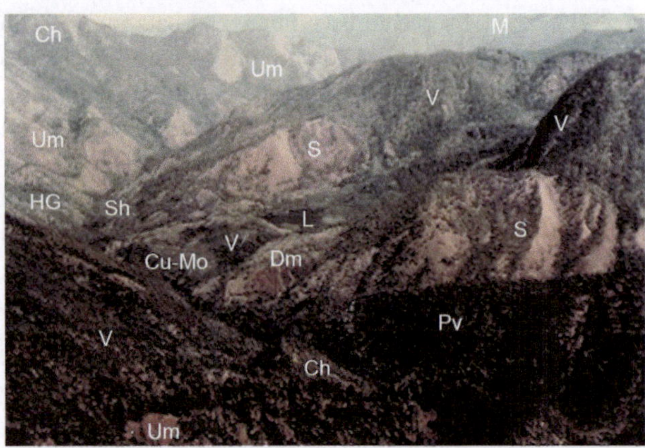

Fig. 3.2 Aerial view of ophiolite, Lacham Lake, Nagaland. *Ch* chert, *Dm* dunite tectonite, *HG* hornblende gabbro, *G* gabbro, *L* Lacham Lake, *V* volcanics, *M* Nimi Formation, *Pv* pelagic-volcanogenic sediments, *S* serpentinite, *Sh* schist, *Um* ultramafic and *Cu–Mo* copper-molybdenum mineralisation

(Chattopadhayay et al. 1983; Acharyya et al. 1986; Chungkham and Jaffar 1998). The ophiolitic rocks in the Chin Hills, located south of the Naga Hills, are unconformably overlain by the Upper Albian limestones (Mitchell 1981). In the southern Chin Hills and further to its south in Arakan Yoma, a homotaxial ophiolite suite of smaller dimension is exposed as tectonised bodies, mélanges and olistostromes associated with Late Cretaceous pelagic sediments (Acharyya et al.1986). However, K–Ar analysis of an ophiolitic basaltic flow from the southeast of Waziho village (25° 38′N, 94° 44′E) yielded a date of 148 ± 4 Ma (Sarkar et al. 1996), while radiolaria in ophiolitic chert from near Salumi belong to the Late Jurassic (Kimmeridgian–Lower Tithonian) age (Baxter et al. 2011). These dates are consistent with a 158 ± 20 Ma date determined by the K-Ar method in hornblende pegmatite within serpentinite in the Chin Hills (Mitchell 1981; Chattopadhyay et al. 1983; Acharyya et al. 1986; Chungkham and Jafar 1998). Therefore, ocean crust generation and marine sedimentation probably began in Late Jurassic. Furthermore, ChRM directions in two-oriented blocks of NHO basalts are consistent with an age of 70 ± 5 Ma (Patil et al. 2012) indicating that the oceanic crust generation may have continued until Late Cretaceous.

3.3 Geological Setting and Stratigraphy

Geological investigation in the Naga Hills ophiolite terrain is beset with several constraints due to a mountainous terrain, remoteness from peninsular India, exposure of outcrops close to the international boundary with Myanmar, rugged topography, lack of infrastructure, high monsoon rain, thick soil/vegetation cover and dismembered lithology (Figs 3.1, 3.2). Salient geological features of different rock types in each tectono-stratigraphic group (Table 3.1), their inter-relationship, deformation pattern and metamorphism are presented below.

3.3.1 Nimi Formation

A thick sequence of folded metasediments dominantly composed of low-grade regionally metamorphosed rocks such as phyllite, feldspathic quartzite, limestone and quartz-sericite schist exposed in the eastern tectonic boundary of the NHO is known as the Nimi Formation after the type section at village Nimi (Figs. 3.3, 3.4, 3.5). The sequence is tightly folded in the form of a major NNE trending overturned anticline (Agrawal and Ghose 1986). This litho-package extends in the west of Nimi to Laluri, and in the east (except the crystalline limestone) to Saramati peak at the India-Myanmar international border. The rocks are vertically inclined and show broad and open warps, often repeated due to folding (Figs. 3.6, 3.7). Graded bedding and cross-bedding are also noted in the quartzite. The quartzites are gritty towards the basal part. Enclaves and tectonic slices of serpentinite and volcanics become less frequent away from the contact. Presence of mylonitic limestone showing flattened quartz and carbonates (see Chap. 5) suggests its formation under extreme pressure as a result of over-thrusting. The presence of garnet-staurolite-biotite-muscovite schist (Chap. 5) belonging to amphibolite facies in the Nimi Formation indicates a temperature exceeding 500 °C in the deeper part of the basin. An eastward increase in the grade of metamorphism is observed in these rocks towards Myanmar.

Extensive occurrence of limestone (140–200 m thickness) trending NNE to NE has been noted with the development of stalactites and stalagmites in caves (Agrawal and Ghose 1986). The limestones are crystalline and associated

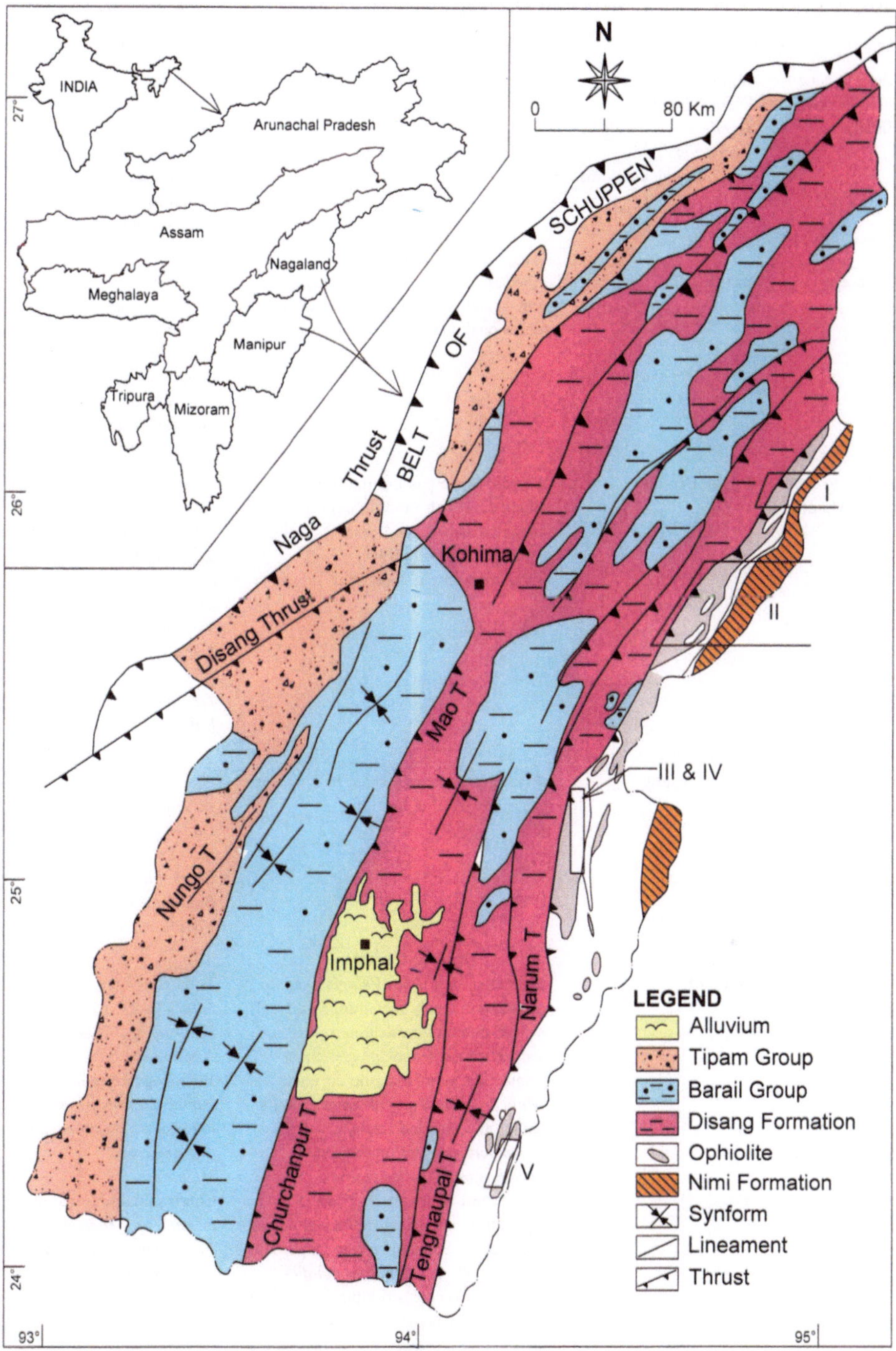

Fig. 3.3 Simplified geological and tectonic map of Nagaland and Manipur states, India (modified after Geological Survey of India and Ningthoujam et al. (2012). Detailed geological maps of the ophiolite terrain (I–V) are given separately

with secondary calcite and quartz veins. The metasediments are affected by tight isoclinal folding with steep axial planes dipping 80–85° towards SSE and show development of slaty cleavage and stripping lineation (Vidyadharan et al. 1986). Tectonically, the metasedimentary rocks can be considered as an accretionary prism complex which has possibly developed on an east-dipping subduction zone (Roy 1989) and has progressively accreted onto the overriding continental margin in western Myanmar.

3.3.2 The Ophiolite Suite

The NHO is exposed as a long arcuate belt with tectonic contacts on either side. It shows evidence of transport into and onto the Disang flysch at the western margin, and is in turn overridden by the Nimi Formation metamorphics from the east (Fig. 3.3). Parallel to sub-parallel tectonic slices of different litho-units of ophiolite are arranged in a NE-SW/N-S trending *en echelon* pattern. Although they are not

Table 3.1 Stratigraphy of the Naga Hills Ophiolite, India (modified after Agrawal and Ghose 1986)

Age	Group/Formation	Lithology	Fossil content	Structure	Environment
Mio-Pliocene (homotaxial to Barail)	**Jopi or Phokphur Formation** *(Patkai Synclinorium) or Surma Formation* (Kohima synclinorium)	Shale and Polymictic tuff breccia Greywacke and arenites Grit Polymictic conglomerate	Gastropods Plant fossils (Angiosperms): *Anthocephalus sp., Aegle sp., Magnifera sp., Psidium sp., Wondlandia sp., etc.* Marine micro-fauna	Mega fold, fracture cleavage	Paralic
		...........***Unconformity***.............			
Oligocene	**Barail**	Shelf sediments: Predominantly sandstone. Alternating with shale and occasional coal seam	*Numsilites chavannesi, Reticulate nummulites, Nummulite sp., Operculina sp.,* Plant fossils: *Dicotilophyllum Garcinia sp., Rhizophera sp., Memecylon sp.,* etc.	Low-plunging open folds with subangular hinge **Schuppen** structure in the northeastern IMR	Shallow marine
Middle to Upper Miocene	**Upper Disang**	**Flysch**: Shale passing into slate alternating with sandstone. Presence of mega cross-bed, load cast, slumping structure and olistoliths. Evaporites in the upper part	*Asmobaculites, Ammudiscus, Gaudr Bathysiphon, Cyclammina sp., Discocyclina dispensa, Discocyclina omphalus,Nummulites margoelari, Taxa nummulites, Venericardia sp., Solen sp., Nemocordium sp., Ammodiscus, Trochamnina,Eggralla, Rubulus and Bullimina*	Three generations of folds: F_1 and F_2 are co-axial and co-planar flexure slip folds formed due to E-W or ESE-WNW compression. F_3 are open type with low to moderate axial plane trending E-W	Shallow marine and Shelf
Upper Cretaceous to Middle Eocene	**Lower Disang**	**Flysch**: Dominantly shale passing into slate and phyllite (turbidite)	*Globotruncana acra, G.Gansseri, G.Stuarti, Globegerina sp., Nodosaria sp., etc.* *Benthonic foraminifera: Ballaina sp., bolivina sp., Noaion sp., etc*		Shallow marine with steep gradient
		..........***Tectonic Contact***............			
Late Jurassic to Eocene K-Ar age of basalt: 148 ± 4 Ma (Sarkar et al. 1996) Paleomagnetic age of basalt: 70 ± 5 Ma (Patil et al. 2012)	**Ophiolite**	Late felsic intrusive; ***Pelagic oceanic sediments***: feebly metamorphosed tuff, pelitic-psammitic-calcareous-carbonaceous sediments and chert containing radiolaria, and coccoliths; ***Mafic/intermediate volcanics***: basalts, minor andesite and trachyte, volcaniclastics and ignimbrite; ***Metabasics*** : zeolite- prehnite-greenschist facies, High-P/low-T blueschists and barroisite eclogite, and Meta-chert: containing blue amphibole and magnetite; ***Cummulate mafic-ultramafic rocks***: peridotite, pyroxenite, gabbroids, plagiogranite and anorthosite; ***dolerite*** (minor); ***peridotite tectonites*** and spinel peridotites; Serpentinite and Rodingite. ***Mineralisations***: Podiform chromitite and Magnetite deposits, Cu-Mo sulfides and minor laterite, native gold and platinum	Tuffs containing gastropods: *Solariola sp., Nerita sp., Littorina sp.,* Panope sp., Assilina sp. Volcanogenic sediments record angiosperm leaf impressions: *Anthocephalina sp., Aegle sp., Magnifera sp., etc* Cherts containing radiolaria*: Kassina sp., Theocyrts sp., Prordiscus sp., Spongoprunum sp., Cenophaere, Conellipasia, Conodiscus, Sethocyrtis, Tricolocapsa, Novixitus, Holocryptocanium sp., Cerops etc.* Cherts containing coccoliths*: Ahmualleralla octoradiata*	D_1 produced strong NE regional schistosity. F_1 folds obliterated by D_2 and D_3. F_2 and F_3 are similar to flysch	Analogy with mid-oceanic ridges (MOR)
		...........***Tectonic contact.........***			
Mesozoic (?)	**Nimi Formation/ Naga Metamorphics**	Dominantly Low-grade metasediments (phyllite, garnet-staurolite schist, limestone, quartzite) with minor serpentinite intrusion (accretionary prism sediments)	Unclassified radiolaria	Major NNE plunging overturned anticline. Five sets of fold structures identified; tight, double plunging, open folds with puckers	shallow to intermediate marine

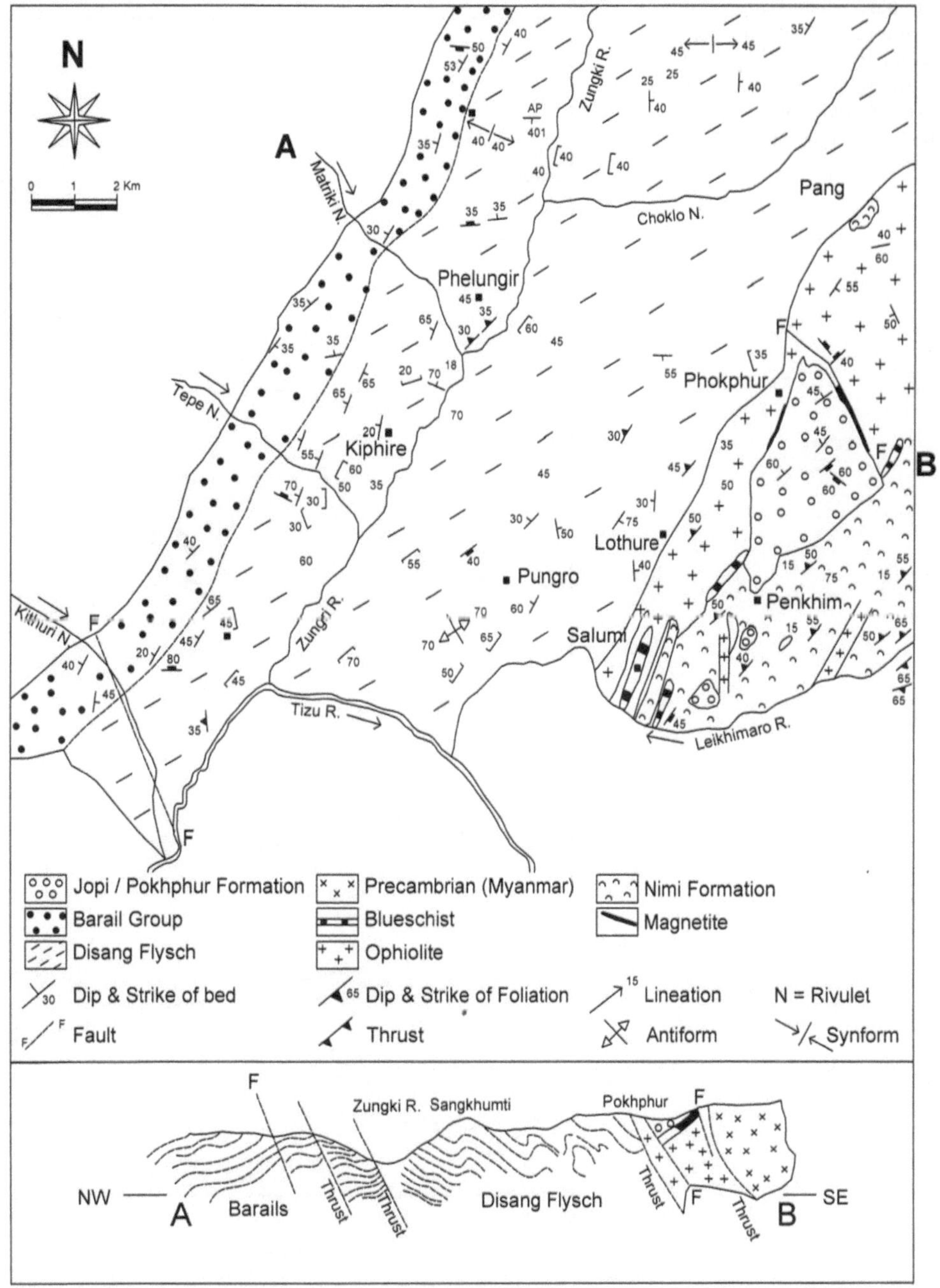

Fig. 3.4 Geological map of the northern part of ophiolite (sector I) around Kiphire-Pokhphur (old Pukphur) (modified after Ghose and Singh 1981). Geological section along *A* and *B* is shown below

preserved in a sequential order typical of other ophiolite sections in the Tethyan domain (e.g. Semail, Oman Mountains), it is evident from field disposition that a systematic order may have existed prior to their emplacement. The ophiolite litho-units may be broadly classified into (i) peridotite tectonite or meta-ultramafics, (ii) spinel peridotite (spinel-bearing dunite/harzburgite/lherzolite), (iii) cumulate mafic–ultramafic sequence (peridotite, pyroxenite, layered and massive gabbroids, plagiogranite and anorthosite), (iv) dolerite and meta-dolerite (minor), (v) volcanics and volcaniclastics and (vi) pelagic sediments. Serpentinisation is ubiquitous, and both very-low (zeolite and chlorite-prehenite schists) to low-grade (greenschists) assemblages and high-P assemblages (blueschists and eclogite) occur in the NHO. Chromitite, nickeliferous magnetite and other metallic and non-metallic minerals also occur in the NHO (Table 3.1).

Some noteworthy petrographic features include the presence of—ocellus of chlorite in basalts, S-C mylonite in spinel peridotite and blueschist, glaucophane-fish in ultra-cataclastic blueschist leading to mortar texture and development of pseudo-tachylite vein, kink carbonate lamellae attached to quartz in pressure shadows in ignimbrite, and late metamorphic breakdown of clinopyroxene to form plagioclase with a rim of orthopyroxene in olivine-websterite. An

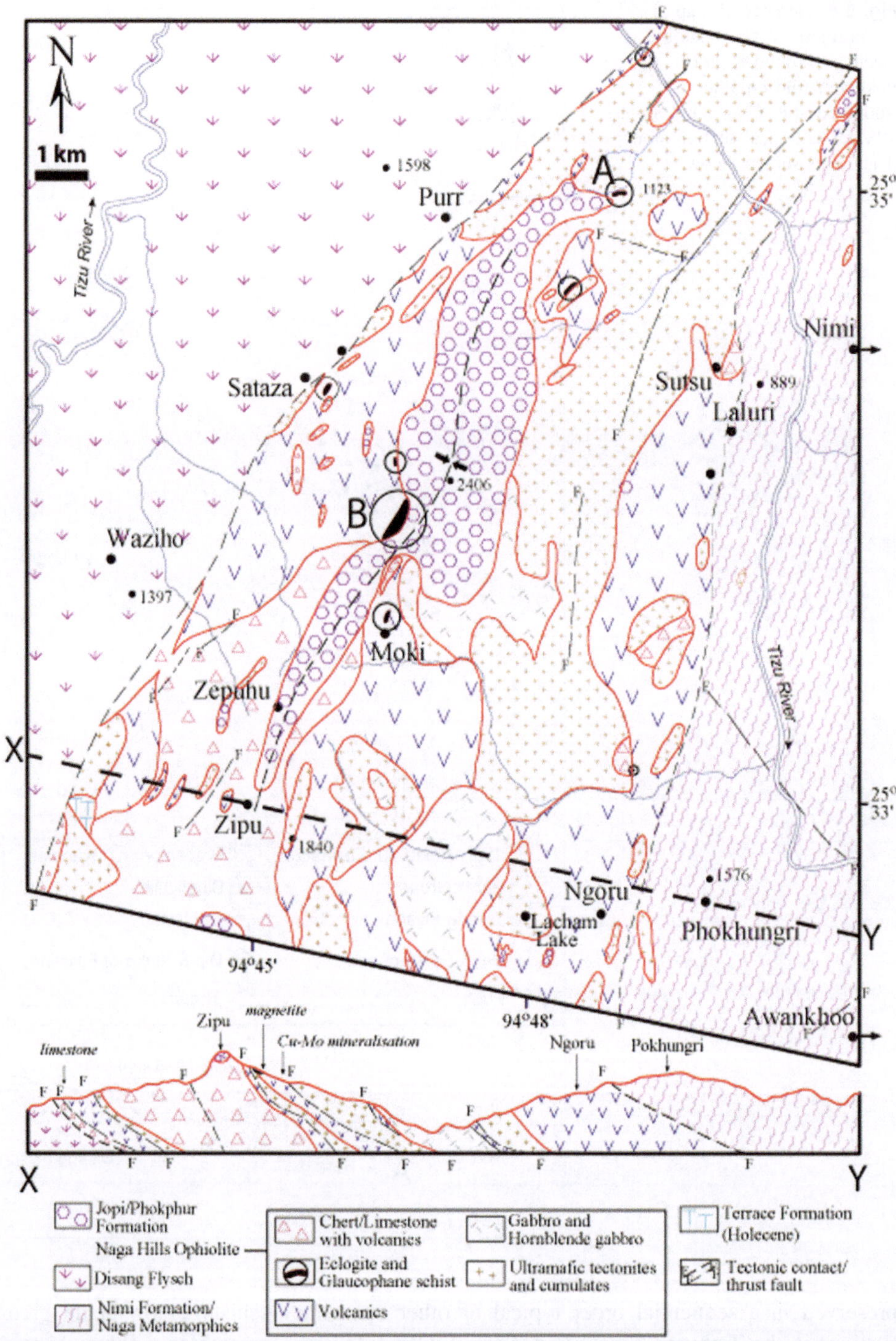

Fig. 3.5 Geological map of the central part of ophiolite belt (sector II) between Zipu and Sutsu (after Agrawal and Ghose 1986). Geological section across the line X-Y is shown below

outstanding feature in a 1.5 cm^2-sized polymictic tuff breccia from the post-ophiolitic sediments of the Jopi Formation shows minerals and rocks representing all components of the Tethyan lithosphere within the scale of a thin-section.

The NHO continues southward into the Indian state of Manipur where it is represented by a dismembered ophiolite suite comprising alternating slices of cumulate ultramafic–mafic differentiates and volcanic rocks (Joshi and Vidyadharan 2008). The ultramafics are dominated by harzburgite with minor dunite and pyroxenite. They are serpentinised and contain magnetite and talc with a typical ribbon and mesh texture. Chromitite bands and pods occur within the ultramafics. Thin, discontinuous lenses of coarse and dark ultramafic segregations with diopside, grossular-almandine garnet (needs verification with the EPMA), magnesio-hastingsite and rutile occur in the northeastern

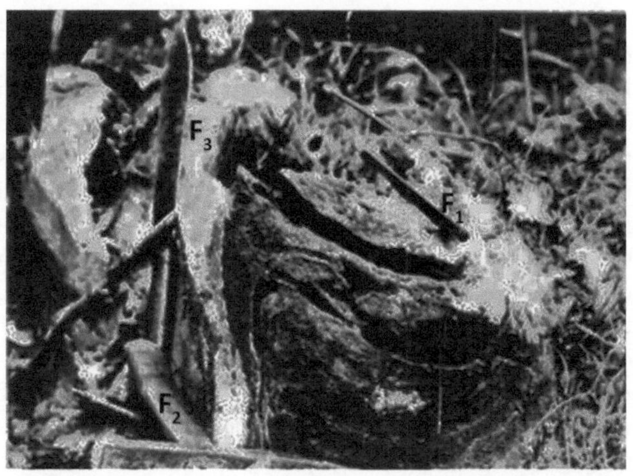

Fig. 3.6 Multiphase deformations in feldspathic quartzite interlayered in limestone of Nimi Formation. Locality: 2 km southwest of Nimi

Fig. 3.7 Quartz veins along axial planes in tightly folded quartzite and phyllite alternation of Nimi Formation. *Locality* Tizu River section, 0.75 km southeast of Sutsu

part in Manipur. They show strong CaO enrichment and severe depletion of MgO from the host peridotites and resemble meta-rodingite formed through extensive Ca metasomatism during serpentinisation of the host rocks followed by high pressure recrystallisation (Evans et al. 1979). The mafic differentiates are represented by rodingitised gabbros and plagiogranite. The volcanic rocks include massive amygdaloidal basalts with low MgO that show intersertal and glomeroporphytitic texture and agglomerate. The eastern margin of ophiolite in Manipur exposes oceanic pelagic cover sediments metamorphosed to a low grade. Dyke swarms are absent. A detailed description of the different lithologies of the NHO is presented in Sects. 3.4–3.9.

3.3.3 Disang Formation (Flysch)

A folded sequence of enormous thickness (>3000 m) of slate, graphitic slate, phyllite, siltstone and fine-grained sandstone, exposed to west of ophiolite is known as the Disang Formation. The NNW–SSE trending ridges and valleys of flysch show a rhythmic alternation of shale/slate and siltstone of distal shelf facies. Slaty cleavage is present often parallel to the bedding. The shale-siltstone sequence is replaced by phyllite-quartzite near the axial planes of folds, faults and thrusts. They exhibit shallow water depositional features like ripple marks, sole marks, graded bedding and cross bedding (Agrawal and Ghose 1986).

The Disang sediments show tight folds, slips, drags, multi-generation quartz veins and salt springs close to the ophiolite contact. These, together with the occurrence of tectonic slivers of serpentinites and its intermixing with sediments (Vaidyanathan et al. 1986), give unequivocal evidence of their emplacement along the deep fractures of basin/continental margin. Despite their large thickness, fossils are rare in the Disang sediments. A limited number of bivalves, gastropods and foraminifera are reported that suggests an Upper Cretaceous-Eocene age of their deposition (Acharyya et al. 1986).

In the southern sector, olistoliths and olistosromal masses occur in abundance, varying in dimensions from a few metres to >0.5 km in the upper part of the Disang Formation (Joshi and Vidyadharan 2008). A mélange zone of folded olistostromal limestone in the upper part of the formation extends from Lambui in the south, through Ukhrul and Paoyi in Manipur (Mitra et al. 1985) to Kiphire in Nagaland (previously referred to as Matriki Formation by Singh and Ghose 1981). Foraminiferal assemblages of different ages such as *Globotruncana Ganseri, Gumbulina striata* of Maestrichtian age, *Nummulites beaumonti, Discocyclina dispensa* of Eocene affinity and *Globorotalia acutespira* of Paleocene age (Mitra et al. 1985; Acharyya et al. 1986) occur in the limestones. Molluscan biozones in the Disang sediments from Manipur reflect an age of Paleocene to Upper Eocene (Lukram and Kachhara 2010).

The turbidite sequences of the Disang Formation grade into coarse shallow marine to fluviatile sediments represented by conglomerate, grit, sandstone and coal streaks of overlying Barail Group. The Disang and Barail sediments were deposited in the distal shelf and on the continental margin, respectively, in an epicontinental sea (Vidyadharan et al. 1989).

3.3.4 Jopi/Phokphur Formation

A clastic shallow marine sedimentary cover ($\sim$1000 m thick) known as the Jopi Formation unconformably overlies

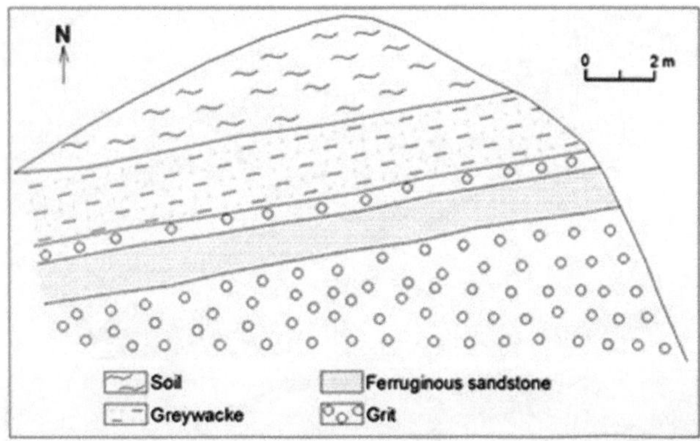

Fig. 3.8 Vertical profile of post-ophiolite cover sediments at Pokhphur-Matungsekien ridge (Jopi Formation)

the NHO with a faulted eastern contact. Primarily derived from ophiolite, this immature and young repeated molassic sedimentary sequence of polymictic boulder—conglomerate—grit—greywacke/sandstone—shale/fine tuff breccia (Fig. 3.8) occupies the higher altitudes of the Mollen-Jopi-Ziphu-Pokphur range that runs along the central part of the ophiolite belt and forms a synclinal ridge (Fig. 3.5). Near Phokphur, the Jopi Formation is referred as Phokphur Formation after the Phokphur-New Basti section by the Geological Survey of India (Vidyadharan et al. 1986). The formation trends NE, shows low to moderate dip and is about 2 km wide along the Zipu-Sangtam Teng ridge. In the type area of the Jopi-Mollen ridge, the formation attains a thickness of 600 m (Agrawal and Ghose 1986). The clast to matrix ratio (>70:30) varies along strike and dip directions with channeling, intertonguing and interfingering relationships between the matrix and the clasts of eroded ophiolite units. Several cyclic sequences of conglomerate-grit-sandstone-tuffaceous shale are also observed (e.g. at Koya rivulet, Purr and Laluri sections). The thickness of the individual cycles varies from <1 to >10 m (Agrawal 1985). The sediments were deposited in the shallow marine waters of an epicontinental sea or lagoon.

The conglomerate contains clasts of serpentinite, gabbro, basalt, chert (with or without radiolaria), crystalline limestone, quartz, jasper, quartzite, sandstone, phyllite and slate embedded in a reworked tuffaceous to siliceous cement. The tuffaceous matrix constitutes up to 50 % by volume of the rock. The sequence could be termed as volcaniclastic (Agrawal 1985), instead of pyroclastic (Chattopadhyay et al. 1983). They grade upward into grit-lithic greywacke-siltstone-sandstone-shale/tuff. A 30-m-thick conglomerate bed is exposed at Pungro-Phokphur road at the cross section with Mukoge rivulet in the northern part of NHO (Singh 1979). It consists of angular to subrounded boulders, cobbles and pebbles embedded in tuffaceous shale.

The sandstone and shale exhibit similar composition of clast and matrix. The shales are tuffaceous, interbedded with fine and coarse tuff, and display various shades of colour such as chocolate brown, reddish brown, purple, dark green and pink. The matrix is red due to the dominance of ferruginous matter. Clasts of serpentinite, pyroxenite, gabbroid and mafic-intermediate-acid volcanics, metabasic, radiolarian chert, arenite, quartzite, greywacke and argillite, containing discrete idiomorphic mineral grains of orthopyroxene, clinopyroxene, plagioclase, glaucophane, ilmenite, garnet, epidote and opaque are present (Fig. 3.9). Thus, all the litho-members of the ophiolite suite are present in the clastic assemblage. Presence of non-volcanics and minor amounts of biogenic constituents may qualify these rocks to be called epiclastic breccia. However, greater abundance of fine tuff and lithic pyroclastics of juvenile and accidental ejectas, qualify the finer material to be named as 'polymictic tuff breccia' (Roger H. Mitchell, personal communication).

The Jopi/Phokphur Formation is characterised by (a) post-orogenic deposition (with post-F_1 folding) followed by late marine transgression, (b) irregular current bedding, (c) dominance of sandstone, (d) rarity of limestone, (e) weak metamorphism at the base and (f) terrestrial and marine fossils of plants, invertebrate and foraminifera. The different tectonic levels of the formation show different degrees of deformation and metamorphism. The weakly metamorphosed units in the lower part of the sequence show alternation of slate and phyllite with quartzite, whereas uppermost units are unmetamorphosed (Vidyadharan et al. 1986). A wide range of invertebrate and plant fossils and microfossils within the formation indicates a subtle change in climate with the advent of the Cenozoic Era. The invertebrate fossils include *Chlamys* sp., *Ostrea* sp., *Trigonia* sp., and *Ficus* sp., whereas leaf impressions of plants include *Graminea, Shorea, Psidium* and *Annona* (Ghose and Singh 1981). The plant fossils *Anthoceplalus* sp., *Litchi* sp., *Syzyginum* sp., etc., indicate a Late Eocene-Oligocene age and a possible homotaxial relationship with the Barail Group deposited over the Disang Formation elsewhere in the Naga Hills (Chattopadhyay et al. 1983). About 5 km south of Phokphur, where the Jopi Formation is known as Phokphur Formation, seven closely spaced, gastropod-dominated, fine tuffaceous breccia horizons with coaly and carbonaceous streaks and plant fossils have been identified (Vidyadharan et al. 1986). Presence of *Assilina* sp. in these horizons suggests an Eocene age (Ranga Rao 1983). The Jopi/Phokphur Formation is thus correlatable with the Mithakhari Group of the Andaman-Nicobar Islands (Bandyopadhyay et al. 1973; Vohra et al. 1989) and the Indus Group of the Indus Suture Zone (Srikantia and Razdan 1980). A simplified stratigraphic sequence of the ophiolite belt is given in Table 3.1.

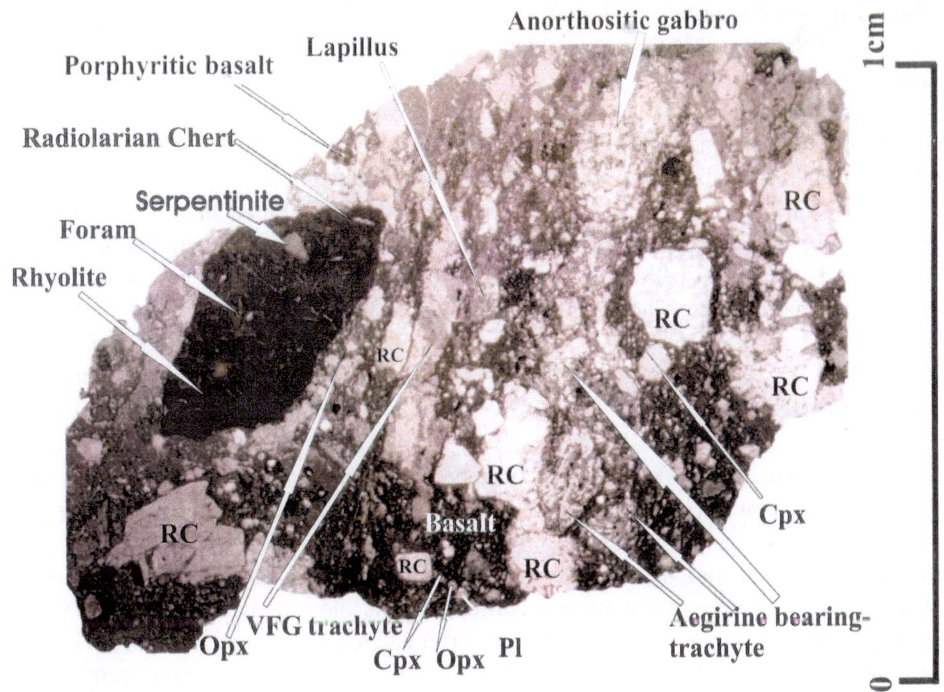

Fig. 3.9 Photograph of the actual size of thin section of polymictic tuff breccia (C21/79) taken under Leitz bionacular microscope containing lithic clasts and minerals of ophiolite lithology from the Jopi Formation

3.4 Ultramafic Rocks and Cumulate Complexes

3.4.1 Peridotite Tectonite

Peridotite, an olivine-pyroxene bearing rock, constitutes an integral part of the NHO. Peridotite tectonite at the base of the ophiolite sequence originated through deformation, plastic flow and shearing as part of the ocean floor was detached from the oceanic lithosphere. It is overlain by strain free cumulate sequence of the layered complexes that formed through fractionation in magma chamber. The tectonites are represented in the NHO by dunite, harzburgite, lherzolite and minor wehrlite, comparable to seismic *Layer-4* of the oceanic crust. One small and two relatively large bodies of peridotite tectonite or meta-peridotite have been identified in the central part of the ophiolite belt: (a) dunite—0.5 km south of Lacham Lake (1.0 × 0.3 km), (b) dunite and harzburgite—2 km southeast of Zipu and (c) a very small body of lherzolite close to Zintang ti rivulet in the central part of the ophiolite belt (Agrawal and Ghose 1986). They contain resorbed or euhedral to anhedral grains of dark coloured opaques or deep brown high-Cr chrome spinel. The mode of deformation varies from brittle to ductile in these rocks. Where serpentinisation is pervasive, distinction between tectonites and cumulate peridotites (*Layer 3*) is difficult in the field.

3.4.2 Spinel Peridotite

The spinel-bearing peridotites occur as thin discontinuous layers, pods and lenses within the peridotite/serpentinite predominantly in the central part of the ophiolite belt, viz., 2.25 km southeast of Purr, Nagaland (Agrawal and Ghose 1986) and east of Tusom Cisi, Manipur (Vidyadharan et al. 1989). The presence of spinel and the absence of plagioclase indicate a deep origin of these rocks. A detailed study necessary for proper tectonic characterisation of these rocks is lacking at present. They may represent peridotite tectonite, mantle xenoliths or both. Three narrow zones of amphibolite-spinel lherzolite-harzburgite have been reported from Tusom Cisi, covering an area of 200–500 m in width and 750 m in length by these authors. There are other occurrences of spinel harzburgite, spinel lherzolite and serpentinite (dunite) that have not been previously reported. Their petrographic features are presented in Table 5.1. The garnet-bearing peridotite reported by previous workers could not be substantiated with EPMA. The reported boulder-like occurrence of garnet lherzolite at Luthur in the north containing garnet, aegirine augite, hornblende, glaucophane, zoisite, chlorite and sphene (Vidyadharan et al. 1986) is likely to be a blueschist. Irregular occurrence of anhedral grains of pale pink coloured chrome spinel intergrown with magnetite as symplectites are often mistaken as garnet. The spinel peridotites are characterised by high-Al chrome spinel as an essential mineral component

(Singh 2013). In contrast, low-Al, high-Cr chromite spinel is commonly encountered in the peridotite tectonites (Chap. 5, Table 5.5).

Rare presence of almost pure andradite garnet as recorded in one sample is probably of metasomatic origin and is related to the formation of rodingite from serpentinite. Occurrence of such Cr- and Fe^{3+}-rich uvarovite-andradite garnet has been reported in serpentinite veins intruding chromitite pods of host peridotite from the Rutland ophiolite in Andaman Islands (Ghosh and Morishita 2011).

3.4.3 Serpentinite

Most of the exposed peridotite bodies (except at the Tizu River section) show pervasive serpentinisation. Serpentinite bodies ranging in size from a few metres to several kilometres is the dominant component of the ultramafic suite of the NHO. They usually form undulating to flat-topped hills and ridges supporting sparse vegetation (e.g. spear grass, palm and pines). The general trend of schistosity is N–S/NE–SW and the dips are subvertical. The absence of mafic cumulates in the large serpentinite bodies, e.g. between Purr and Laluri (Fig. 3.5), indicates that the serpentinites are derived from an ultramafic/peridotite precursor. Serpentinites also occur as small slices, lenses, rafts, breccias and slivers in the other ophiolitic units. Presence of talc-serpentine schist at the tectonic contacts of the NHO on both eastern and western margins may indicate extreme pressure during their emplacement. The serpentinite bodies show intertonguing relationship with enveloping sediment that may owe their origin to plastic flow in a late ductile shear regime.

Megascopically, serpentinite is a green or dark grey coloured, massive or schistose rock with almost complete obliteration of the original igneous fabric. The massive type exhibits a mesh texture even in hand specimen with a fine network of dark veins composed of serpentine and magnetite. In the schistose type, the schistosity is produced by polygonal fracturing and shearing. Secondary veins of chrysotile or fibrous antigorite/lizardite are common in sheared and fractured zones, and at the periphery of large serpentinite bodies. Lenses or beds of quartz-chlorite-sericite schist and feldspathic schists occur within the serpentine bodies, possibly representing tectonic inclusions of oceanic sediments.

Serpentinite is formed by metamorphism and/or hydration of peridotite, chiefly affecting olivines and to a lesser extent, pyroxenes. X-ray diffraction studies have established antigorite as the major phase and lizardite and chrysotile as the minor phases in serpentinite (Ghose et al. 1986; Ningthoujam et al. 2012). Secondary quartz, calcite and magnetite veins are also common. Alteration of ultramafic to talc-serpentine schist at the tectonic contacts occurred through addition of CO_2, and removal of H_2O and a very small amount of O_2 (Naldrett 1966).

3.4.4 Rodingite

Small *in situ* outcrops or boulders of rodingite commonly occur along the hill slopes and river courses in the southern part, e.g. east of Kudengthabi and Kwatha (Figs. 3.10, 3.11, 3.12) (Shukla 1989), and less commonly in the north (Agrawal and Ghose 1986). It is a metasomatic rock of igneous or sedimentary parentage showing sharp or gradational contact with the ultramafics. These rocks are genetically related to serpentinisation of the ultramafics. Calcium released by pyroxene is the primary source of metasomatism, chiefly affecting gabbros and other surrounding rocks like greywacke, pelagic sediments and volcanics. They occur as discontinuous lenses, pockets, stringers, veins and pods. Rodingites are relatively hard (H = 5.5–7.0) and dense (sp. gr. 3.0–3.5) rocks, and the translucent variety is used as semi-precious gem stones. The colourless variety is common in the NHO, and the milky white and pale green varieties are rare.

3.4.5 Cumulate Complexes

In many Tethyan ophiolites, viz., Semail massif in the Oman Mountains, the basal peridotite tectonites merge upward into a cumulate zone comparable to continental-layered gabbroic complexes (viz., Bushveld, South Africa; Stillwater, Montana, U.S.A.). A number of mafic-ultramafic cumulate sequences are exposed in the NHO which appear to be a result of high degree of tectonic slicing during emplacement. The base of the cumulate sequence consists of peridotite and pyroxenite, which constitute early fractionates of basaltic magma. These are overlain by a mafic sequence of gabbroids represented by olivine gabbro-norite-gabbronorite-hornblende gabbro-plagiogranite-anorthosite, which are probably products of progressive fractional crystallisation. Good exposures of these rocks occur in the Pang and Tizu River sections in the southern part of the belt, where cumulates of peridotite, serpentinite, pyroxenite, gabbro and spilite are exposed (Fig. 3.13). A brief description of different cumulate rocks is given below.

3.4.5.1 Peridotite Cumulates

The peridotite cumulates are green to olive green or black in colour, medium- to coarse-grained. Due to limonitisation, the outcrops show reddish or brick-red colouration. Partly serpentinised peridotites have an olive green amorphous groundmass with shiny prisms of pyroxene giving rise to

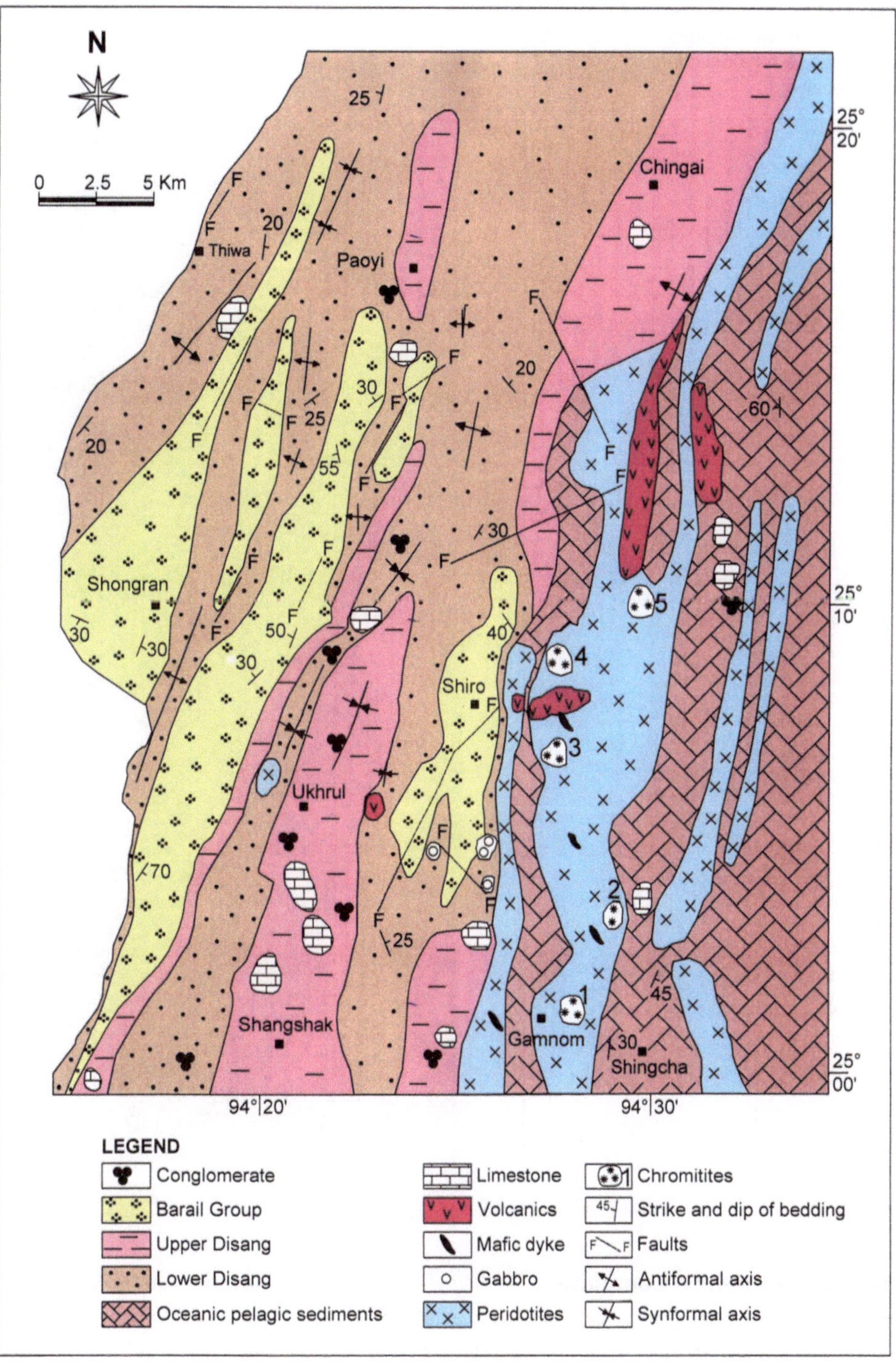

Fig. 3.10 Geological map of south-central part of ophiolite (sector III) between Gamnom and Chingai (modified after Vidyadharan et al. 1989)

hob-nail structure on weathering. Serpentine and talc are developed along the fractures. They occur both in the basal part of the layered cumulate gabbros (e.g. Tizu river), as well as isolated bodies of large dimension showing indistinct or gradational contacts with pyroxenite (Agrawal and Ghose 1986). Veins of wehrlite, pyroxenite, pegmatitic gabbro, antigorite and crysotile traverse the dunite cumulates (Figs. 3.13, 3.14, 3.15). The contact between the gabbro and the dunite is devoid of any chilled margin indicating their comagmatic nature (Fig. 3.16). The dunite grades into harzburgite, lherzolite and wehrlite. The peridotites contain lenses or disseminations of chromitite. However, workable deposits of chromitite are found only in the southern part of the ophiolite belt in Manipur (Ghosh et al. 1984; Ghosh and Goswami 1986; Ghose and Shrivastava 1986; Singh et al. 2012a). Nickeliferous magnetite bodies within cumulate lherzolite have also been noted in Nagaland.

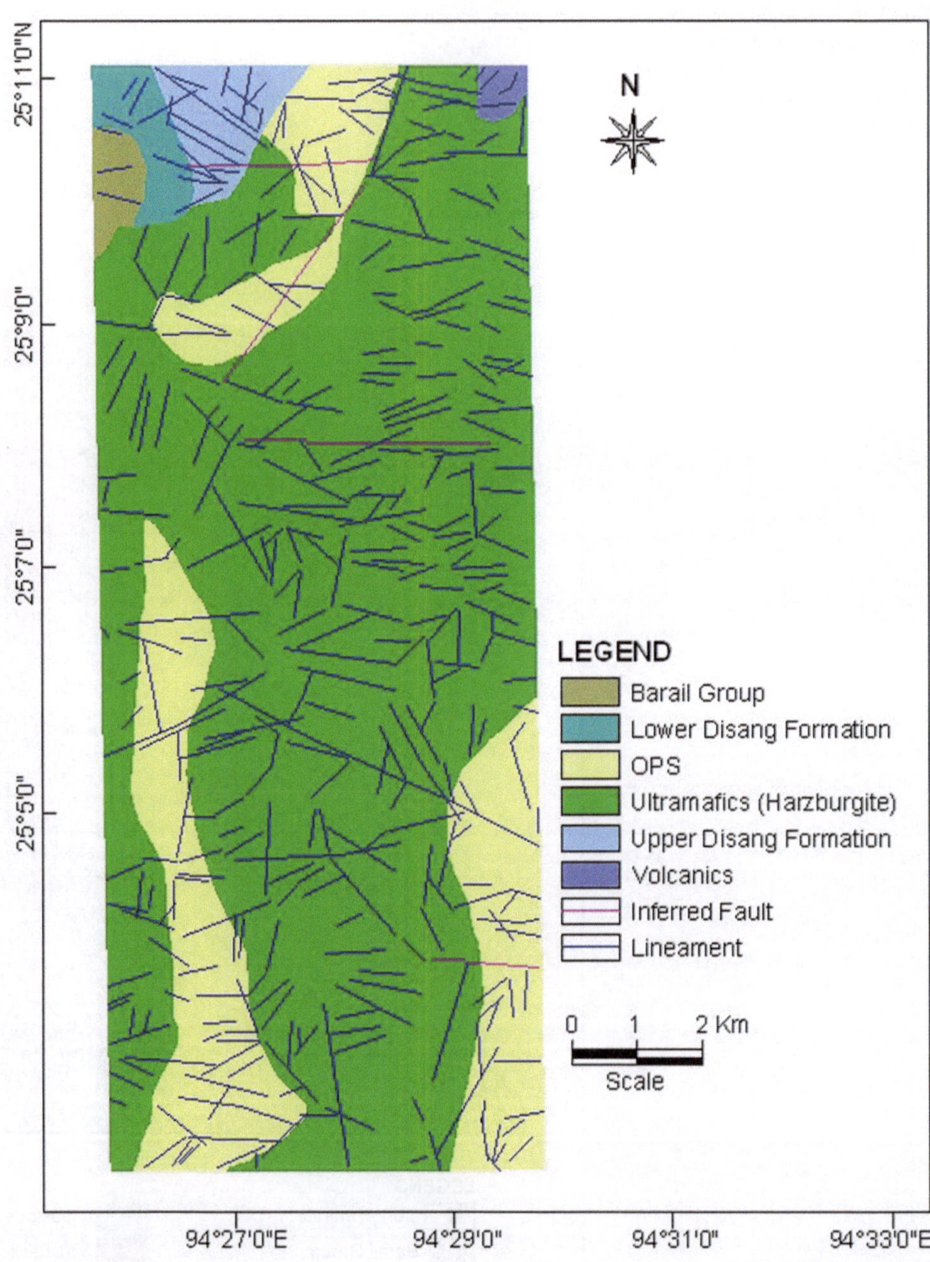

Fig. 3.11 Lineament map of the Chandel Block of Manipur ophiolite (Sector IV, Courtesy: R.N. Singh)

3.4.5.2 Pyroxenite Cumulates

The pyroxenite cumulates occur as large bands and lenses cutting across the peridotite cumulate sequence, and also as a part of the layered sequence, viz., at Tizu river and Zipu (Fig. 3.17). They are distinguished from peridotites by their coarse grain size, shiny resinous lustre, shades of green colouration and pale yellow or cherry red weathered crusts. They are mainly composed of orthopyroxene and clinopyroxene with or without olivine and minor opaques.

3.4.5.3 Mafic Cumulates (gabbroids)

The mafic cumulates include gabbro and its derivatives, viz., plagiogranite and anorthosite. Discontinuous lenses of layered gabbroids showing cryptic and rhythmic layering (Fig. 3.18) are exposed at five different locations in the thickest central part of the ophiolite belt (Agrawal and Ghose 1986) as described below:

(i) Tizu River gorge (4.5–5 km northeast of Purr): Layered gabbro about 200 m thick is exposed in contact with dunite (Fig. 3.16). Veins of leucogabbro and diorite cut across harzburgite/serpentinite (Figs. 3.19, 3.20). Two prominent sets of joints are displayed by the dunite.

(ii) Upper reaches of Zintang-ti rivulet (4 km southeast of Purr): A well-developed mafic-ultramafic cumulate sequence composed of harzburgite at the base, followed upward by websterite/clinopyroxenite, and gabbro with

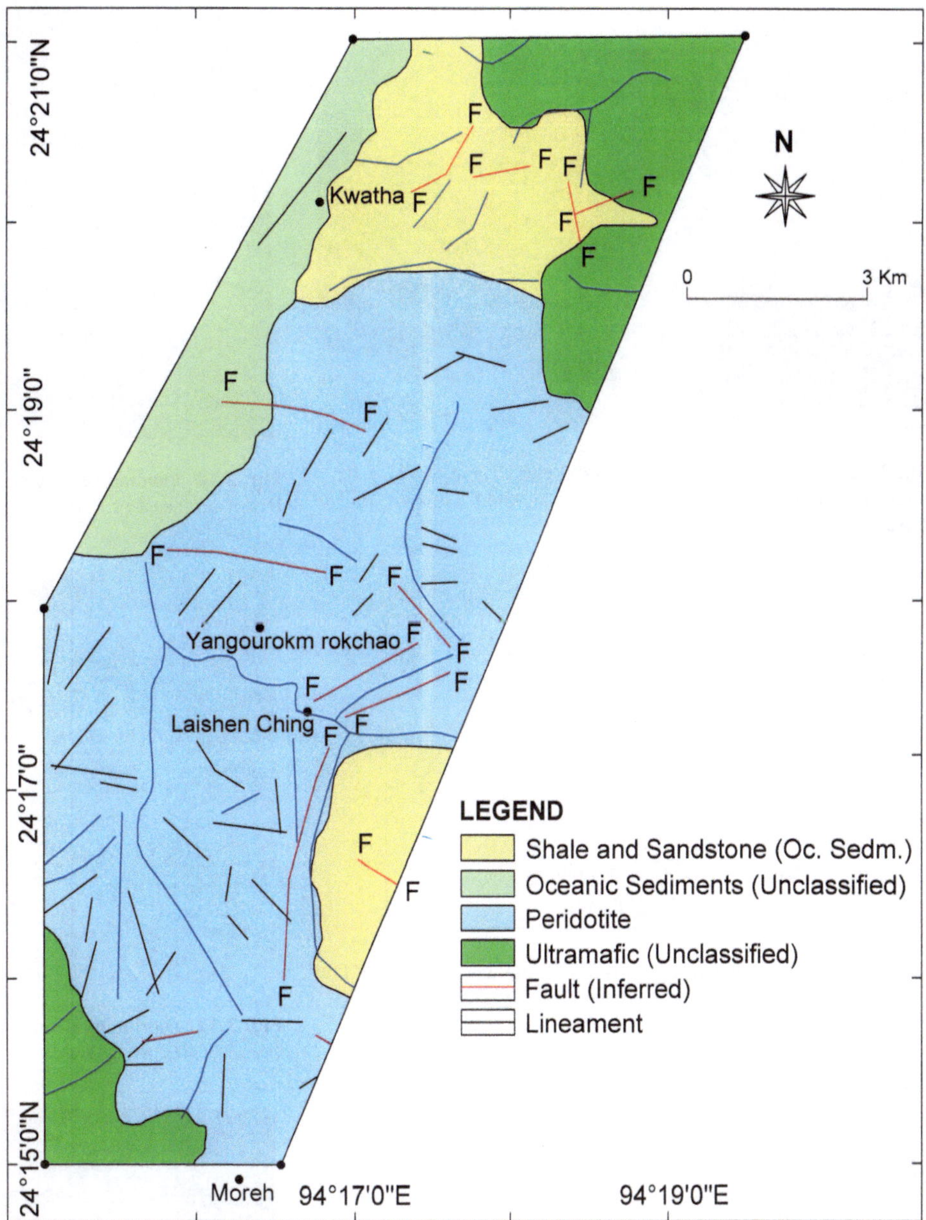

Fig. 3.12 Lineament map of the Ukhrul Block (sector V), southernmost ophiolite belt (sector V, Courtesy: R.N.Singh)

veins and layers of plagiogranite and anorthosite at the top (Fig. 3.21, 3.22). Slips and fractures in the body indicate that it is tectonically disturbed.

(iii) East of Moki: A large body of layered gabbro (3 × 2.5 km, >300 m thick) shows multiple fractures, joints and slips with abundant scree material on slopes. It is overlain by low-dipping cover sediments of the Jopi Formation.

(iv) 2 km ESE of Zipu: A layered sequence of lherzolite, olivine clinopyroxenite and websterite at the base, and feebly layered highly fractionated gabbroids at the top. The gabbroids include olivine gabbro, norite, leucogabbro and plagiogranite, showing gradational contacts. At places, pegmatitic gabbro occurs as discordant veins in the Fe-rich gabbro with abundant opaque minerals.

(v) West of Lacham Lake (Loya ti rivulet): An extensive body (2 × 4 km) of medium-grained hornblende gabbro with crude mineral lineation and banding that impart gneissic foliation to the rock. Ubiquitous presence of hornblende (up to 1 cm in size) distinguishes it from other occurrences.

High-level massive isotropic gabbros are exposed in the central part of the NHO, notably on the road section between Waziho and Zipu. They occur as dikes and veins showing discordant relationship with the host volcanics and serpentinites. They are traversed by veins of plagiogranite. In the southern sector, they occur as oval-shaped bodies of smaller dimensions (5 × 10 m, 15 × 25 m) within the pelagic sediments (Singh et al. 2012a).

Fig. 3.13 Dunite (*D*) overlain by harzburgite (*Hz*) in the Tizu River gorge. Disang flysh (*Df*) is seen in the background. *Locality* NNE of Sutsu

Fig. 3.16 Outcrop of layered gabbro (*G*) and dunite. *Locality* Tizu River, northeast of Purr

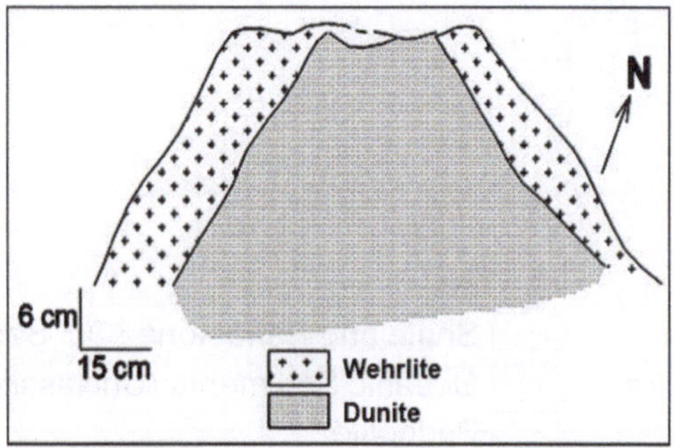

Fig. 3.14 Dunite is flanked by veins of wehrlite. *Locality* East of Purr

Fig. 3.17 Diorite intrusion in clinopyroxenite showing displacement. *Locality* Tizu River, Sutsu

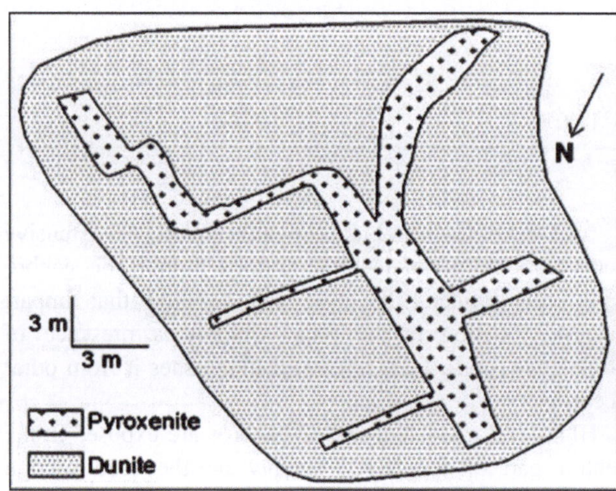

Fig. 3.15 Veins of pyroxenite in dunite. *Locality* Tizu River

Fig. 3.18 Sample of gabbro showing rhythmic layering. *Locality* Tizu River, northeast of Purr

Fig. 3.19 Folded diorite veins in serpentinite. *Locality* Tizu River

Fig. 3.20 Displacement of diorite (*Dr*) intrusion in harzburgite (*H*). *Locality* Tizu River section, NE of Purr

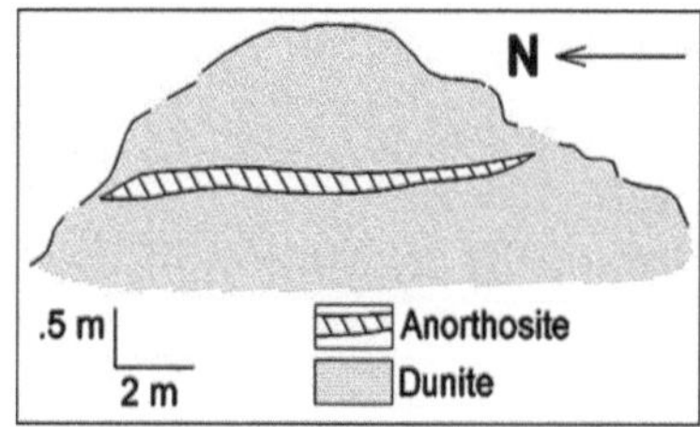

Fig. 3.21 Anorthosite vein in dunite. *Locality* East of Purr

Fig. 3.22 An ophiolitic melange, consisting of blocks of serpentinite (*S*), volcanics (*V*), limestone (*Lst*) and chert (*Ch*). *Locality* Sataza, western tectonic contact

3.4.6 Mafic Dikes

Sheeted mafic dike complexes are an integral part of many Tethyan ophiolites. They are considered to be feeders to the overlying volcanism. However, these are rarely encountered in the NHO. Their rarity in the NHO has been attributed to high-level emplacement of the gabbroids in the oceanic crust. Minor occurrence of dolerite dike is noted at the river section ENE of Zipu in association with volcanics and elsewhere in the cumulate ultramafics. They show discordant and chilled margin relationships with their host rocks (Agrawal and Ghose 1986).

3.5 Mafic Volcanics and Volcaniclastics

Mafic volcanics constitute the second most dominant component after ultramafics in the NHO. The volcanics are more abundant in the northern sector and are chiefly exposed along the tectonic contacts on both the eastern and western margins of the NHO. The basalts occurring close to thrust contacts are highly brecciated, fractured, silicified and sheared, forming a tectonic mixture with the serpentinites and the pelagic sediments (Figs. 3.22, 3.23). There is an apparent lack of continuity between the peridotites and the basalts. These exotic basaltic blocks form an ophiolitic melange (Gansser 1976). Transported blocks of greenschist, blueschist and cherts have also been observed in the Matungse-kien Range near magnetite deposits of Phokphur (formerly Pukphur; Singh 1979).

Notable mafic volcanic outcrops occur in the east of Sataza, Waziho-Zipu road, between Lacham Lake and Sutsu, and Ngazu and Mokiin the central part of the NHO. The volcanics are interlayered, intercalated and intermixed with pelagic sediments, viz., radiolarian cherts, green and grey cherts and crystalline limestone (Fig. 3.24). Multiple flows differing in colour, glass content, vesicularity, granularity, texture, structure, degree of metamorphism and alteration are noted. They are mostly basaltic in composition. At least three flows can be distinguished in the fresh outcrops of Zipu road sections (Fig. 3.25). The thickness of the flows between Waziho and Zipu vary between 3 and 6 m. Megascopically, the rocks show variable textures including massive, aphanitic to medium-grained, porphyritic, vesicular, amygdaloidal and schistose. The mafic volcanics occur at different topographic as well as stratigraphic levels. They

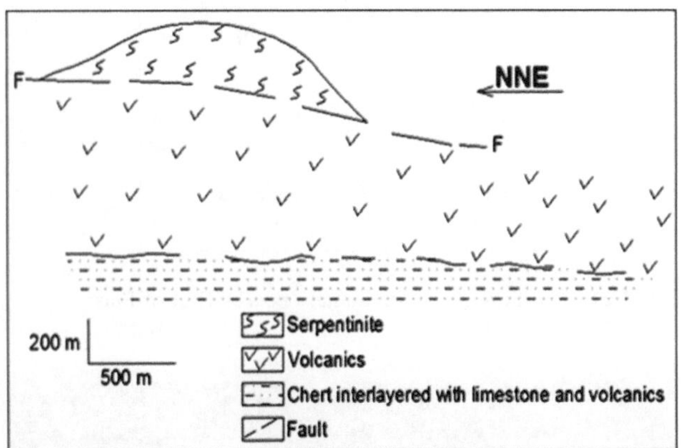

Fig. 3.23 Ophiolitic melange comprising schistose serpentinite in tectonic contact with volcanics and pelagic sediments. *Locality* Sataza

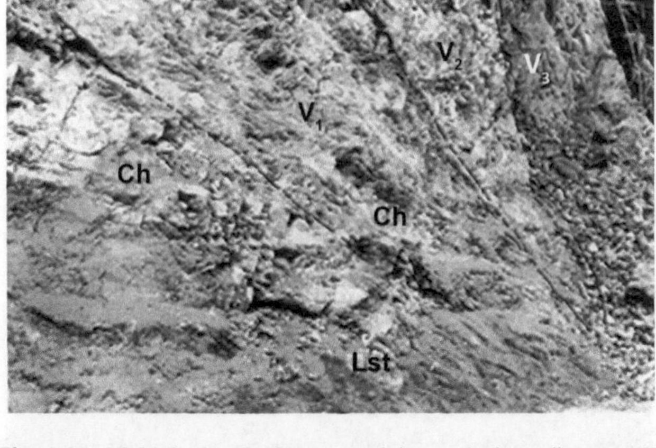

Fig. 3.25 Multiple basalt flows overlying pelagic sediments. V_1 glassy basalt, V_2 aphyric basalt, V_3 massive basalts, *Ch* chert and *Lst* limestone. *Locality* Waziho-Zipu road

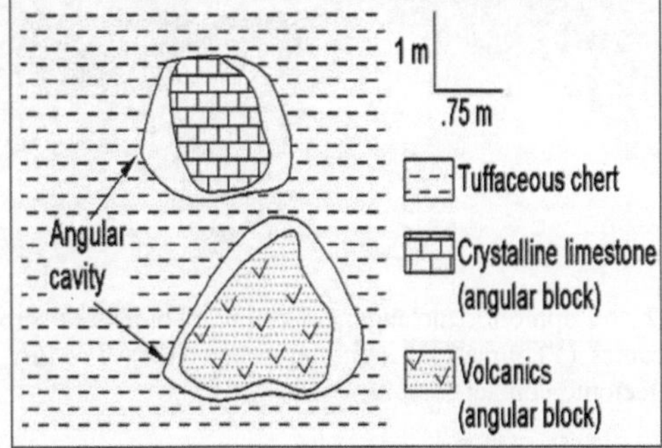

Fig. 3.24 Angular fragments of limestone and volcanics in tuffaceous chert rimmed by cavities. *Locality* Northeast of Waziho

Fig. 3.26 Pillow lava. *Locality* NNW of Zipu on Zipu road

have a minimum-estimated thickness of 800 m, although the actual thickness may be greater if tectonic dismemberment is taken into account (Agrawal and Ghose 1986).

The volcanics are tectonically overlain and underlain by serpentinite bodies in the east of Sataza. Excellent pillow structures with glassy chilled margins and a massive core are exposed at Salumi and New Basti, Tuensang district in the northern part (Singh and Ghose 1981), Waziho-Zipu road, Phek district in the central part (Fig. 3.26, Agrawal and Ghose 1986), and in the southern part (Singh et al. 2008). Eruption underwater in submarine conditions and subsequent interaction with seawater transformed the basalts into greenish grey spilite or sodic basalt composed essentially of chlorite, albite/oligoclase, epidote and calcite. In the partially altered basalts, the plagioclase has been transformed to albite/oligoclase and epidote retaining original igneous texture, and the pyroxene and olivine have been partially transformed to K-rich illite, chlorite and serpentine (Table 5.2). In view of variable alteration, the mafic volcanics have been classified in this book mainly on the basis of texture. On the other hand, the spilites show complete alteration of the mafic volcanics to albite/oligoclase, chlorite, epidote, ±ilmenite, ±calcite assemblages.

Amygdaloidal basalts with vesicles filled with calcite, quartz, rare feldspar and chlorite occur in repeated sequences in scarp sections close to the tectonic contacts and presence of lithic tuff at Sataza (western), Lagela-Sutsu and Lacham Lake-Ngazu (eastern) and Wazeho-Zipu road section (central) indicate the existence of volcanic vents. Collapse structures generated by draining away of lava near the central eruptive vents are not uncommon. Outcrops of hyaloclastite showing quenched texture, volcanic breccia, agglomerate, scoria, welded crystal-vitric tuff, lithic tuffs and amygdaloidal basalt are extensively exposed to the east and north of Lacham Lake at low elevations. Light pink and green hyaloclastites composed of basaltic fragments and pyroclastics are exposed at the eruptive centres. Pyroclastic flow deposits like ignimbrite of rhyolitic composition are encountered in an ephemeral rivulet (Leikimoro) west of Lacham Lake. Volcanic glass is either altered to palagonite or shows devitrification. Predominance of basalts and minor basaltic andesite and andesite may indicate origin at a spreading centre. However, the presence of minor andesite and tuff/ash beds intercalated with chert beds indicates volcanism and sedimentation in a shallowing marine environment perhaps with the development of a nascent volcanic arc.

3.6 Metabasic Rocks

The basalts are metamorphosed very-low to low-temperature zeolite-prenhite-greenschist facies assemblages, and high-pressure blueschist facies and type-C eclogite, indicating low geothermal gradients (Singh and Ghose Ghose and Singh 1980; Ghose et al. 1984, 1986, 2011; Vidyadharan et al. 1986; Chatterjee and Ghose 2010). Linear, isolated and narrow detached lenses, bands and tectonic slices of metabasics/blueschists trending N–S to NE–SW are developed mainly along the tectonic contacts in the northern part of NHO (Fig. 3.27). They show sharp tectonic contacts with their surrounding rocks. Schematic sections displaying the occurrence of blueschist/eclogite within two different surrounding lithologies, viz., ultramafic and volcanic, are shown in Fig. 3.28. Two sets of lineation in the fabrics of the metabasalts and oceanic sediments are possibly related to tectonic overpressures corresponding with before and after the continent-continent collision.

Several narrow slivers of blueschist have been reported from the northern and central parts of the ophiolite belt, parallel to the western mega-lineament and some other lineament (Vidyadharan et al. 1986). Important occurrences (Fig. 3.27) of these are listed together with associated rocks below:

(i) Chipur-Chokla: in the northernmost part the glaucophane-bearing rocks are associated with metavolcanic and meta-arenite.
(ii) East of Phokphur: blue amphiboles are associated with metachert striking N–S and NE–SW. Re-examination of the rock reveals presence of albite and calcite instead of lawsonite and aragonite respectively (Ghose and Singh 1980)
(iii) Moki-Kamku-Moya: associated with gabbros and serpentinites.
(iv) Lacham-Zipu: northwest of Lacham Lake along the Zipu road in contact with gabbro and volcanics.
(v) Luthur: garnetiferous glaucophne schists occur as lenses within sheared serpentinites.
(vi) Shilloi rivulet (*nala*) and Zipu ridge: in association with volcanics.
(vii) Akhen: lenses of garnetiferous serpentinite and spinel lherzolite within metaharzburgite south of Akhen close in contact to Disang flysch.
(viii) Tusom Sisi: three xenoliths of garnetiferous metabasic and lherzolite are delineated in the east of Tusom Cisi, in association with harzburgite, gabbro, metavolcanic and pelagic sediments.

The low-temperature metabasic rocks containing zeolite (natrolite), chlorite-prehnite schist and greenschists are massive to schistose bodies of small dimension, occurring as bands, lenses and pods. They also occur as small slivers within ultramafics and cherts (viz., at Sataza). They show crenulation cleavage, striations and multiple sets of quartz veins as in the blueschists. The very low-grade assemblages of zeolite and chlorite-prehnite facies were recrystallised below temperatures of 400 °C, and are distinguished from the low-grade greenschist facies rocks that contain albite, chlorite, actinolite and epidote formed at higher temperatures of ~500 °C (Ghose and Fareeduddin 2011).

Eclogite has been recorded at two places: (i) 3.5 km southeast of Sataza, and (ii) 3 km ENE of Purr. It is dark green in colour and contains uniformly distributed garnet porphyroblasts. It is genetically related to the glaucophane-bearing blueschists as indicated by interlayering and intermixing of the two rock types.

3.7 Oceanic Sediments

The pelagic oceanic and volcanogenic sediments of the NHO include tuffaceous/pelitic/psammitic/calcareous and iron-rich sediments, and cherts. The sedimentary suites have undergone low-grade greenschist facies metamorphism and deformation. Minerals commonly developed in these rocks are chlorite, epidote, sericite, phengite, muscovite, biotite, quartz, albite, potash feldspar, calcite, actinolite, tremolite, serpentine, grunerite, graphite and opaque, depending on the composition of the host rock and provenance, viz., clastic, volcanic or chemical. Late fluids derived by low-temperature metamorphic reactions and hydrothermal activity facilitate alteration and recrystallisation of the original minerals. Deformation features include folds and faults in the microscopic to macroscopic scale.

Volcanic tuff formed through submarine eruptions occurs as an admixture with lutite or muds (silts and clays) in tuffaceous sediments. Low-grade metamorphism develops a

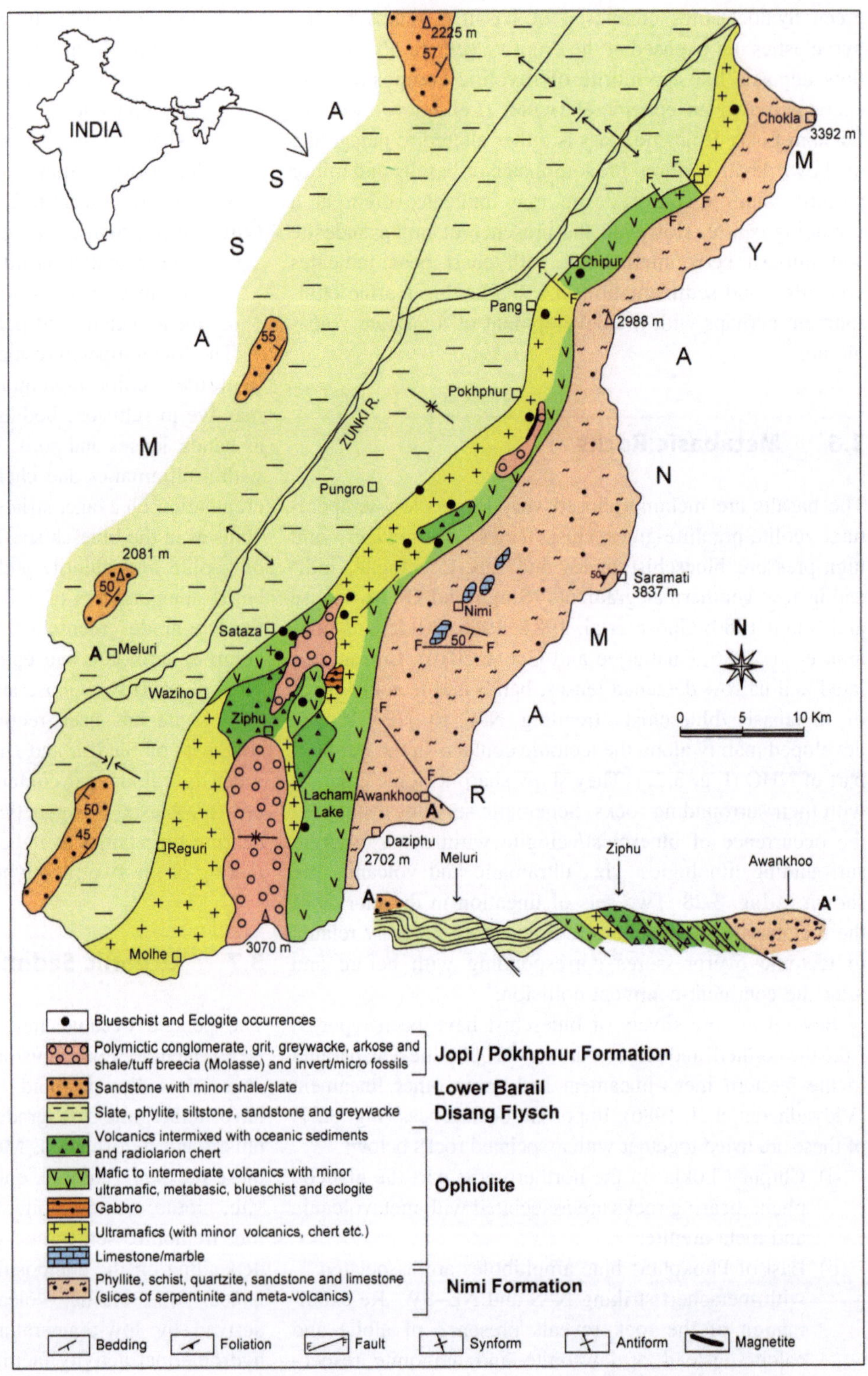

Fig. 3.27 Occurrences of blueschist and eclogite (solid circles) in the ophiolite belt of Nagaland. The geological map is modified after Agrawal (1985). A geological section along *A-A'* is shown below

mineral assemblage that includes tremolite, actinolite, chlorite, biotite, epidote, opaque and albite.

The pelitic sediments are represented by variegated shale, slate and phyllite, and rarely, schist. Phyllite shows convolute banding due to minor disturbances during sedimentation and load cast structure.

The psammitic sedimentary rocks include greywacke, arkose, subarkose and quartzite. Lithic greywacke occurs as small wedges within shale, as boudins along the foliation planes of phyllite, and as slivers and rafts within the volcanics. Arkose shows vague foliation as a result of weak metamorphism and deformation. Feebly metamorphosed

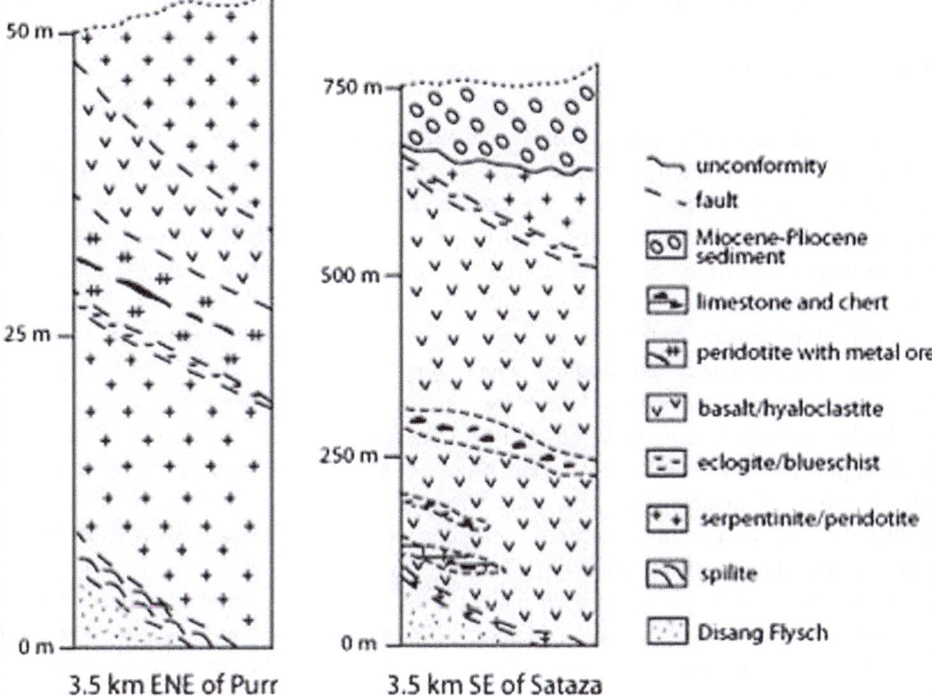

Fig. 3.28 Schematic sections of the occurrence of eclogite and blueschist schist: *A*—a lens within ultramafic cumulate, and *B*—a tectonic slice of blueschist within volcanics. Height is measured from the base of the ophiolite section (after Ghose et al. 2010)

subarkose represented by sericite-quartz schist contains mineralisation of magnetite and sulphides, e.g. pyrite and sphalerite. Banded quartzite shows alternate bands of thin dark-coloured minerals comprising chlorite-magnetite-ilmenite-sulphides-epidote and lighter bands of quartz.

The chemically precipitated (authigenic) impure siliceous limestones and marl occur as small to very large detached bands and lenses in association with chert, phyllite, serpentinite and volcanics. They commonly form steep and sheared contacts with their host.

The silica-rich cryptocrystalline cherts are extensively preserved among the oceanic sediments because of their resistance to weathering. They show a variety of colours including light grey, bluish grey, light green, dull to bright red and brown. They occur as thin beds and lenses (1–10 m) intermixed and interlayered with volcanics, ash and limestone in large outcrops (3 × 5 km, 250 m thickness). Slices and lenses of cherts are also found in serpentinites. They are massive, highly jointed, fractured and sheared (Fig. 3.29) with the development of microscopic folds and faults, and breccia with networks of recrystallised quartz veins. They are often recrystallised to coarse quartzite in contact with volcanics due to thermal effect.

Red cherts are exposed in a rivulet 0.5 km southeast of Zipu. Deep brown or reddish iron-rich siliceous sediments associated with chert beds are exposed east of Reguri. Microfossil-rich red cherts are interstratified with volcanics between Waziho and Zipu. They contain coccoliths, viz., *Zygolithus penticulus* and *Ahmullerella octoradiata*, and radiolarians, viz., *Cenozphaera* sp., *Spongodiscus* sp., etc., belonging to the Maastrichtian age (Agrawal 1985). Planktonic foraminifera and coccoliths from the pelagic limestones of Ukhrul in the southern part of the NHO also indicate the same age (Chungkham and Jafar 1998). However, radiolarian chert from the northern sector of the NHO yielded a Late Jurassic age (Baxter et al. 2011) indicating sedimentation in the early Neotethys.

Fig. 3.29 Outcrop of sheared chert. *Locality* Zipu road

3.8 Economic Minerals

3.8.1 Chromite

A number of chromitite pods, lenses and streaks occur within the NHO peridotites (dunite and harzburgite) and serpentinites. These occurrences may be compared with the Alpine-type occurrences. They are small in dimension (1–5 m in length, 1–3 m in width), have steep dips and plunge 15–20° towards north or south (Chattopadhyay et al. 1983). The chromite-bearing bodies are classified into massive, granular, disseminated and nodular types (Ghose and Shrivastava 1986). The nodular type is rare. Potential and workable chromite-bearing rocks (20 × 10 m) are confined to Manipur sector in south, viz. Sirohi, Moreh, Gamnom and Kwatha (Venkataramana and Bhattacharyya 1989). Distribution of the chromitite lenses is irregular and they show a high degree of alteration. At Gamnom, the chromitite lenses occur in an *en echelon* pattern with N–S to NE–SW elongation within closure of folds that plunge 50° towards SSW (Ghosh and Goswami 1986).

Investigation under reflected light shows that massive chromitite contains large amounts of interstitial silicate minerals. However, sintered (recrystallised) massive chromitite shows negligible interstitial silicate minerals. Chromite grains are extensively fragmented. Brecciation is confined to grain boundaries and sometime form microbrecciated zones. Chromite crystals in nodular and disseminated ores are often zoned with unaltered cores and an outer rim of Cr-magnetite (Singh et al. 2012a). Textural relationship suggests that hematite, goethite, sulfide and serpentinite were formed after crystallisation of chromite and magnetite (Ghose and Shrivastava 1986).

The nodular chromitites contain rounded chromite grains with high interstitial silicates which is greater than other textural types. Large grains of nodular chromitite containing inclusions of BMS (pentlandite), rounded silicates and polygonal grains of laurite (PGE mineral which requires confirmation by EPMA, Sisir Mondal, person. commun.). Laurite (RuS_2) is a primary magmatic phase of PGM entrapped in chromite at a high temperature.

3.8.2 Magnetite

Multiple metal-bearing magnetite occurrences have been reported from several localities in the northern part of the NHO. They contain high concentrations of Ni, Cr and Co (Nayak et al. 2010). The Phokphur magnetite deposit, the largest among seven, occurs along the inner rim of a broad, low-plunging synform truncated by high-angle faults on either side to form a graben-like structure (the Molhen-Jopi-Zipu-Sangtam-Teng-Phokphur ridge, Agrawal and Ghose 1989). The strata-bound, sheet-like deposit of magnetite overlies a serpentinite/pyroxenite of cumulate ultramafic basement (Fig. 3.30), and is unconformably overlain by younger paralic sediments of the Jopi Formation. The tabular magnetite body trends N–S to NNE–SSW, and it dips 25–40° towards west. It can be traced intermittently for about 1 km along strike. The thickness varies between 5 m and 15 m. A thin chromite layer (0.5–1.0 cm thickness) with inter-granular magnetite occurs at the base of the magnetite layer (Fig. 3.31). The origin of magnetite is attributed to serpentinization in bringing out substitution of Cr^{3+} by Fe^{3+} (Ghosh and Goswami 1986).

3.8.3 Sulphide Mineralisation

A number of occurrences of syngenetic sulphide mineralisation are associated with the volcanics, cherts, quartz-chlorite rock, occasionally with gabbro, serpentinite, igneous breccia and late felsic intrusives throughout the ophiolite belt (Fig. 3.32). Fe and Cu-bearing sulfides, viz., pyrite, chalcopyrite, bornite, cuprite, arsenopyrite, sphalerite and chalcocite together with the Cu-bearing carbonates azurite and malachite are commonly noted as specs and dissemination. The minerals in the gossan zone include Fe–Mn oxides (goethite, hematite and pyrolusite) and quartz (Ghosh and Goswami 1986). The sulphides largely occur as dissemination, fracture filling, veins, banded rims and gossans (Agrawal and Ghose 1989). The syngenetic sulphide mineralisation is synchronous with the evolution of mafic-ultramafic rocks of the ophiolite suite. These were formed as a result of basalt-sea water interaction at the accretionary ridge of the oceanic crust.

The epigenetic hydrothermal Cu–Mo sulphide mineralisation marks the second phase which is related to I-type acidic magma intrusion either during collision or post-collision period. The mineralisation is caused by transportation and concentration of Cu–Mo base-metals by late granitoids that intruded the volcanics through pre-existing fractures or shear zones. The origin of the felsic intrusives has been related to partial melting of basalt in the contact aureole of an oceanic spreading centre (Ghose and Chatterjee 2011).

3.8.4 Laterite

Thin, isolated and scattered laterite cappings are developed at favourable physiographic sites (undissected topography and gentle slope aided by warm and dry climate) over

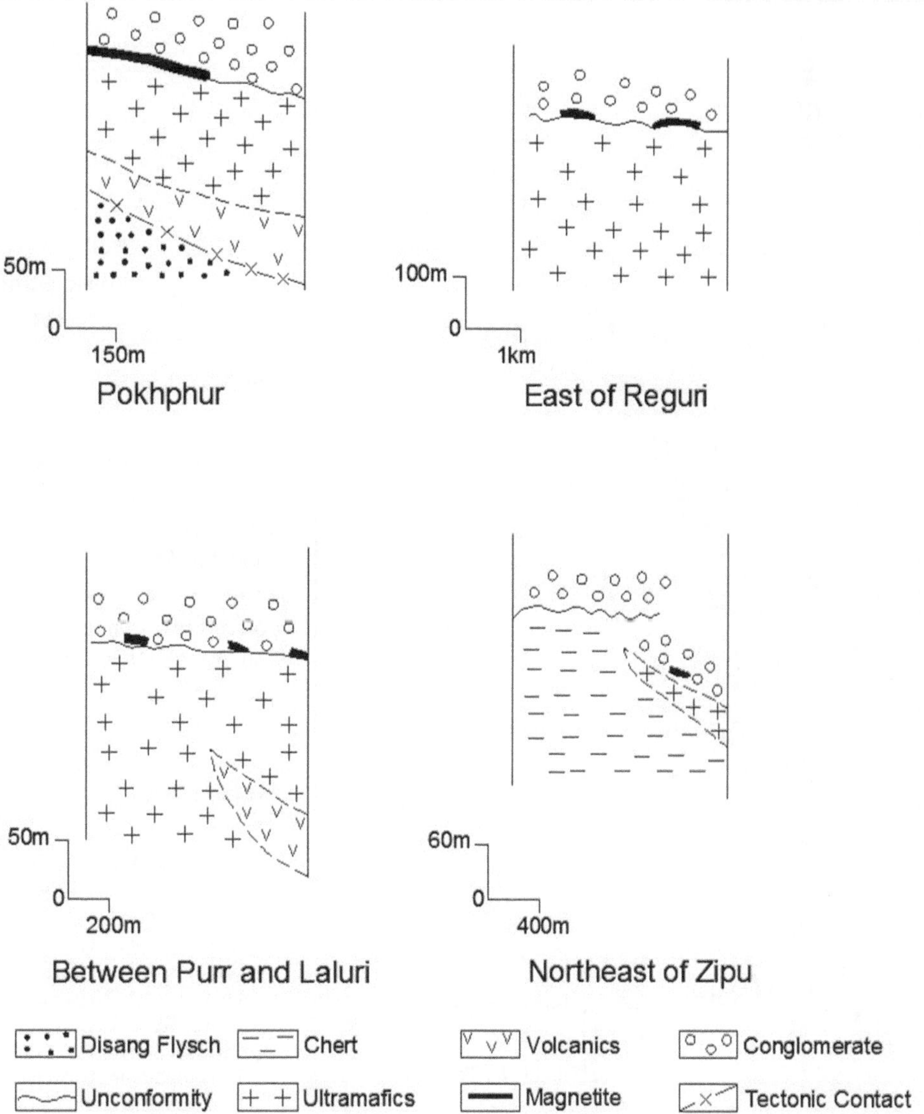

Fig. 3.30 Schematic diagrams showing occurrence of nickeliferous magnetite in the Naga Hills ophiolite

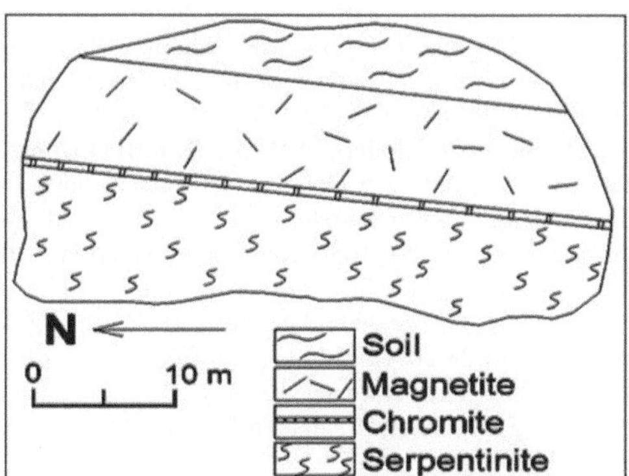

Fig. 3.31 Lamination of chromitite in nickeliferous magnetite at Matungsekien ridge, Pokhphur

serpentinised peridotite and the cumulate ultramafics between elevations of 1200 and 1800 m (Agrawal and Ghose 1989). They are rich in goethite or garnierite with high Ni abundances (Ghosh and Goswami 1986). Exquisitely preserved laterite profiles (sections) are exposed on the southern side of the Tizu River, particularly between Purr and Sutsu in the central part of the NHO. They are between 0.1 and 10 m^2 in size, and range in thickness between 3 cm and <2 m. The laterite profiles have been divided into four zones from bottom to top: (a) bed rock, (b) zone of granular goethite-limonite, (c) zone of pisolitic goethite and cryptocrystalline quartz and (d) top soil (5–20 cm) with granules of laterite (Venkataramana et al. 1986).

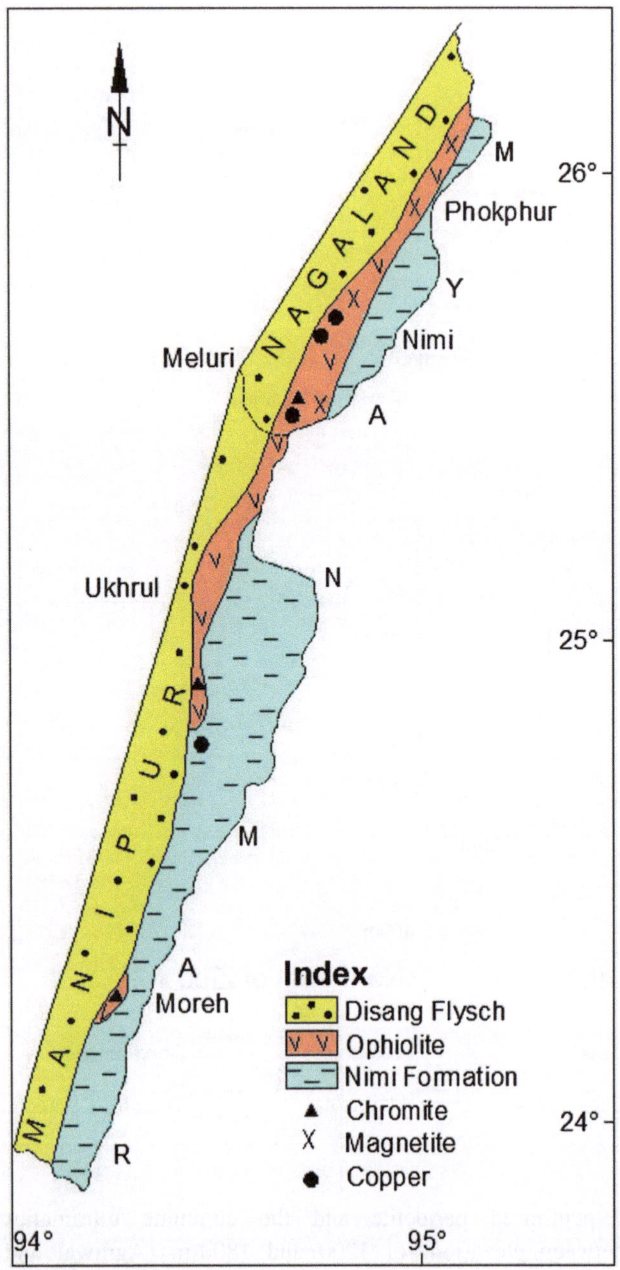

Fig. 3.32 Occurrence of chromite, nickeliferous magnetite and base metals in the Naga Hills ophiolite

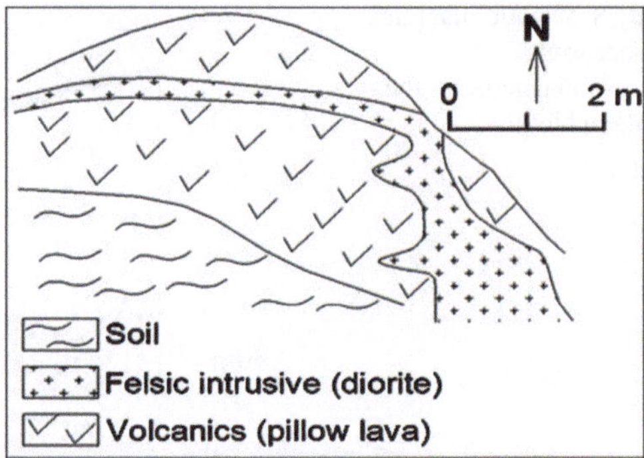

Fig. 3.33 Late felsic granite cut-across pillow lava. *Locality* Salumi

3.9 Late Felsic Intrusives/Late Tertiary Granitoids

Minor, leucocratic veins and intrusions of felsic/granitoid composition and variable thickness cut through the cumulate sequences and the volcanics (Fig. 3.33) of the NHO, and its cover sediments - Jopi Formation (Ghose and Chatterjee 2011). Mineralogically, they range in composition from quartz porphyry to granite and granodiorite. The late granitoids are distinguished from the plagiogranites of layered cumulates by higher abundance of modal quartz and presence of potash feldspar. Emplacement of these late felsic granitoids were controlled by two macrofractures, trending N–S to NE–SW and E–W in the central part of NHO. The Cu–Mo mineralisation in Zipu and Lacham Lake are associated with these late intrusives.

References

Acharyya SK, Roy DK, Mitra ND (1986) Stratigraphy and palaeontology of the Naga Hills ophiolite belt. Mem Geol Surv India 119:64–79

Agrawal OP (1985) Geology and geochemistry of the mafic-ultramafic complex of Indo-Burman ranges between Meluri and Awankhoo, Phek district, Nagaland, India. Unpublished Ph.D. thesis, Patna University, Patna

Agrawal OP, Ghose NC (1986) Geology and stratigraphy of the Naga Hills ophiolite between Meluri and Awankhoo, Phek district, Nagaland, India. In: Ghose NC, Varadarajan S (eds) Ophiolites and Indian plate margin. Sumna Publishers, Patna, pp 163–195

Agrawal OP, Ghose NC (1989) Mineral resources in the ophiolite belt of Nagaland, N.E.India. In: Ghose, (ed) Phanerozoic ophiolites of India. Sumna Publishers, Patna, pp 245–280

Bandyopadhyay S, Subramanyam MR, Sharma PN (1973) The geology and mineral resources of the Andaman and Nicobar Islands. Rec Geol Surv India 105:25–68

Baxter AT, Aitchison JC, Zyabrev SV, Ali JR (2011) Upper Jurassic radiolarians from the Naga Ophiolite, Nagaland, northeast India. Gondwana Res 20(2–3):638–644

Chatterjee N, Ghose NC (2010) Metamorphic evolution of the Naga Hills eclogite and blueschist, Northeast India: implications for early subduction of the Indian plate under the Burma microplate. J Metamorph Geol 28:209–225

Chattopadhyay B, Venkataramana P, Roy DK, Bhattacharyya S, Ghosh S (1983) Geology of Naga Hills ophiolites. Rec Geol Surv India 112(2):59–115

Chungkham P, Jafar SA (1998) Late Cretaceous (Santonian-Maestrichtian) integrated Coccolith-Globotruncanid biostratigraphy of pelagic limestone from the accretionary prism of Manipur, northeast India. Micropalaeontology 44:68–83

Das JN, Gupta KS, Setty SRK and Mohanty SN (2008) Geology, structure and mineral potential of mélanges of Phungyar-Kamjong areas of Manipur – part of an ancient subduction-margin terranes. Seminar on Indo-Myanmar Ranges in the tectonic framework of the Himalaya and Southeast Asia, Abstract, 96–97

Evans BW, Trommsdorff V, Richter W (1979) Petrology of an eclogite metarodingite suite at Cima di Gagnone, Ticino, Switzerland. Am Mineral 64:15–31

Gansser A (1976) The great suture zone between Himalaya and Tibet: a preliminary account. Colloques internationaux du Centre national de la recherche scientifique 168:181–191

Ghose, NC (2013) Occurrence of native gold and gold-silver alloy in the olivine gabbro of layered cumulate sequence of Naga Hills ophiolite, India. Curr Sci, Subjudice

Ghose NC, Chatterjee N (2011) Diorite vein in quenched basalt and its implication for the origin of late-granitoid intrusives in Naga Hills Ophiolite, Northeast India. In: Srivastava RK (ed) Dyke swarms: keys for geodynamic interpretation. Springer, Berlin, pp 315–330

Ghose NC, Fareedudeen (2011) Textural fingerprints of magmatic, metamorphic and sedimentary rocks associated with the Naga Hills Ophiolite, northeast India. In: Ray JS, Sen G, Ghosh B (eds) Topics in Igneous petrology: a tribute to professor Mihir K. Bose. Springer, Berlin pp 321–351

Ghose NC, Shrivastava MP (1986) Podiform chromite of Naga Hills ophiolite, N.E. India. In: Petrascheck W, et al. (eds) Chromites: Theophrastus Publication, Athens, pp 263–284

Ghose NC, Singh RN (1980) Occurrence of blueschist facies in the ophiolite belt of Naga Hills, east of Kiphire, N.E.India. Geologiche Rundschau 69:41–43

Ghose NC, Singh RN (1981) Structure of the Naga Hills ophiolites and associated sedimentary rock in the Tuensang district of Nagaland, N.E.India. Ofioliti 6:237–254

Ghose NC, Agrawal OP, Windley BF (1984) Geochemistry of the blueschist-eclogite association in the ophiolite belt of Nagaland, India. Cenozoic crustal evolution of the Indian plate margin. Seminar abstracts, Patna University, Patna, pp 27–30

Ghose NC, Agrawal OP, Singh RN (1986) Geochemistry of the ophiolite belt of Nagaland, N.E. India. In: Ghose NC, Varadarajan S (eds) Ophiolites and Indian plate margin. Sumna Publishers, Patna, pp 241–294

Ghose NC, Agrawal OP, Chatterjee N (2010) A geological and mineralogical study of eclogite and glaucophane schists in the Naga Hills Ophiolite, Northeast India. Island Arc 19:336–356

Ghosh S, Goswami HK (1986) Mineral occurrences and metallogenetic aspect. Mem Geol Surv India, Geol Nagaland, 119:80–93

Ghosh B, Morishita T (2011) Andradite-uvarovite solid solution from hydrothermally altered podiform chromitite, Rutland ophiolite, Andaman, India. Can Mineral 49:573–580

Joshi A, Vidyadharan KT (2008) Litho stratigraphy of the Naga-Manipur Hills (Indo-Burma range) ophiolite belt from Ukhrul District, Manipur, India. Extended abstracts: 23rd Himalayan-Karakoram-Tibet Workshop, India. Himal J Sci 5/7(special Issue):73–74

Lukram JS, Kachhara RP (2010) Molluscan biostratigraphy of the Disang Group of rocks in parts of Manipur, India. Mem Geol Soc India 75:149–163

Mitchell AHG (1981) Phanerozoic plate boundaries in mainland SE Asia, the Himalayasand Tibet. J Geol Soc Lond 138:109–122

Mitra ND, Vidyadharan KT, Gaur MP, Singh SK, Mishra UK, Joshi A, Khan IK, Ghosh S (1985) A note on the olistromal deposits of Manipur. Rec Geol Surv India 114:61–76

Naldrett AJ (1966) Talc-carbonate alteration of some serpentinized ultramafic rocks south of Timmins, Ontario. J Petrol 7:489–499

Nayak B, Vaish AK, Singh SD, Bhattacharyya KK (2010) Petrography, chemistry and economic potential of the magnetite ores of Pokphur area, Nagaland. Mem Geol Soc India 75:341–348

Ningthoujam PS, Dubey CS, Guillot S, Fagion A-S, Shukla DP (2012) Origin and serpentinization of ultramafic rocks of Manipur Ophiolite Complex in theIndo-Myanmar subduction zone, Northeast India. J Asian Earth Sci 50:128–140

Patil SK, Thong GT, Watitemsu, Temjenrenla, Rao BV (2012) Geochemistry and paleomagnetism of the basalt of the ophiolite suite in parts of Phek district, Nagaland. National symposium on recent advances in applied geochemistry: current status and future trends, Atomic Min Dept, Hyderabad, Indian Soc Appl Geochemists, Abstract Vol, 20–21

Ranga Rao A (1983) Geology and hydrocarbon potential of a part of Assam-Arakan basin and its adjascent region. In: Bhandari LL et al (eds) Petroliferousbasins in India. Petrol Asian J, Dehra Dun 127–158

Roy RK (1989) Meso-Cenozoic accretionary prism on the margin of Indo-Burman range ophiolite and its implications. In: Ghose NC (ed) Phanerozoic ophiolites of India. Sumna Publishers, Patna, pp 145–164

Sarkar A, Datta AK, Poddar BC, Bhattacharyya BK, Kollapuri VK, Sanwal R (1996) Geochronological studies of Mesozoic igneous rocks from eastern India. In: Ghose NC, Kent RW, Saunders AD (Guest eds) Mesozoic magmatism of the Eastern Margin of India. J Southeast Asian Earth Sci 13(special Issue):77–81

Shukla R (1989) Occurrence of rodingite in the ophiolite belt of Manipur. In: Ghose NC (ed) Phanerozoic ophiolites of India. Sumna Publishers, Patna, pp 189–196

Singh AK (2013) Petrology and geochemistry of Abyssal Peridotites from the Manipur Ophiolite Complex, Indo-Myanmar Orogenic Belt, Northeast India: Implication for melt generation in mid-oceanic ridge environment. J Asian Earth Sci 66:258–276

Singh AK, Singh NI, Debala Devi L, Singh RKB (2008) Pillow basalts from the Manipur ophiolite complex, Indo-Myanmar Range, Northeast India. J Geol Soc India 72:168–174

Singh AK, Singh NB, Debala Devi L, Singh RKB (2012a) Geochemistry of Mid-Ocean Ridge mafic intrusives from the Manipur ophiolite complex, Indo-Myanmar orogenic belt, N.E.India. J Geol Soc India 80:231–240

Singh, AK, Devala Debi L, Ibotombi Singh N, Subramanyam KSV (2012b) Platinum group of elements and gold distributions in the peridotites and associated podiform chromitites of the Manipur ophiolite complex, Indo-Myanmar orogenic belt, NorthEast India. Chemie der Erde, http://dx.doi.org/10.1016/j.chemer.2012.07.004

Singh RN (1979) Geochemistry of the ophiolite suite east of Kiphire, Tuensang district, Nagaland. Unpublished Ph.D. thesis, Patna University, Patna

Singh RN, Ghose NC (1981) Geology and stratigraphy of the ophiolite belt of Naga Hills, East of Kiphire. N.E. India. Recent Res Geol 8:359–381

Srikantia SV, Razdan ML (1980) Geology of part of central Ladakh Himalaya with particular reference to the Indus Tectonic Zone. J Geol Soc India 21:523–545

Venkataramana P, Datta AK and Acharyya SK (1986) Petrography and petrochemistry. In: Mitra, N.D. and 12 others, (eds.) Geology of Nagaland ophiolite, Memoir Geological Survey of India 119, 33–63

Venkataramana P, Bhattacharyya S (1989) Mode of occurrence and origin of chromite, magnetite, and Ni-bearing laterite in the Naga

Hills ophiolite, N.E.India. In: Ghose NC (ed) Phanerozoic ophiolites of India. Sumna Publishers, Patna, pp 213–234

Vidyadharan KT, Shrivastava RK, Bhattacharyya S, Joshi A, Jena SK (1986) Distribution and description of major rock types. Geol Nagaland Ophiolite, Mem Geol Surv India 119:18–27

Vidyadharan KT, Joshi A, Ghosh S, Gaur MP, Shukla R (1989) Manipur ophiolites: its geology, tectonic setting and metallogeny. In: Ghose NC (ed) Phanerozoic ophiolites of India. Sumna Publishers, Patna, pp 197–212

Vohra CP, Haldar D, Ghosh Roy AK (1989) The Andaman-Nicobar ophiolite complex and associated mineral resources—current appraisal. In: Ghose NC (ed) Phanerozoic ophiolites of India. Sumna Publishers, Patna, pp 381–315

4 Structure and Tectonics of the Naga Hills

4.1 Introduction

The evolution of the NHO is linked with the fragmentation and dispersion of Gondwana (protocontinent of the southern hemisphere) and its subsequent collision and amalgamation with Eurasia during Meso-Cenozoic time. The post-fragmentation history of Gondwana and its unification with Eurasia is responsible for shaping the present physiographic configuration of southern Europe and Southeast Asia, including larger tracts of the Russian Federation and China. The collisional boundary, called the geosuture, demarcates the convergence of the two protocontinents that witnessed the emplacement of ophiolite between the two continental plates. A segment of the Tethyan oceanic crust of Southeast Asia, obducted onto the eastern continental margin of India, now stands exposed as the Naga Hills ophiolite.

Structurally, the Naga Hills form a part of the Indo-Myanmar Range (IMR) that has a parallel geological history to that of the Himalayan Mountain Range in the northwest and the Indonesian Arc in the southeast. The deformations that resulted in making the loftiest mountain chains at one end and the deepest trench to the opposite end provide a better understanding of the mechanism of crustal plate movements in Southeast Asia. As originally suggested by Chibber (1934), the IMR and Andaman arcs, together provide an important transitional link between the Himalayan collision zone and the Indonesian Arc, which is connected with the Western Pacific arc system. The Indus-Yarlung suture between the Tibet and the Himalaya, joining southward through Lohit Himalaya (Lushai and Patkai hills, Ghosh and Ray 2003; Ghosh et al. 2007)-Naga-Chin-Arakan Yoma Hill Ranges between India and Myanmar to the Andaman-Nicobar Islands forming an arc, are known for the occurrence of a series of dismembered ophiolites (Agrawal and Kacker 1980; Roy and Kacker 1980; Chattopadhyay et al. 1983; Mitchell et al. 2010).

Mathur and Evans (1964) gave a lucid account of the regional geology of northeastern India including parts of the IMR within Indian territory. They considered the Assam and the Arakan basin of Myanmar as a contiguous geosyncline with a geanticline along the IMR as the dividing line between two sub-basins. The earliest sediments of this basin, similar to Disangs of the IMR, were deposited in Upper Cretaceous and continued up to Eocene. These are contemporaneous with the shelf sediments (Jaintia Group) of Assam (Das Gupta and Biswas 2000).

4.2 Indo-Myanmar Range

The Indo-Myanmar Mountain Range provides evidence of strong orogenic deformation that is characteristic of Cenozoic fold mountain belts; in particular, the Himalayas. The 1,200-km-long IMR, trending NNW–SSE to NE–SW, has been sub-divided into three segments each of about 400 km in length from south to north, as Arakan Yoma and Chin Hills (in Myanmar) and Naga Hills (in India and Myanmar, Brunnschweiler 1974). The IMR is bounded in the east by the Central Lowland (CL) of Myanmar (Neogene-Quaternary) and in the west by the Belt of Schuppen (Evans 1964). It is separated from the Proterozoic Shillong Plateau-Mikir Hills of the eastern Indian shield margin by the Dawki tear fault and the Brahmaputra valley by the Naga thrust (Figs. 3.3, 4.1). The IMR and CL, 250–700 km wide, is a Tertiary orogenic belt, forming a broad convex arc towards the Indian foreland (Brunnschweiler 1974). The CL is divided into two troughs, eastern and western, by a volcanic arc (Mitchell and McKerrow 1975). The eastern trough containing Tertiary marine and fluviatile sediments is juxtaposed with the Proterozoic Shan plateau (Shan-Tenasserium block). The western trough sediments are thrust westward over the IMR and comprise turbidites and serpentinites (Evans 1932).

The shelf areas of the Indian shield overlying the Proterozoic basement of Shillong-Meghalaya-Mikir Hills, plunges eastward. The 'Belt of Schuppen' (Late Paleogene-Neogene), a molasses basin known for imbrication into a series of thrust faults, tectonically overrides the Indian

N. C. Ghose et al., *A Petrographic Atlas of Ophiolite*,
DOI: 10.1007/978-81-322-1569-1_4, © Springer India 2014

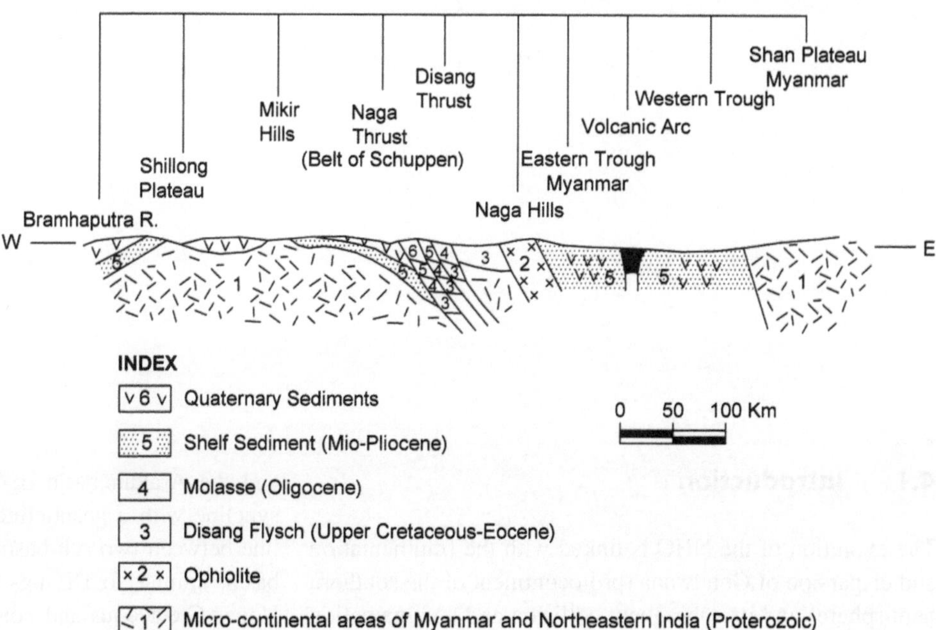

Fig. 4.1 Geological section along Latitude 25° N, across northeast India and northern Myanmar

foreland spur towards northwest (Evans 1932; Mathur and Evans 1964). The Tertiary rocks have an early marine facies and a late fluviatile facies. These consist of thick argillaceous sediments at the base (Disangs) passing upward into sandstone inter-bedded with coal seams and conglomerate (Barail), separated by unconformity. The Cretaceous-Paleogene rocks constituting the IMR are dissected by easterly dipping thrusts (Clegg 1941).

The mafic–ultramafic assemblages occurring between Myanmar and the IMR likely constitute three discrete ophiolite belts possibly of successively younger age from east to west (Fig. 4.1). One ophiolite belt is exposed along the western margin of Shan Plateau up to the north of Mandalay, consisting of glaucophane- and jadeite-bearing schists, serpentinite and pre-Eocene basalts of Early-Mid-Jurassic age (Mitchell 1981; Curray et al. 1979). The second belt (middle) occurs along the western margin of the Central Myanmar magmatic arc of Mid-Jurassic to Early Cretaceous age indicated by presence of Mid-Cretaceous and Eocene granodiorite plutons (Acharyya 2010). The ophiolites of IMR (Naga Hills-Chin-Arakan Yoma) of Late Cretaceous to Paleocene are possibly the youngest of these (Patil et al. 2012).

4.3 Geophysical Signature

The Bouguer anomalies over Myanmar and adjoining Indian territories (Verma and Mukhopadhyay 1977) show two distinct contour trends reflecting an E–W structural lineation over Shillong plateau and N–S over Bengal basin and Myanmar (Burmese) arc. These were attributed to the eastern Himalayan fore-deep and the periphery of Indian shield margin and Myanmar arc respectively. An E–W transact has brought out a distinct low gravity anomaly over the Central Belt (CB) molasse basin of Myanmar (Mukhopadhyay and Dasgupta 1988).

Active subduction of the Indian plate is currently occurring beneath the Myanmar arc in an east-dipping Benioff zone that extends to a depth of about 180 km (Mukhopadhyay and Dasgupta 1988). The over-riding Myanmar plate forms an inland seismic slab that is deflected downwards near the Benioff zone. Based on gravity data, the thickness of the Central Lowlands of Myanmar has been estimated to be 10–15 km by these authors.

GPS data indicate that more than half of the present-day convergence (36–40 mm/yr) between the Indian and Eurasian plates, forming western China and surrounding regions, is one of the widest diffuse plate-boundary zones. The subduction zones around the eastern margins of the Asian continent contribute little to the intra-continental deformation in the Asian continent at present but may have played a major role in the early Cenozoic, when most of the Tibetan Plateau was not fully uplifted and back-arc spreading was intense along the eastern margins of the Asian continent (Liu et al. 2007).

Inhomogeneous tectonic set-up associated with varying subduction rate of Indo-Myanmar subduction margin is associated with varying seismicity distribution. Higher seismicity with occasional incidences of large earthquakes is related to higher degree of deformation and stress accumulation in the subducting lithosphere (Khan et al. 2010).

Analysis of focal mechanism solutions for the Himalayas and the Arakan-Yoma shows that in addition to thrusting, strike-slip faulting is also taking place along some of the transverse features which behave like steeply dipping fracture zones (Verma and Krishna Kumar 1987). The deformation coincides with the western and eastern N–S directed major lineaments in the northern Indian Ocean manifested on land into the Quetta Line (west) and Arakan Yoma (east), culminating into western and eastern syntaxial bends of the Himalayas respectively (Gansser 1966). Orientation of P-axes for all thrust solutions in the two regions shows a sharp change from predominantly E–W along the Myanmar arc to N–S and NE–SW along the Himalayas. The direction further changes to NW–SE along the Baluchistan arc. It gives an impression that the Indian lithosphere is under compression from all sides (Verma and Krishna Kumar 1987).

Seismotectonic studies of the collision zones around the Indian plate in the Himalayas and the IMR present a contrasting view concerning the tectonic development and stress regime in the Myanmar arc in support of the above inference. The earthquakes in the Northeast Himalayan Collision Zone are deeper (>30 km) compared to that of the Western Himalayan Seismic Zone (<30 km), the former being related to transverse tectonics (Kayal 2010). The earthquakes in IMR on the other hand, are caused by continent–continent subduction; the shallower (<90 km) events show normal and strike-slip faulting, whereas the deeper earthquakes (90–200 km) are generated by thrust faulting within the Benioff zone. Different segments along Myanmar-Andaman-Sumatra subduction margin experience widely varying seismicity with maximum concentration in Sumatra (Khan et al. 2010).

4.4 Structure

The ophiolite belt is classified into three distinct structural domains separated by two major tectonic zones. The three domains from west to east (Fig. 3.4) are:

(i) Disang and Barail folded sequences
(ii) Ophiolite suites and its cover sediments - Jopi/Phokphur Formation
(iii) The metamorphic basement (Nimi Formation/Naga Metamorphics)

The ophiolite is an allochthonous body of oceanic crust ancestry. It has been transported to present sites through tectonic slicing, imbrication, overthrusting, fracturing, mylonitization and tectonic agglomeration in the formation of ophiolitic melange. It shows polyphase deformation and metamorphism. Phases of upliftment of Himalayan orogeny and subduction of the Indian plate against Eurasia are a continuous process bringing changes in the distribution of land and composition of the mantle. The major and minor structures of the ophiolite and its relationship with the associated rocks are presented below and summarised in Table 4.1.

4.4.1 Folds

A systematic study of structural history of the ophiolite belt is yet to be made. Early studies have shown three generations of fold systems broadly correlatable with the major folding episodes of the Himalayan orogeny (Ghose and Singh 1981; Roy et al. 1982; Chattopadhyay et al. 1983; Acharyya et al. 1986a, b; Nandy 1986; Bhattacharyya and Venkataramana 1986; Agrawal and Ghose 1986; Shukla 1989; Roy 1989; Ghose and Agrawal 1989). Combining dislocation episodes with folding, five phases of deformation have been recognised by Bhattacharjee (1991). According to him the first deformation (F_1-folds) on regional scale, has affected the pre-flysch and the flysch sediments. These have developed tight isoclinal to recumbent folds with steep axial planes with low plunging axes. The schistosity is parallel to axial planes of these folds in areas of low-grade metamorphites. The second deformation (F_2-folds) is marked by a series of reverse faults trending NNE–SSW in the north (Nagaland) and N–S in the south (Manipur). The thrust blocks moved westward on easterly dipping fault planes, bringing older rocks from west to east. The third deformation (F_3-folds) has produced a number of asymmetrical folds with axial planes steeply dipping east. The western limbs are dismembered by reverse faults. The fourth phase of deformation has produced faults, viz., (i) gravity faults which occur all along the belt, (ii) conjugate sets of strike-slip faults, confined to the foreland of the orogenic belt and (iii) gravity faults striking parallel to the hinges of the anticlinal folds of foreland areas of the orogen. The fifth phase has produced a series of asymmetrical folds outside the ophiolite belt. The (iii) fault could be associated with this late folding phase. The F_1 and F_2 folds are ascribed to be comparable with the early and late Alpine-Himalayan deformations and the F_3-fold with the Pliocene-Quaternary movements (Roy and Kacker 1980).

Statistical interpretation of S_1 poles to foliation and cleavage planes of the flysch indicates 25° plunge of F_2-axis towards N 195° (Fig. 4.2a). Similarly, S_1 poles to the bedding planes of the cover sediments (Jopi Formation), shows distribution along a great circle girdle, the ß-axis of which dips 25° towards N 220° (Fig. 4.2b). The F_1 fold axis shows a significant amount of rotation as a result of superimposition of later folds.

Table 4.1 Relationship between deformation/structural elements and metamorphic episodes in the Naga Hills Ophiolite, India (modified after Ghose and Agarwal 1989)

Phase	Major structure	Minor structure	Resultant impact	Metamorphic effects
F_1-phase	Major regional folds and small-scale first fold with reclined and inclined geommetry	1. Major fold on S_0 parallel type with varied geometry 2. Well-developed axial plane schiostosity/cleavage (S_1) marked by NNE–SSW orientation of flaky minerals and flattened quartz grains. Intersections of S_0 with S_1 developed as L_1 on S_1 plane 3. Mineral lineation L_1	1. Development of folds and axial plane foliation (S_1), broadly synchronous and developed by flexure slip as a result of layer parallel compression from WNW–ESE 2. Variation in the attitude and orientation of S_1 mainly due to superposed folding	Pelitic sediments were converted to phyllite and mica schist, chert to meta-chert, and basalts to zeolite-prehnite-greenschist facies, blueschists and eclogite (M_1 episode)
F_2-phase	Superposed puckers/ mesocopic folds with low dipping fracture cleavage/slip schiostosity with variable amount and direction of plunge	1. Second fold on S_0 and S_1, similar type and tightly appressed 2. Development of old pucker 3. Development of younger pucker sub-parallel or oblique to earlier pucker 4. Crenulation foliation formed parallel to axial plane of F_2 folds	1. Development of second fold and broadly synchronous axial plane fracture cleavage/ slip schistosity 2. Developed slightly later than F_2 folds by progressive deformation parallel to axial plane of F_2 fold on S_1	Recrystallisation of sericite, quartz and antigorite along S_2
F_3-phase	Superposed open folds with sub-horizontal axial plane trending E–W	Superposed symmetrical folds with low dippping axial planes	Reorientation of earlier fabric elements	

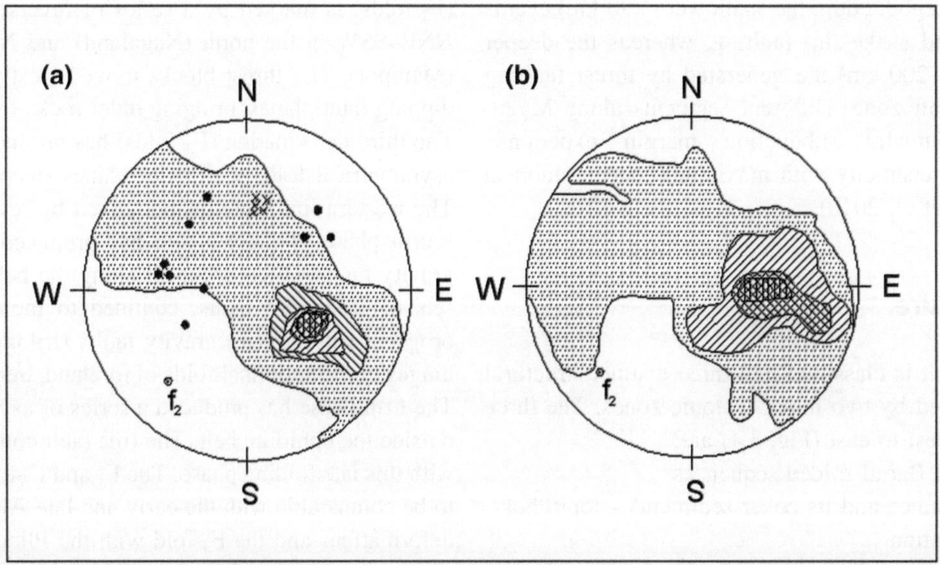

Fig. 4.2 **a** S_1 diagram for 112 S_1 poles (parallel S_2) from the Jopi Formation. Contours 16, 12, 8, 4 and 2 % per 1 % area. Solid circles and crosses represent attitude of f_1 and f_3 axes respectively. **b** S_1 diagram for 110 S_1 poles from the Disang Flysch. Contours 6, 4, 2 and 1 per 1 % area

4.4.2 Faults

Three generations of macro-fractures consisting of longitudinal and transverse faults have occurred in the NHO (Agrawal 1985). The most conspicuous of these is the longitudinal faults that trend NNE–NE, constituting high-angle reverse faults with steep easterly dips (Singh and Ghose 1981; Bhattacharyya and Venkataramana 1986). This has resulted in juxtaposition of the ophiolite with the Nimi Formation in the east and the Disang flysch in the west (Fig. 3.3).

The WNW–ESE transverse faults are dextral or sinistral gravity-cum-wrench faults. A third set of faults trending northwest is also encountered as reverse-cum-wrench faults. Both the WNW and NW trending faults are also high-angle faults with moderate to steep dips (45–60°, Ghose and Agrawal 1989). Courses of major rivers in the Naga Hills are guided by the thrust planes. These lineaments cause abrupt termination, juxtaposition and off-setting of the litho-units. It is best displayed in the 'Belt of Schuppen' (Mathur and Evans 1964) which caused westward thrusting of the orogenic belt against the Indian continent.

4.4.3 Contact Between Disang Flysch *and Ophiolite*

The boundary thrust in the western contact between the Disang Flysch and the NHO is marked by brecciation, silicification, presence of fault gouge and intermixing of lithologies (Ghose et al. 2010). The Disangs are tightly folded and faulted with ferrugination along the shear zones. The Upper Disangs and the oceanic pelagic sediments (OPS) preserve E–W or NW–SE trending minor folds which are often recumbent with W-, M- and S-type (Vidyadharan et al. 1989).

Slivers of serpentinite within the Disang phyllite and slate and *vice versa* are common. Discordant, imbricate tectonic lenses and bands of ophiolite occur as sub-vertical rootless folds along the Tizu and Arachu Rivers in the central part of the belt (Agrawal and Ghose 1986). Ophiolite mélange (Figs. 3.22, 3.23) so produced indicates emplacement of the ophiolite through uplift onto and into the marine sediments of the Disang Flysch. The high-pressure litho-assemblages of glaucophane schist and eclogite are emplaced close to the boundary thrust between Disang Flysch and ophiolite (Fig. 3.27).

4.4.4 Contact Between Nimi *Formation/Naga Metamorphics and Ophiolite*

The eastern tectonic contact of ophiolite with the metamorphosed-Nimi Formation also trends NE–SW with moderate to steep dip towards SE. The Nimi Formation has thrust over the ophiolite from the east which is marked by imbrication, mylonitization, silicification, brecciation, truncation of beds and formation of sheared mass of agglomeration of litho-units of the two at the confluence of Turati–Tizu Rivers (Agrawal 1985). Fragmented klippe of metamorphics overlie the ophiolite as well as the sedimentary cover of Jopi/Pokhphur Formation (Bhattacharyya and Venkataramana 1986).

4.4.5 Relationship Between Intra-ophiolite Litho-Units

The intra-ophiolite structural fabric is characterised by intensive tectonic slicing, shearing (e.g. spinel peridotite, blueschist and eclogite), brecciation (e.g. gabbro, serpentinite and glaucophane schist), disruption, mylonitization (e.g. arenaceous sediments, blueschist), compressional and extensional features and intermixing of lithologies in chaotic tectonic setting to form ophiolitic melange (Agrawal and Ghose 1986). The individual member of ophiolite suite maintains tectonic contact with each other with limited strike length. Their wedge-shaped linear pattern of litho-units indicate emplacement along high-angle faults. They show large-scale development of fault breccia, crush conglomerate, imbrication, stretching, micro-shearing and S–C mylonite (Plate SP-13, 15) on microscopic scales (Ghose and Fareeduddin 2011). A common sight in ophiolite terrain is the presence of fault scarp comprising chert and volcanics. Some N–S to NE–SW trending isoclinal folds have been preserved in the chert and limestone beds. Absence of intrusive bodies, contact metamorphism and apophyses and other field evidences give unequivocal support to a mechanical displacement or emplacement of ophiolite as a result of tectonic uplift and exhumation.

4.4.6 Ophiolite and Sedimentary Cover

The ophiolite derived clastic to volcaniclastic rocks formed along the margin of the sea and continental deposits (paralic, lagoonal and fluvial type) of Jopi/Phokphur Formation and got folded and imbricated with the ophiolite basement. The sedimentary envelope is variably deformed, shows incipient metamorphism in the basal part and is overridden by the Nimi Formation from the east (Nandy 1986, 2000). Overturned to isoclinal folds (F_1) with axial plane parallel to N–S to NE–SW are common (Bhattacharyya and Venkataramana 1986). The F_2-fold is encountered where F_1 is refolded coaxially. Based on fossil assemblages, the Jopi Formation has been ascribed Eocene (Ranga Rao 1983; Acharyya et al. 1986a, b). In consequence, F_1 and F_2 deformations may be considered as post-Eocene (Bhattacharyya and Venkataramana 1986).

4.5 Eastern Indian Plate Margin

The validity of Naga Hills ophiolite as eastern suture of the Indian plate has been questioned of late. The eastern ophiolite belt located along the magmatic arc of Central Myanmar Lowland, a zone of gravity high, is said to have been overthrusted westward as rootless, flat-lying nappes and juxtaposed with the flyschoid sediments now constituting the Naga Hills ophiolite (western ophiolite belt), lying close to a zone of negative gravity anomaly (Sengupta et al. 1990, Acharyya 2010). This is contradicted by presence of S–C mylonite in NHO (Ghose et al. 2010; Ghose and Fareeduddin 2011). Geophysically, large negative isostatic anomalies of the order of a few hundred milligals in the western belt have been related to flexure of the crust and mantle as a result of collision between two plates (Watts and Talwani 1974). It is corroborated that negative isostatic anomaly in the Assam-Arakan basin is due to flexure of the lithosphere and focal mechanism solutions for earthquakes taking place in the region (Verma 1985 p.167). Therefore, occurrence of ophiolitic melange and S–C mylonite provide unequivocal support for ductile shear zone along which the ophiolite was mechanically emplaced and defines the eastern Indian plate margin.

References

Acharyya SK (2010) Tectonic evolution of Indo-Burma range with special reference to Naga-Manipur Hills. Mem Geol Soc India 75:25–43

Acharyya SK, Mitra ND, Nandy DR (1986a) Regional geology and tectonic setting of northeast india and adjoining region. Mem Geol Surv India 119:6–12

Acharyya SK, Roy DK, Mitra ND (1986b) Stratigraphy and palaeontology of the Naga Hills ophiolite. Mem Geol Surv of India 119:64–74

Agrawal O.P (1985) Geology and geochemistry of the mafic-ultramafic complex of Indo-Burman ranges between Meluri and Awankhoo, Phek district, Nagaland, India. Unpublished Ph.D. thesis, Patna University, Patna

Agrawal OP, Ghose NC (1986) Geology and stratigraphy of the Naga Hills ophiolite between Meluri and Awankhoo, Phek district, Nagaland, India. In: Ghose NC, Varadarajan S (eds), Ophiolites and Indian Plate Margin. Sumna Publishers, Patna, pp 163–195

Agrawal O.P and Kacker R N (1980) Nagaland ophiolites, India - A subduction zone ophiolite complex in a Tethyan orogenic belt. In: Panayiotou A (ed.) "Ophiolites", proceedings International Ophiolite symposium, Cyprus Geol Surv Dept, 1979, 454–461

Bhattacharjee CG (1991) The ophiolites of northeast India – a subduction zone ophiolite complex of the Indo-Burman orogenic belt. Tectonophysics 191:213–222

Bhattacharyya S, Venkataramana P (1986) Structure and metamorphism of Naga ophiolite. Mem Geol Surv India 119:28–32

Brunnschweiler R O (1974) Indoburman ranges. In: Spencer, A M (ed.), Mesozoic-Cenozoic orogenic belts. Geol Soc London Sp Pub 4:279–299

Chattopadhyay B, Venkataramana P, Roy DK, Bhattacharyya S, Ghosh S (1983) Geology of Naga Hills ophiolites. Rec Geol Surv India 112(2):59–115

Chibber HL (1934) Geology of Burma. MacMillan, London 538p

Clegg ELG (1941) The Cretaceous and associated rocks of Burma. Mem Geol Surv India 74:1–101

Curray JR, Moore DG, Lawver LA, Ehmel FJ, Baitt EW, Henry M, Kielhefer R (1979) Tectonics of Andaman Sea and Burma: geological and geophysical investigations of continental margins. Am Assoc Petrol Geol Mem 29:189–198

Das Gupta AB (1977) Geology of the Assam-Arakan region. Q J Geol Min Metall Soc India 49:1–54

Das Gupta AB, Biswas AK (2000) Geology of Assam. Geol Soc India, Bangalore, 169 p

Evans P (1932) Tertiary succession in Assam. Trans Min Geol Inst India 27:155–250

Evans P (1964) The tectonic framework of Assam. J Geol Soc India 5:80–96

Gansser A (1966) Indian Ocean and the Himalayas: A geological interpretation. Ecol Geol Helv 59:831–848

Ghose NC, Agrawal OP (1989) Geological framework of the central part of Naga Hills ophiolite, Nagaland. In: Ghose NC (ed) "Phanerozoic ophiolites of India". Sumna Publishers, Patna, pp 165–188

Ghose N C and Fareeduddin (2011) Textural fingerprints of magmatic, metamorphic and sedimentary rocks associated with the Naga Hills Ophiolite, northeast India.In: Ray J S, Sen G and Ghosh B (eds.) "Topics in Igneous Petrology: A Tribute to Professor Mihir K. Bose", Springer-Verlag, Berlin- Heidelberg, pp 321–351

Ghose NC, Singh RN (1981) Structure of the Naga Hills ophiolites and associated sedimentary rock in the Tuensang district of Nagaland, N.E.India. Ofioliti 6:237–254

Ghose NC, Agrawal OP, Chatterjee N (2010) A geological and mineralogical study of eclogite and glaucophane schists in the Naga Hills Ophiolite, Northeast India. I Arc 19:336–356

Ghosh B, Ray J (2003) Mineral chemistry of ophiolitic rocks of Mayodia-Hunli area of Dibang valley district, Arunachal Pradesh, Northeastern India. Mem Geol Soc India 52:447–471

Kayal JR (2010) Seismotectonics of Northeast India: A review. Mem Geol Soc India 75:55–68

Khan PK, Mukherjee G, Chakraborty PP (2010) Seismotectonic overview of the Burma-Andaman-Sumatra subduction margin preceding the 2004 off Sumatra mega event. Mem Geol Soc India 75:81–95

Liu M, Yang Y, Shen Z, Wang S, Wang M and Wan Y (2007) Active tectonics and intracontinental earthquakes in China: the kinematics and geodynamics. In: Stein S and Mazzotti S (ed) Continental Intraplate Earthquakes: Science, Hazard, and Policy Issues. Geol Soc of Am Sp Paper 425:209–318

Mathur L P and Evans P (1964) Oil in India. In: 22 Intern Geol Cong, New Delhi, p 86

Mitchell AHG (1981) Phanerozoic plate boundaries in mainland SE Asia, the Himalayas and Tibet. J Geol Soc Lond 138:109–122

Mitchell AHG, Hlaing T, Nyunt H (2010) The Chin Hills segment of the Indo-Burman Ranges: not a simple accretionary wedge. Mem Geol Soc India 75:3–24

Mitchell AHG, Mckerrow WS (1975) Analogous evolution of the Burma orogen and the Scottish Caledonides. Bull Geol Soc Am 86:305–315

Mukhopadhyay M, Dasgupta S (1988) Deep structure and tectonics of the Burmese arc: constraints from earthquake and gravity data. Tectonophysics 149:299–322

Nandy DR (1986) Tectonics, seismicity and gravity of northeastern India and adjoining region. Geol Nagaland Ophiolite. Mem Geol Surv India 119:13–17

Nandy DR (2000) Tectonic evolution of northeastern India and the adjoining area with special emphasis on contemporary geodynamics. Indian J Geol 72:175–197

Patil SK, Thong GT, Watitemsu, Temjenrenla, Rao BV (2012) Geochemistry and paleomagnetism of the basalt of the ophiolite suite in parts of Phek district, Nagaland. In: National Symposium on Recent Advances in Applied Geochemistry: Current Status and Future Trends, Atomic Min Dept , Hyderabad, Indian Soc of Appl Geochemists, Abstract Vol, 20–21

Ranga Rao A (1983) Geology and hydrocarbon potential of a part of Assam-Arakan basin and its adjascent region. In: Bhandari L L et al. (eds) "Petroliferous basins in India", Petroleum Asian Journal, Dehra Dun, 127–158

Roy RK, Kacker RN (1980) Tectonic analysis of Naga hills orogenic belt along eastern peri-Indian suture. Himal Geol 10:374–402

Roy RK (1989) Meso-Cenosoic accretionary prism on the margin of Indo-Burman range ophiolite and its implications. In: Ghose NC (ed) Phanerozoic ophiolites of India. Sumna Publishers, Patna, pp 145–164

Roy RK, Kacker RN, Chattopadhyay B (1982) Geochemical characteristics and tectonic setting of the Naga hills ophiolite volcanics, India. Ofioliti 2(3):479–498

Sengupta S, Ray KK, Acharyya SK, de Smeth JB (1990) Nature of ophiolite occurrences along the eastern margin of the Indian plate and their tectonic significance. Geol 18:439–442

Shukla R (1989) Occurrence of rodingite in the ophiolite belt of Manipur. In: Ghose NC (ed) Phanerozoic ophiolites of India. Sumna Publishers, Patna, 189–196

Verma RK (1985) Gravity field, seismicity and tectonics of the Indian peninsula and the Himalayas. Allied Publications, New Delhi 213p

Verma RK, Krishna Kumar GVR (1987) Seismicity and the nature of plate movement along the Himalayan arc, Northeast India and Arakan Yoma: a review. Tectonophysics 134:153–175

Verma RK, Mukhopadhyay M (1977) An analysis of gravity field in northeastern India. Tectonophysics 42:283–317

Vidyadharan KT, Joshi A, Ghosh S, Gaur MP, Shukla R (1989) Manipur ophiolites: its geology, tectonic setting and metallogeny. In: Ghose NC (ed) Phanerozoic Ophiolites of India. Sumna Publishers, Patna, 197–212

Watts AB, Talwani M (1974) Gravity anomalies seaward of deep-sea trenches and their tectonic implications. Geophys J R Astron Soc 36:57–90

5 Petrography

5.1 Introduction

The textural relations among minerals are shown in optical photomicrographs. Unless specified otherwise, the photomicrographs were collected in the plane-polarised, transmitted light mode. A few backscattered electron images were also collected with the electron microprobe to supplement the optical microscope images. Minerals belonging to solid solution series such as sodic amphibole, calcic-amphibole, garnet, opaques, pyroxene, plagioclase feldspar and spinel are identified by their group names in the photomicrographs. Phases labelled as Gln and Amp/Act in the photomicrographs broadly represent sodic amphibole and calcic-amphibole respectively. Texture, mineral association and relative abundance of minerals were used in the identification of a rock. Since sea water interaction alters the primary constituents of sub-marine basalts, a combination of factors such as freshness of sample, presence of critical minerals, morphology (shape) of minerals, relative abundance and texture were taken into account in classifying the basalts. Limited EPMA data on samples from our recently published works and new data collected exclusively for this study (Tables 5.1, 5.2) were also used to confirm mineral identification. The new EPMA data collected from 14 representative samples comprising 90 spot analyses of minerals are listed in Table 5.2. In order to project an overall petrological account, relevant work from published literature have been used. The dimension of this work demands consultation of a wide range of text and reference books which could not be cited in the reference for want of space and scope of the book.

5.2 Peridotite Tectonite/Meta-ultramafics

The peridotite tectonites also designated as meta-peridotites are dominated by dunite followed by harzburgite, lherzolite and wehrlite. These rocks are classified on the basis of abundance of olivine (>40 volume %), ortho- and clinopyroxenes, which are variably altered to serpentine. Olivine shows preferred orientation and together with serpentine develops a strong foliation. Olivine occurs in two distinct modes: (a) large porphyroclasts (sheared type) and protogranular (>2 mm) olivine and (b) small neoblasts (sheared type) and fine-grained (<1 mm) olivine (Plate PT-2, 5). The olivine porphyroclasts are flattened or curved, and often brecciated (Plate PT-2 to PT-8). Unlike their extrusive equivalent, viz., komatiite, that shows spinifex texture of radiating/acicular olivine grains due to quenching, these plutonic rocks show flattened olivine grains under severe stress. Plastic strain in the form of strain shadow in olivine porphyroclasts (Plate PT-1) and kink bands in pyroxenes (Agrawal and Ghose 1986) represent post-igneous features. The neoblasts show uniform extinction. These rocks possibly originated at the base of lithosphere.

An unusual feature in the olivine of dunite tectonite is the presence of an amorphous greyish core (Plate PT-10). An enlarged view reveals fine streaks of opaque minerals traversing diagonally across the core and confined within the crystal outline (Plate R-16-2). The outer border of olivine is altered to red-brown Fe-oxide due to post-emplacement oxidation.

The tectonites are essentially spinel-bearing and devoid of plagioclase. This suggests derivation from a deeper region of the lower crust beyond the stability field of plagioclase. The rocks have undergone at least two phases of deformations (Plate 107/80-2). The spinel grains are usually octahedral in shape, sometimes lensoid, and equidimensional or skeletal. The cores of the serpentine veins often entrap trails of fine magnetite granules. Spinel containing olivine inclusions indicates that it crystallised after olivine. Abundance of spinel is greater in the tectonites than in the peridotites of the layered sequence (seismic Layer 3) overlying the tectonites (Fig. 1.1a).

The dunite tectonite (100/79) is composed of olivine (Fo_{93}), enstatite orthopyroxene with low alumina (Al_2O_3 0.6 wt.%, $En_{92}Fs_7Wo_1$), magnesio-chromite (Cr# 80) and secondary magnetite and serpentine (Table 5.2). The lherzolite tectonite (96/80A) also contains clinopyroxene (Al_2O_3 1.6

N. C. Ghose et al., *A Petrographic Atlas of Ophiolite*,
DOI: 10.1007/978-81-322-1569-1_5, © Springer India 2014

Table 5.1 Mineral paragenesis of selected samples of the NHO studied by EPMA (this study and from publication elsewhere[a])

Sample No	Rock type	Texture	Mineralogical composition/association	Remarks
Ultramafic (tectonite)				
100/79	Dunite	Porphyroclastic	Ol (Mg# 85), Opx (Mg# 84), Cr-Spl (Cr# 75), Srp	Flattened porphyroclast and brecciated neoblast of Ol
Basalt				
27/79	Hyaloclastite	Porphyritic, intergranular	Ab, Aug, Ep, Chl, Ttn, Ap, Cal (late veins of Ep-Ab and Chl-Cal)	Primary Aug phenocrysts partially chloritised; primary Pl phenocrysts pseudomorphed by Ab + Ep
90/80	Basalt with plagioclase megacryst	Porphyritic, intergranular	Pl-megacryst, Ab, Zo/Ep, Chl, Ttn, Hem, Ap (intergrowths of Zo-Chl and Ttn-Hem; late veins of Chl lined with Ep/Zo)	Zo and Chl are rimmed by Hem; Pl altered to pure Ab + Zo
371/79	Basalt with Pl glomerocrysts and Chl-ocelli	Intergranular, amygdaloidal	Ab + Ep/Zo pseudomorph after Pl megacryst, Aug, Chl, Ttn (Chl-ocelli with Qz)	Pl megacrysts in glomerocrystic aggregates; base metal—Py veins in megacrysts and groundmass
330/90	Augite-phyric basalt	Porphyritic, amygdaloidal	Ab, Aug, Ph, Ep, Cal, Aln, Opq (vesicles lined with Ph and filled by Cal with minor Qz, Cct and Cu)	Base metals—metallic Cu, Cct
567/78	Amygdaloidal basalt	Intergranular, amygdaloidal	Ab, Chl pseudomorph after Cpx, Ep, Ab-Chl groundmass with fine Aeg-Aug + Ttn intergrowth (Cal-filled vesicles lined with Ep)	Primary Aug pseudomorphed by zoned Chl; Pl laths replaced by pure Ab; late Aeg-Aug + Ttn intergrowth and euhedral Ep in groundmass
230/100	Olivine basalt	Intersertal/ porphyritic	Ab, Cpx ($En_{50}Fs_{29}Wo_{21}$), Chl, Cr-Spinel, Ep	Chl pseudomorph after Ol, Chl shows sectorial twinning
580/79	Augite-phyric trachybasalt	Porphyritic, amygdaloidal	Aug, Ab-Ph pseudomorph after Fs, Ed-Hbl + Ab groundmass with minor Ep (ocelli with spherulitic Chl and minor Ep + Aln rimmed by Ab)	Resorbed primary Aug phenocrysts; primary Fs pseudomorphed by Ab + Ep; minor Aln within ocelli
Blueschist				
Z-74	Garnet-glaucophane schist	Schistose	resorbed Alm-Sps Grt, zoned Na-Amp (core-Mrbk, rim-Wnc), Ab/Kfs, Ph, Qz, Ap, Zrn	two generations of Grt: resorbed porphyroblasts and small idioblastic; accessory Zrn + Ap inclusion in Ph
212/80	Omphacite-glaucophane schist	Schistose	Omp, Gln (Mg# 76), Ph, Chl (Mg# 70), Ep, minor Cal, accessory Rt, Ttn, late-vein Ab	Omp with Chl alteration veins; late veins and patches containing Ab, Ph, Cal
139/80	Glaucophane garnet-barroisite rock	Granoblastic	Brs, Gln, Grt (core richer in Adr-Sps, poorer in Grs-Pyr), Ep, Ph, Ab, Rt, Qz, Ttn, Hem	
139/80b	Garnet-glaucophane schist	Granoblastic	Brs, Gln, (Amp 49 vol%), Alm-Grs Grt (3 vol%), Ep (23 vol%), Qz (21 vol%), Ph (4 vol%), accessory Rt	Resorbed Grt porphyroblasts with Ph, Ep, Qz and Rt in alteration veins; Brs with Gln rims; Ab only in late veins
ML/70	Glaucophane schist	Schistose	Grt, Brs, Gln, Ep, Chl, Ph, Bt, Qz, Rt, Ttn, Mag	Qz contains fluid inclusions
ML/69	Glaucophane schist	sub-ophitic	Grt, Gln (Mg# 84), Ed-hbl, Ep, Chl (Mg# 75), Ph, Bt, Qz, Rt	
2/70	Glaucophane schist	Schistose, sheared	Grt, Gln, Amp (core-Act, rim-Cum), Ep, Ph, Qz, Rt, Ttn, Cal, Hem, Aln	Ab occurs as inclusion in Gln
ML/225	Glaucophane schist	Schistose, S–C mylonite	Ab, Gln, Act, Ep, Ph, Spn, Qz, Mag, Hem	presence of 'mica fish'
N-7	Glaucophane schist	Schistose	Ab, Gln, green Amp, Ep, Chl, Ph, Qz	Ep dominates (>50 % by vol.), minor Gln
M/234	Glaucophane schist	Schistose	Gl, Ph, Rt, Qz, Cal, Mag	Highly carbonated
76/80	Glaucophanite	Schistose with micro-shear	Gln (Mg# 82), Act (Mg# core—86, rim—77)	>70 % Gln by vol., Gln two generations—prismatic and rhombic

(continued)

Table 5.1 (continued)

Sample No	Rock type	Texture	Mineralogical composition/association	Remarks
Greenschist				
22/80	Greenschist	Schistose	Ab, Mhb, Ep, Opq	
Eclogite				
77/80	Barroisite eclogite	Granoblastic	Grt (8 vol%), Omp (2 vol%), Brs, Gln, (Amp 47 vol%), Ep (30 vol%), Ph (4 vol%), Chl (1 vol%), Qz, Rt, Ttn	poikiloblastic Grt with Omp, Gln, Ep, Rt and Qz inclusions; Omp porphyroblasts contain Ep inclusions; Brs is replaced by Gln during retrogression; Ep porphyroblasts have veins of Ph and Qz; late Ab vein cuts through sample
Meta-chert/meta-arenite				
855	Meta-chert	Very fine-grained	Mn-bearing Gru, inclusion of Gru in Qz, Chl, Ap, Opq	Gru needles at coarse-grained Qz grain boundaries; base metals—Gn, Ccp
49/70	Meta-arenite	Mylonite	Grt (Alm > Grs > Prp), Mrbk (Mg# 66), Ph, Ap, Bt, Rt	

*Abbreviations**Act* actinolite, *Aeg* aegirine, *Ab* albite, *Aln* allanite, *Alm* almandine, *Amp* amphibole, *Adr* andradite, *Ap* apatite, *Aug* augite, *Brs* barroisite, *Bt* biotite, *Cal* calcite, *Ccp* chalcopyrite, *Cct* chalcocite, *Chl* chlorite, *Cu* metallic copper, *Ed* edenite, *En* enstatite, *Ep* epidote, *Fs* feldspar, *Gn* galena, *Grt* garnet, *Gln* glaucophane, *Grs* grossular, *Gru* grunerite, *Hem* hematite, *Hbl* hornblende, *Kfs* K-feldspar, *Mag* magnetite, *Mhb* magnesio hornblende, *Mrbk* Magnesio riebeckite, *Omp* Omphasite, *Opq* Opaque, *Ph* Phengite, *Qz* Quartz, *Prq* Pyrope, *Rt* rutile, *Spl* spinel, *Spn* sphene, *Sps* spessartine, *Ttn* titanite, *Wnc* winchite, *Wo* wollastonite, *Zrn* zircon, *Zo* zoisite

[a] *Sources*: EPMA analysis of samples from Chatterjee and Ghose (2010) and Ghose et al. (2010)

wt.%, Mg# 92, $En_{51}Fs_4Wo_{45}$) and minor magnesio-hornblende (Mg# 88). High forsterite content of olivine is typical of mantle peridotite. Serpentine (Mg# 92-93 in harzburgite tectonite, 98 in dunite tectonite) occurs as an alteration product. The Cr_2O_3 (61–62 wt.%) and Al_2O_3 (9–11 wt.%) contents of high-Cr spinel are almost uniform, whereas the MgO:FeO ratio varies between 0.8 and 1.0 ($Mg \leq Fe^{2+}$). In magnetite, Cr^{3+} is replaced by Fe^{3+}. The chromium content of the magnetite inclusions in chromite is lower ($Cr_2O_3 = 0.21$ wt.%) compared to the magnetite occurring in fracture filling veins ($Cr_2O_3 = 2.91$ wt.%).

Micro-shear planes associated with step-faulting (S–C mylonite) in lherzolite (K-72) suggest a ductile to brittle tectonic environment (Plate SP-13). A thin layer of magnetite within serpentine is displaced by the step-fault in the shear plane (Plate SP-13). The rock contains olivine, non-pleochroic orthopyroxene, clinopyroxene, serpentine and magnetite. In another lherzolite sample, brecciated and pulverised clinopyroxene grains and serpentine fill the micro-shear zone (96/80b, Plate SP-15).

5.2.1 Spinel Peridotite

The peridotite tectonites are often associated with spinel, some of which are veined by magnetite. Three types of spinel peridotites have been distinguished. In the order of abundance, these are: (a) Spinel harzburgite, (b) Spinel lherzolite and (c) Serpentinite/dunite. Salient petrographic features of spinel peridotites are given in Table 5.3. The spinel compositions reflect solid solutions between chromite ($Fe^{2+}Cr_2O_4$) and spinel *sensu stricto* ($MgAl_2O_4$) as determined by the EPMA. These Mg–Al-bearing chrome-spinels (Al_2O_3 51 wt.%, MgO 18 wt.%, Mg# 72 and Cr# 19 in spinel harzburgite ML-77-12, Table 5.2) are typical of those found in ultramafic rocks of mantle origin. They occur as hypidiomorphic, xenomorphic large reddish brown grains with veinlets of magnetite. Tabular, granular and trihedral grains of magnetite released during serpentinisation form fracture fillings, lenticular patches in micro-shears trails along serpentine (Mg# 92-93) veins, and granules around relict olivine (Fo_{91}), and across the cleavage planes of high-alumina clinopyroxene (Al_2O_3 5.7 wt.%, Mg# 92, $En_{52}Fs_4Wo_{44}$) and non-pleochroic orthopyroxene (Al_2O_3 0.69 wt.%, $En_{93}Fs_6Wo_1$).

Two types of Cr-spinel have been distinguished in the NHO peridotites/serpentinites: (a) *high-Cr spinel* with variable FeO:MgO ratio ($Mg \leq Fe^{2+}$) and low Al_2O_3 associated with the tectonite or metamorphic peridotite of mantle origin (this study, see above) and (b) *high-Al spinel* with variable FeO/MgO ratio ($Mg > Fe^{+2}$), and low Cr_2O_3 associated with the serpentinites derived from the cumulate peridotites (equivalent to abyssal peridotites) of Manipur (Singh 2013; Singh et al. 2013) in the southern part of NHO. The Cr content of spinel in peridotite is a good indicator of the degree of partial melting of sub-crustal rocks. Low Cr-spinel indicates less depleted peridotite, whereas high-Cr-spinel indicates more depleted peridotite (Dick and Bullen 1984). The serpentinites and the abyssal peridotites of NHO with high-Al chromite is related to low degree of partial melting and MORB extraction in a mid-oceanic ridge environment (Ningthoujam et al. 2012; Singh 2013).

Table 5.2 Composition of minerals in the NHO rocks

	Peridotite tectonite										
Sample No.	96/80A		100/79				384/79	402/79		ML-77-12	
n	1	1	3	1	5	3	1	3	1	2	1
	Cpx	Mhb	Ol	Opx	Spl	Mag	Srp	Opx	Srp	Ol	Cpx
Weight percent											
SiO_2	53.98	48.22	41.48	56.99	0.34	0.45	44.63	57.23	39.65	40.18	51.32
TiO_2	0.16	0.47	0.02	0.01	0.04	0.01	0.02	0.01	0.02	0.04	0.27
Al_2O_3	1.58	10.15	0.04	0.64	9.95	0.04	0.14	0.69	0.06	0.00	5.71
Cr_2O_3	0.18	0.33	0.14	0.42	60.97	2.74	0.17	0.31	0.00	0.01	1.11
Fe_2O_3					0.00	65.46					
FeO	2.71	4.30	6.82	4.76	18.49	31.35	1.45	4.27	5.97	9.27	2.74
MnO	0.00	0.09	0.18	0.13	0.30	0.03	0.00	0.12	0.10	0.13	0.12
MgO	17.69	17.86	51.59	35.29	9.32	0.27	41.18	36.20	42.08	51.31	18.10
CaO	22.02	12.29	0.04	0.60	0.06	0.01	0.03	0.53	0.14	0.04	21.03
Na_2O	0.08	1.12		0.14			0.00	0.01	0.06		0.63
K_2O	0.02	0.02		0.03				0.01	0.01		0.10
NiO										0.33	
Total	98.42	94.85	100.31	99.01	99.47	100.36	87.62	99.38	88.09	101.30	101.13
Atomic proportions											
O	6	23	4	6	4	4	7	6	7	4	6
Si	1.981	6.930	0.999	1.975	0.011	0.017	2.050	1.970	1.882	0.974	1.843
Ti	0.004	0.051	0.000	0.000	0.001	0.000	0.001	0.000	0.001	0.001	0.007
Al	0.068	1.719	0.001	0.026	0.390	0.002	0.008	0.028	0.003	0.000	0.242
Cr	0.005	0.037	0.003	0.012	1.603	0.083	0.006	0.008	0.000	0.000	0.032
Fe^{3+}					0.000	1.880					
Fe^{2+}	0.083	0.517	0.137	0.138	0.516	1.001	0.056	0.123	0.237	0.188	0.082
Mn	0.000	0.011	0.004	0.004	0.008	0.001	0.000	0.003	0.004	0.003	0.004
Mg	0.968	3.826	1.853	1.823	0.462	0.015	2.820	1.858	2.978	1.853	0.969
Ca	0.866	1.892	0.001	0.022	0.002	0.000	0.001	0.020	0.007	0.001	0.809
Na	0.006	0.312		0.009			0.000	0.001	0.006		0.044
K	0.001	0.004		0.001			0.000	0.000	0.001		0.005
Ni			0.000		0.000	0.000				0.006	
Sum	3.982	15.299	2.998	4.011	2.993	3.000	4.942	4.012	5.118	3.026	4.037
Mg#	92.1	88.1	93.1	93.0	47.3	1.5	98.1	93.8	92.6	90.8	92.2
Cr#					80.4	97.7					
Fo			93.1							90.8	
Fa			6.9							9.2	
En	50.5			91.9				92.9			52.1
Fs	4.3			7.0				6.2			4.4
Wo	45.2			1.1				1.0			43.5
An											
Ab											

(continued)

Table 5.2 (continued)

	Peridotite tectonite		Peridotite cumulate									
Sample No.	ML-77-12		160/80			131/80					K-14	
n	4	3	2	1	1	2	3	2	1	1	2	1
	Spl	Srp	Ol	Cpx	Srp	Ol	Opx	Mhb	Spl	Srp	Ol	Cpx
Weight percent												
SiO_2	0.13	41.1	39.09	53.36	41.75	40.53	54.09	45.39	0.23	39.15	41.15	53.75
TiO_2	0.06	0.04	0	0.22	0.09	0.08	0.11	0.58	0.46	0.09	0.01	0.18
Al_2O_3	51.14	1.77	0.02	2.75	2	0.13	2.25	11.54	28.54	3.01	0	2.46
Cr_2O_3	17.49	0.34	0.06	0.53	0.27	0.06	0.29	0.81	27.03	0.33	0.07	0.73
Fe_2O_3	0.16								9.96			
FeO	12.45	5.66	20.01	4.05	9.92	14.28	10.07	5.4	26.06	6.67	10.26	2.72
MnO	0.12	0.13	0.25	0.07	0.21	0.18	0.25	0.07	0.36	0.25	0.18	0.02
MgO	17.9	38.17	39.78	16.53	33.11	44.02	31.14	18.42	6.87	36.16	48.64	17.1
CaO	0.03	0.5	0.04	22.26	0.33	0.16	0.93	12.1	0.06	0.07	0.03	22.85
Na_2O		0.06		0.18	0.1		0.05	2.1		0.2		0.25
K_2O		0.05		0.04	0.07		0.06	0.09		0.21		0.04
NiO	0.29											
Total	99.76	87.84	99.23	99.99	87.85	99.43	99.25	96.48	99.57	86.14	100.33	100.10
Atomic proportions												
O	4	7	4	6	7	4	6	23	4	7	4	6
Si	0.003	1.942	1.012	1.944	2.004	1.017	1.922	6.521	0.007	1.9	1.006	1.948
Ti	0.001	0.002	0	0.006	0.003	0.001	0.003	0.063	0.011	0.003	0	0.005
Al	1.617	0.099	0.001	0.118	0.113	0.004	0.094	1.953	1.057	0.172	0	0.105
Cr	0.371	0.013	0.001	0.015	0.01	0.001	0.008	0.092	0.671	0.013	0.001	0.021
Fe^{3+}	0.003								0.235			
Fe^{2+}	0.279	0.223	0.433	0.124	0.398	0.3	0.299	0.649	0.685	0.271	0.21	0.082
Mn	0.003	0.005	0.005	0.002	0.009	0.004	0.008	0.009	0.01	0.01	0.004	0.001
Mg	0.716	2.688	1.535	0.898	2.369	1.647	1.65	3.945	0.322	2.616	1.772	0.924
Ca	0.001	0.025	0.001	0.869	0.017	0.004	0.036	1.862	0.002	0.004	0.001	0.887
Na		0.006		0.013	0.009		0.004	0.584		0.019		0.018
K		0.003		0.002	0.004		0.003	0.016		0.013		0.002
Ni	0.006		0			0			0		0	
Sum	3	5.006	2.987	3.991	4.938	2.979	4.027	15.693	3	5.02	2.993	3.993
Mg#	71.9	92.3	78	87.9	85.6	84.6	84.6	85.9	32	90.6	89.4	91.8
Cr#	18.7								38.9			
Fo			78			84.6					89.4	
Fa			22			15.4					10.6	
En				47.5			83.1					48.8
Fs				6.5			15.1					4.4
Wo				46			1.8					46.9
An												
Ab												

(continued)

Table 5.2 (continued)

	Peridotite cumulate		Gabbro								Anorthosite	
Sample No.	K-14		155/80					150/80			T-74	
n	1	1	4	1	2	6	2	1	2	2	1	3
	Spl	Mag	Cpx	Pl	Ed	Mhb	Chl	Cpx	Mhs	Ep	Pl	Mhb
Weight percent												
SiO_2	0.32	0.57	52.53	44.89	46.42	47.17	34.75	53.33	45.17	37.88	47.05	46.35
TiO_2	0.2	0.01	0.13	0.01	0.5	0.57	0.02	0.15	0.62	0.01	0.01	1.31
Al_2O_3	25.5	0.05	2.31	34.95	12.15	11.23	14.22	2.31	12.86	30.06	34.03	10.66
Cr_2O_3	33.78	0.25	0.13	0	0.1	0.21	0.04	0.03	0.13	0.04	0.03	0.13
Fe_2O_3	9.49	66.78								5.06		
FeO	14.69	31.09	5.06	0.13	8.07	8.21	14.67	4.17	7.09	0	0.42	7.45
MnO	0.24	0.01	0.16	0	0.18	0.11	0.1	0.11	0.09	0.1	0.1	0.11
MgO	13.66	0.2	16.55	0.03	14.44	16.1	23.98	16.42	17.5	0.27	0.6	16.83
CaO	0.04	0.1	22.56	18.39	12.54	11.7	0.4	24.07	12.49	23.45	16.81	11.89
Na_2O			0.29	0.74	1.87	2.04	0.15	0.3	2.52	0.08	1.5	1.93
K_2O			0.06	0.1	0.18	0.26	0.16	0.07	0.12	0.04	0.04	0.22
NiO												
Total	97.91	99.06	99.77	99.24	96.42	97.59	88.46	100.96	98.57	96.97	100.58	96.87
Atomic proportions												
O	4	4	6	8	23	23	28	6	23	12.5	8	23
Si	0.01	0.022	1.934	2.085	6.717	6.741	6.782	1.937	6.408	2.952	2.149	6.67
Ti	0.005	0	0.004	0	0.054	0.062	0.002	0.004	0.066	0.001	0	0.141
Al	0.927	0.002	0.1	1.913	2.072	1.891	3.271	0.099	2.149	2.762	1.832	1.809
Cr	0.824	0.008	0.004	0	0.011	0.024	0.006	0.001	0.015	0.002	0.001	0.014
Fe^{3+}	0.22	1.945								0.372		
Fe^{2+}	0.379	1.006	0.156	0.005	0.976	0.981	2.394	0.127	0.841		0.016	0.897
Mn	0.006	0	0.005	0	0.021	0.013	0.017	0.003	0.011	0.006	0.004	0.014
Mg	0.628	0.012	0.908	0.002	3.114	3.43	6.976	0.889	3.701	0.031	0.041	3.61
Ca	0.001	0.004	0.889	0.915	1.943	1.792	0.083	0.937	1.898	1.958	0.823	1.833
Na			0.021	0.067	0.525	0.566	0.055	0.021	0.692	0.011	0.133	0.539
K			0.003	0.006	0.032	0.047	0.039	0.003	0.021	0.004	0.002	0.041
Ni	0	0										
Sum	3	3	4.023	4.994	15.466	15.546	19.624	4.021	15.801	8.1	5.001	15.567
Mg#	62.4	1.1	85.4		76.1	77.8	74.5	87.5	81.5			80.1
Cr#	47.1	77										
Fo												
Fa												
En			46.5					45.5				
Fs			8					6.5				
Wo			45.5					48				
An				93.2							86.1	
Ab				6.8							13.9	

(continued)

Table 5.2 (continued)

	Anorthosite		Basalt								Tuff
Sample No.	T-74		230/79		585/79						439
n	1	3	1	2	3	1	1	1	2	1	3
	Zo	Chl	Cpx	Ep	Cpx	Ab	Aln[a]	Chl	Ilt	Ttn	Ttn
Weight percent											
SiO_2	38.5	32.26	49.04	39.1	50.91	67.04	36.11	29.02	53.52	33.08	32.01
TiO_2	0.01	0.05	1.55	0.16	1.37	0.05	0.03	0.04	0.06	29.83	30.77
Al_2O_3	31.55	15.83	6.52	21.59	3.7	19.37	20.32	17.33	22.19	2.43	3.81
Cr_2O_3	0.06	0.1	0.19	0.04	0.4	0	0.07	0.08	0.05	0	0.00
Fe_2O_3	1.8			13.35			13.41				
FeO	0	3.99	7.95	1.29	5.82	0.82	2.3	16.98	5.64	2.26	2.54
MnO	0.02	0.81	0.17	0.13	0.17	0	0.06	0.22	0.05	0	0.08
MgO	0.25	31.85	13.34	0.14	15.29	0	0.08	24.25	5.02	0.25	0.09
CaO	23.97	0.27	19.95	21.5	22.15	0.15	20.33	0.19	0.12	24.69	25.80
Na_2O	0.12	0.1	0.36	0.56	0.44	11.35	0.08	0.08	0.29	0.4	0.28
K_2O	0.03	0.09	0.05	0.04	0.05	0.13	0.06	0.38	9.9	0.12	0.24
NiO											
Total	96.31	85.35	99.12	97.88	100.31	98.91	98.35	88.57	96.82	93.06	95.63
Atomic proportions											
O	12.5	28	6	12.5	6	8	12.5	28	22	5	5
Si	2.987	6.244	1.832	3.119	1.875	2.974	3.062	5.808	7.12	1.145	1.086
Ti	0.001	0.008	0.044	0.01	0.038	0.002	0.002	0.006	0.006	0.776	0.785
Al	2.885	3.611	0.287	2.03	0.16	1.013	2.031	4.088	3.479	0.099	0.152
Cr	0.004	0.016	0.006	0.003	0.012	0	0.005	0.013	0.005	0	0.000
Fe^{3+}	0.185		0	0.801			0.856				
Fe^{2+}	0	0.646	0.249	0.086	0.179	0.03	0.163	2.842	0.628	0.065	0.072
Mn	0.001	0.133	0.005	0.009	0.005	0	0.004	0.037	0.005	0	0.002
Mg	0.029	9.191	0.743	0.016	0.84	0	0.01	7.236	0.996	0.013	0.005
Ca	1.993	0.057	0.799	1.837	0.874	0.007	1.847	0.041	0.016	0.915	0.937
Na	0.018	0.038	0.026	0.087	0.032	0.976	0.013	0.031	0.074	0.027	0.018
K	0.003	0.021	0.002	0.004	0.002	0.007	0.006	0.097	1.68	0.005	0.01
Ni											
Sum	8.06	19.965	3.993	8	4.018	5.01	8	20.199	14.009	3.046	3.0638
Mg#		93.4	74.9		82.4			71.8	61.3		
Cr#											
Fo											
Fa											
En			41.5		44.4						
Fs			13.9		9.5						
Wo			44.6		46.2						
An						0.7					
Ab						99.3					

Determined by EPMA/WDS at GSI, Bangalore, using a Cameca SX-100 electron microprobe operating at 15 kV accelerating voltage and 15 nA beam current
n = number of analyses; Mg# = 100.Mg/(Mg + Fe^{2+}); Cr# = 100.Cr/(Cr + Al); Fe^{2+} and Fe^{3+} in spinel and epidote/zoisite/allanite calculated through charge balancing

Locality: 96/80A Lherzolite tectonite, Southeast of Purr; 100/79 Dunite tectonite, East of Zipu; 384/79 Dunite tectonite, South of Lacham Lake; 402/79 Harzburgite, South of Lacham Lake; ML-77-12 Spinel harzburgite; 160/80 Olivine websterite, East of Sataza; 131/80 Lherzolite cumulate, East of Purr; K-14 Harzburgite cumulate; 155/80 Olivine gabbro, 2.7 km northwest of Sutsu; 150/80 Olivine gabbro, 2.7 km northwest of Sutsu; T-74 Anorthosite; 230/79 Olivine basalt, Northeast of Zipu; 585/79 Alkali basalt, tectonic contact, Laluri; 439 Lithic tuff, SSW of Ngazu

Mineral abbreviations (after Whitney and Evans 2010): *Ab* albite, *Aln* allanite, *An* anorthite, *Chl* chlorite, *Cpx* clinopyroxene, *Ed* edenite, *En* enstatite, *Ep* epidote, *Fa* fayalite, *Fs* ferrosilite, *Fo* forsterite, *Ilt* Illite, *Mag* magnetite, *Mhb* magnesio-hornblende, *Mhs* magnesio-hastingsite, *Ol* olivine, *Opx* orthopyroxene, *Pl* plagioclase, *Spl* spinel, *Srp* serpentine, *Ttn* titanite (sphene), *Wo* wollastonite, *Zo* zoisite

[a] Contains 5.5 wt.% Ce_2O_3

5.3 Serpentinite

The serpentinites display a wide variety of textures, viz., massive, mesh or reticulate, ribbon, fibrous, bladed, sheared and brecciated (Ghose and Fareeduddin 2011). Besides serpentine and relict olivine and pyroxene, other minor phases are opaque/chrome-spinel, talc, tremolite, chrysotile, calcite, biotite, muscovite and chlorite. Three structural types of serpentinite have been recognised; e.g. (a) massive or blocky, (b) sheared and (c) cross-fibred veins. The first is dominantly composed of antigorite with minor lizardite, whereas the last one is mainly composed of chrysotile (Ghose et al. 1986; Ningthoujam et al. 2012). Mesh (honeycomb) texture is common, consisting of primary, secondary and tertiary chords (veins) of serpentine enclosing fractured grains of olivine and pyroxene. In some cases, serpentine exhibits a ribbon texture composed of parallel and undulating bands. The bladed/mat texture shows blended and inter-felted mass of fine-to-medium size serpentine blades belonging to more than one generation. Thin chrysotile veins occur as fracture fillings. Both magnetite and chromite occur as disseminations, fracture fillings and also along cleavage traces. Al-spinels (Al_2O_3 = 46–52 wt.%) are common in serpentinites derived from ultramafic cumulates (Ningthoujam et al. 2012) and in abyssal peridotites (Singh 2013).

5.4 Rodingite

Rodingites of diverse composition are developed during serpentinisation through calcium metasomatism. They are characterised by low silica (SiO_2 ~ 35 wt.%) and high lime contents (CaO ~ 35 wt.%, Shukla 1989). A variety of lithology of ultramafic, volcanic, gabbro and pelagic sediments were affected by this process. X-ray diffraction studies indicate the presence of grossularite, vesuvianite, diopside, idocrase, tremolite, actinolite, talc, albite, chlorite, prehnite, pyrophyllite, opaque and sphene in these rocks (Shukla 1989). Grossularite and vesuvianite form the major constituents of the rock. Metasomatised gabbronorite containing Pl-Opx-Cpx-Ep-Zo-Chl and serpentinite containing Tlc-Grt-Opq-Srp show replacement of Cpx by Pl, Grt by Opq, Opx by Cpx, Cpx by Ep and Pl by Zo. The ferromagnesian minerals show alteration to chlorite.

5.5 The Cummulate Complexes

5.5.1 Ultramafics

The cumulate litho-assemblages of the NHO may be classified as (1) peridotite (dunite, harzburgite lherzolite and minor wehrlite), (2) pyroxenite (olivine websterite, websterite, olivine clinopyroxenite and clinopyroxenite), (3) gabbroids (olivine gabbro, olivine gabbronorite, norite, gabbro and hornblende gabbro), (4) plagiogranite and (5) anorthosite. These are emplaced as tectonic slices. Late plagiogranite intrusions cut through the entire magmatic sequence. Based on mutual field relationship and mineral paragenesis, a sequence similar to those encountered in the Precambrian layered magmatic complexes may be envisaged prior to emplacement (Agrawal and Ghose 1986). The peridotites of cumulate sequence host the stratabound nickeliferous magnetite deposits in the NHO.

The cumulate peridotites show a protogranular texture with medium- to fine-grained, granular, tabular or smoothly curved olivine crystals. High forsterite content of olivine ($Fo_{84\text{-}90}$) is comparable to that of the alpine peridotite (Ghose et al. 1986; Singh et al. 2010, this study). The composition of orthopyroxene is fairly homogeneous $En_{89\text{-}87}$ (enstatite), whereas the clinopyroxene is diopsidic augite ($En_{62\text{-}47}Fs_{6\text{-}4}Wo_{49\text{-}32}$) (Singh et al. 2010). Absence of any strain effect, foliation and lineation characterise these peridotites. Spinel grains are subhedral to anhedral, skeletal and less frequently octahedral. The abundance of opaque minerals is usually dependent on the degree of serpentinisation of parent minerals. Cracks in olivine and grain boundaries are commonly filled with magnetite and sepentine. Orthopyroxene contains inclusions of olivine and spinel, whereas poikilitic clinopyroxene has inclusions of olivine and orthopyroxene. The cumulate lherzolite (131/80) consists of olivine (Fo_{85}), orthopyroxene (Al_2O_3 2.3 wt.%, $En_{83}Fs_{15}Wo_2$), clinopyroxene, chrome-spinel (Cr# 40, Mg# 32) and minor magnesio-hornblende (Mg# 86) around clinopyroxene (Table 5.2). Serpentine (Mg# 91) is present as alteration of olivine. The cumulate harzburgite (K-14) contains olivine (Fo_{89}), orthopyroxene, minor clinopyroxene (Al_2O_3 2.5 wt.%, Mg# 92, $En_{49}Fs_4Wo_{47}$), chrome-spinel (Cr# 47) and secondary magnetite (Table 5.2).

5.5.2 Pyroxenite

The pyroxenites (olivine <40 volume %) are usually coarse to very coarse-grained and exhibit hypidiomorphic granular texture. They range in composition from olivine websterite, hornblende websterite, websterite, olivine clinopyroxenite and clinopyroxenite. Olivine occurs as discrete grains in websterite or as inter-cumulus grains in olivine clinopyroxenite. Based on optical properties (2 V angle), the composition of olivine in websterite is $Fo_{93\text{-}83}$ (Ghose et al. 1986). The composition of clinopyroxene varies from diopsidic augite to magnesian diopside (X-ray diffraction). The clinopyroxene shows polysynthetic twinning in clinopyroxenite. EPMA analysis of the olivine websterite (160/80) reveals that it contains olivine (Fo_{78}), clinopyroxene

(Al_2O_3 2.8 wt.%, Mg# 88, $En_{48}Fs_7Wo_{46}$) and secondary serpentine (Mg# 86) (Table 5.2). The orthopyroxene contains inclusions of chalcopyrite.

Olivine occurs as inclusions both in the ortho- and clinopyroxenes. Orthopyroxene forms a reaction rim around plagioclase in olivine websterite (160/80-1, 160/80-2, Plates Px-5, 6). Such mineralogical transformation in an otherwise ultramafic milieu composed of olivine and pyroxenes is unusual and points to a change in P–T condition. A plausible mechanism for the formation of olivine websterite is that a part of the crystallised magma chamber collapsed due to the weight of the overlying volcanic edifice after being isolated from the centre of volcanic activity, and clinopyroxene subsequently recrystallised into plagioclase and orthopyroxene at high P–T condition (Mei-Fu Zhou, person. commun.).

5.5.3 Gabbroids

The mafic cumulates occurring both as massive and layered gabbroid complexes are represented by olivine gabbro, hornblende-olivine (serpentinised) gabbro, norite, gabbronorite, gabbro, hornblende gabbro and pegmatitic gabbro with veins and small bodies of plagiogranite and anorthosite (Ghose et al. 1986). The gabbros are medium- to coarse-grained, commonly showing hypidiomorphic texture, and occasionally porphyritic, poikilitic, equigranular, sub-ophitic and brecciated. Poikilitic clinopyroxene (Al_2O_3 2.3 wt.%, Mg# 85-58, $En_{46\text{-}47}Fs_{7\text{-}8}Wo_{46\text{-}48}$, Table 5.2) in olivine gabbro contains inclusions of both olivine and orthopyroxene (155/80-2, 155/80-3) and exsolution lamellae of orthopyroxene. Plagioclase is rich in anorthite component ($An_{93}Ab_7$, Table 5.2). Edenite (Mg# 76), magnesio-hornblende (Mg# 78) and magnesio-hastingsite (Mg# 82) together with chlorite occur as inclusions within clinopyroxene and represent metamorphic alteration products. The post-crystallisation transformation of clinopyroxene to chlorite (Mg# 75) can be attributed to feeder channels connecting the inside to the outside of the clinopyroxene grain (155/80-4, Plate G-4, p. 116). Post-crystallisation transformation of plagioclase to epidote is also noted. The hornblende-olivine gabbro contains abundant magnetite due to a change in oxidation state of the magma with progressive fractionation. Minor presence of ilmenite, magnetite and pyrrhotite ($Fe_{1\text{-}n}S$) as minute flecks and blebs are detected by EPMA.

The orthopyroxene-bearing members of the gabbroid family include norite and gabbronorite. Norite is highly deformed showing effects of strain and brecciation. Clinopyroxene shows subsolidus exsolution of magnetite lamellae. The gabbronorite shows transformation of pleochroic orthopyroxene and green amphibole without disruption of cleavage. The orthopyroxene contains exsolution lamellae of inverted clinopyroxene. The plagioclase grains are commonly altered to sericite, epidote and clay minerals.

Preferred orientation of hornblende suggests post-crystalline deformation as evidenced by bent twin lamellae of plagioclase and pyroxene, and granulation of minerals (Ghose et al. 1986). The effect of deformation and brecciation is found to be of a lesser degree compared to the serpentinites on account of massive nature and toughness rendered by feldspar. At least two phases of deformations associated with mineralization of Fe–Ti oxides are recorded.

5.5.4 Plagiogranite

The plagiogranites range in mineralogical composition from diorite, tonalite, trondhjemite and albite granite (Agrawal and Ghose 1986). They are coarse-grained, showing panidiomorphic to hypidiomorphic, granular and porphyritic or gneissic texture. Laths of plagioclase are associated with quartz, green amphibole, muscovite rimmed by biotite, epidote, opaque and secondary chlorite. K-feldspar is involved in myrmekitic intergrowth. Plagioclase occurs as laths, plates or inter-stellar form with multiple twins. They show normal zoning and are altered to sericite or a mixture of quartz, calcite and epidote. Greenish brown amphibole occurs in clusters or as intercumulus grains. Breakdown of brown amphibole to green chlorite, opaque and epidote suggests late metamorphic recrystallisation. Late carbonate veins cuts across plagioclase and amphibole.

5.5.5 Anorthosite

The anorthosite is characterised by multiple growths of amphibole and minor presence of corundum. The anorthosite (T-74) is dominantly composed of plagioclase ($An_{86}Ab_{14}$, Table 5.2) with minor amounts of magnesio-hornblende (Mg# 80), zoisite, chlorite (Mg# 93), apatite, titanite and calcite.

5.6 Dolerite Dikes

The mafic diabase dikes are rare and occur as intrusives in the ultramafic cumulates. They are medium-to fine-grained, showing ophitic to sub-ophitic texture, and consist of plagioclase, clinopyroxene, magnetite and secondary chlorite. Prismatic or sub-rounded grains of clinopyroxene are altered to fine needles of radiating tremolite, epidote and iron oxides.

5.7 Basalt and Andesite

A wide variety of sub-alkaline and alkaline basalts are encountered in the NHO. The basalts have been classified here on the basis of texture, type and abundance of mafic minerals. The sub-alkaline basalt dominates (Ghose et al. 1986; Srikant et al. 2004) over less commonly occurring alkali basalt (Sengupta et al. 1990) and intermediate volcanics, viz.., andesite and trachyte. Basalts with oriented microlites of plagioclase, K-feldspar, clinopyroxene, opaques and glass are described here as trachybasalt. These are equivalent to reported trachybasalts that display trachytic texture, have low Mg-number (Mg# 23-49), and contain discrete K-feldspar, plagioclase and palagonite in the groundmass (Venkataramana and Datta 1987). Outcrop of trachyte has not been reported so far in the field despite its occurrence as clasts in the polymictic tuff breccia (Fig. 3.9) of the ophiolite derived cover sediments (Jopi Formation). The volcaniclastic rocks, though far more expansive in the central part of ophiolite than hitherto considered (Ghose et al. 1986; Ghose and Fareeduddin 2011), have received very little attention until now. These rocks are characterised by their wide range of granularity, mode of origin and variable concentration of crystal, vitric and lithic components.

The basalts are classified into (1) olivine basalt (olivine altered to chlorite), (2) porphyritic basalts (plagioclase/clinopyroxene-phyric and with plagioclase megacrysts), (3) crater facies basalt, (4) porphyritic andesite, (5) trachybasalt, (6) spilite and (7) hyaloclastite. The contact between successive flows is marked by chilled margins or by a thin sinuous layer of silica (584/79-5) or tuff (37/79-10). The plagioclase-phyric basalt overlies trachybasalt and contains fragmented xenoliths of the latter (584/79-5). The basalts are associated with a wide range of pyroclastic rocks including ignimbrite containing zircon. The volcaniclastics are dealt in a separate Sect 5.8.

The basalts display a wide range of textures; e.g. quenched (radiating laths of plagioclase), vitrophyric, variolitic, intersertal, spherulitic, intergranular, interpenetration, porphyritic, vesicular and amygdaloidal. These are developed under varied P–T-fluid conditions and partial pressure of oxygen, different degrees of fractional crystallisation and different source rock compositions and degrees of partial melting. Additionally, they were influenced by sea water interaction and metamorphism before and after emplacement. Most of the basalts are devoid of primary olivine and pyroxene as a result of pronounced crystal fractionation and hydrothermal alteration.

The vesicles in basalts are of different sizes and shapes, sometimes merging into fine fissures. They are also involved in deformation and oriented along flow direction. They have chilled margins and are often corroded due to reaction with secondary minerals. The vesicles are filled with secondary calcite, zeolite, chlorite, quartz and albite. The glassy materials are devitrified.

The basalts and trachybasalts of the NHO are metamorphosed to greenschist facies. They commonly contain amygdules and veins filled with late-stage calcite, quartz, chlorite, epidote and albite. Ocellus and spherulitic structures are rarely observed. In the basalts, plagioclase is altered to pure albite + zoisite/epidote which pseudomorph the primary phenocrysts. Some samples contain large altered megacrysts that are also pseudomorphed by albite + zoisite/epidote. Some of the megacrysts contain secondary pyrite veins. Primary clinopyroxene is pseudomorphed by chlorite, although some samples contain unaltered augite phenocrysts. The groundmass contains new growths of aegirine-augite, titanite and hematite. Fine-grained disseminated Fe-oxide occurs in the groundmass of some samples. The calcite-filled vesicles occasionally contain chalcocite and Cu-mineralisation. The trachytes contain phenocrysts pseudomorphed by albite + phengite. The groundmass of trachyte is composed of edenitic hornblende, albite and epidote. Fine- to medium-grained chlorite and epidote also occur in the groundmass. Some of the vesicles contain chlorite associated with epidote and REE-rich allanite.

Greenschist facies metamorphism of the volcanic rocks probably occurred in the oceanic crust and was aided by sea water circulation. High Na-content of the basalts, alteration of primary plagioclase to pure albite, recrystallisation of primary clinopyroxene to chlorite and aegirine-augite reflect interaction of newly erupted basalt with sea water. Presence of hematite indicates that alteration happened in an oxidising environment.

5.7.1 Types of Basalt

5.7.1.1 Olivine Basalt

Fine-grained olivine basalt shows intersertal texture and contains phenocrysts of plagioclase and highly altered olivine pseudomorphed by chlorite. The groundmass shows quenched texture with criss-cross-acicular crystals of clinopyroxene and remnants of primary clinopyroxene phenocrysts with high-Al content (Al_2O_3 6.52 wt.%, Mg# 75, $En_{41}Fs_{14}Wo_{45}$, Table 5.2). Secondary minerals such as chlorite and epidote are common. Olivine is commonly

Table 5.3 Petrographic features of spinel peridotites (harzburgite/lherzolite/serpentinised dunite) from the Naga Hills Ophiolite

Sample No.	Rock type	Locality	Texture	Characteristic mineralogical features	Nature of spinel
K-72	Spinel lherzolite		Porphyroclastic, micro-shear with pulverised olivine grains indicating ductile brittle environment. Step-faulting along the shear resulted in offset of magnetite vein	Stout prismatic grain of clinopyroxene containing inclusions of opaques along cleavage traces. Olivine grains in the shear zone are brecciated and associated with serpentine	Colourless, flattened and anhedral spinel, devoid of any fracture, contains inclusions of colourless orthopyroxene
R-16	Spinel harzburgite		Protogranular	Relatively fresh rock containing large plates of non-pleochroic orthopyroxene, serpentine with relict olivine, minor clinopyroxene, spinel and opaque. Orthopyroxenes are surrounded or veined by serpentine. Clinopyroxene contains inclusion of orthopyroxene	Skeletal brown spinel (~2 vol%) contains inclusions of serpentine
202/80	Spinel-bearing serpentinite (dunite)	Tizu river, 5.4 km northeast of Purr	Brecciated blocks of serpentinite, schistose	Dominated by serpentinised olivine. Variable orientation of serpentine and spinel in different blocks. Porphyroclasts of clinopyroxene are altered to chlorite. Chrysotile veins cut across foliation planes of serpentine	Thin rectangular spinels oriented along foliation
N179	Spinel harzburgite		Protogranular	Prophyroclasts of colourless orthpyroxene (25 vol%) and olivine dominate. Olivine is altered to serpentine. Clinopyroxene constitutes about 8 % by volume	Small anhedral red spinels (<1 vol%)
96/80b	Spinel lherzolite	2.25 km southeast of Purr	Protogranular, micro-shear plane shows crushing, granulation and brecciation of minerals, producing angular fragments of clinopyroxene and serpentine set in brown glassy groundmass	Olivines are altered to green serpentine both outside and inside the shear plane. Towards one end of the shear plane, both clino- and ortho (non-pleochroic)-pyroxenes are set in a groundmass of green serpentine, developing protogranular texture in ductile brittle crust	Small rounded/octahedral grains of brown spinels are developed on either side of shear plane (<1 vol%)
M77/19	Spinel lherzolite	Tizu river section, Meluri	Protogranular	Serpentinised lherzolite containing olivine, serpentine, non-pleochroic orthopyroxene, unaltered clinopyroxene and Cr-spinel. The spinel contains inclusions of both serpentine and clinopyroxene endorsing its late crystallisation. The spinel and clinopyroxene are intricately veined by chrysotile	Skeletal red spinel
M77/20	Spinel harzburgite	Tizu river section, Meluri	Protogranular	Olivine shows alteration to serpentine. Clinopyroxene constitutes about 20 % by vol. Resorbed Cr-spinel contains inclusions of serpentine and clinopyroxene	Skeletal spinels occur as discrete grains
ML77/8	Spinel Peridotite	Tizu river section, Meluri	Medium-grained	Serpentinised peridotite with relict olivine, non-pleochroic orthopyroxene, clinopyroxene and Cr-spinel	Vermicular spinel

(continued)

Table 5.3 (continued)

Sample No.	Rock type	Locality	Texture	Characteristic mineralogical features	Nature of spinel
M	Spinel harzburgite	Tizu river section, Meluri	Coarse-grained, protogranular	Fresh olivine, prismatic non-pleochroic orthopyroxene dominates over clinopyroxene, spinel ~0.5 vol%	Fractures in red spinel is filled with magnetite
ML77/11	Spinel Lherzolite	Tizu river section, Meluri	Medium-grained, fresh looking rock	Olivine altering to serpentine, both clinopyroxene and non-pleochroic ortho-pyroxene and rutile	Red spinel with fracture filling magnetite
ML77/12	Spinel harzburgite	Tizu river section, Meluri	Medium-grained	Olivine (Fo_{91}) shows alteration to serpentine. Porphyroclasts of non-pleochroic orthopyroxene are closely associated with red spinel. Minor clinopyroxene ($En_{52}Fs_4Wo_{44}$)	Resorbed and fractured spinel (Cr#19)
M77/49	Spinel lherzolite (carbonate rich)	Tizu river section, Meluri	Fine-grained, brecciated	Dominantly composed of carbonates ($\sim$45 vol%) and serpentine ($\sim$40 vol%). Other constituents are clinopyroxene, minor colourless ortho-pyroxene, opaque and secondary veins of chrysotile. Contains a xenolith (?) with chilled margin. Presence of almost pure andradite garnet (EPMA), developed in the formation of rodingite during serpentinisation	Anhedral red spinel

altered to chlorite and serpentine. Some of the chlorite pseudomorphs after olivine show a rim of epidote (230/79-1).

5.7.1.2 Porphyritic Basalt

The porphyritic basalts contain abundant phenocrysts of plagioclase and clinopyroxene which are commonly altered to albite, zoisite, epidote and chlorite. The matrix shows intergranular to intersertal texture with abundant quenched plagioclase and clinopyroxene. Basalt containing skeletal plagioclase phenocrysts (W11-1) with spherules of glass having radiating fibers (W11-3) indicates high temperature of crystallisation. Some other plagioclase phenocrysts are rounded due to resorption.

Two types of clinopyroxene-phyric basalts are identified. One contains highly fractured, sub-hedral phenocrysts of clinopyroxene showing sector zoning or hourglass structure (330/79-2). The other contains hexagonal prisms of clinopyroxene (585/79-2) with high-Al content (Al_2O_3 3.7 wt.%, Mg# 82, $En_{44}Fs_{10}Wo_{46}$, Table 5.2). They show little sign of reaction with magma. These late-formed euhedral clinopyroxene phenocrysts (585/79-2, V-17, p. 132) contrast with the early-formed clinopyroxene phenocrysts in 330/79-2 that shows partial resorption (Plate V-13). Thus the early- and late-formed clinopyroxene phenocrysts can be easily distinguished on the basis of their morphology representing two different phases of eruptions. Both types of clinopyroxene are marginally altered to chlorite (Mg# 72). Primary plagioclase is replaced by almost pure albite (Ab_{99}) and epidote. Presence of vesicles rimmed by phengite and chlorite, and filled with calcite and minor quartz, REE-rich allanite (Ce_2O_3 5.5 wt.%) and titanite suggests metamorphic fluid infiltration. Presence of significant amounts of illite (Mg# 61, K_2O $\sim$ 10 wt.%, Table 5.2) perhaps indicates an alkaline source for basalt 585/79.

5.7.1.3 Basalt Containing Plagioclase Megacryst

Porphyritic basalts containing large megacrysts of plagioclase (altered to albite + zoisite) (90/80 and 371/79) have been distinguished from other porphyritic basalts on account of their formation through multiple processes in the magma chamber including crystallisation, flotation and coarsening of the plagioclase phenocrysts by interaction with a continuous supply of hot magma. The groundmasses of the two basalts are different; one showing vitrophyric texture (371/79-3) crystallised at very high temperature, while the other is intergranular texture (90/80-3). Hematite is associated with titanite in the groundmass and pyroxenes are altered to chlorite.

The basalt showing vitrophyric texture (371/79-3) is composed largely of short prismatic clinopyroxene, skeletal plagioclase (altered to albite + zoisite), opaque, pyrite, sphene and possibly devitrified glass. Resorbed clinopyroxenes (371/79-1) show variable orientation along with plagioclase laths. Large altered albite + zoisite pseudomorphs after plagioclase megacrysts occur in glomerocrystic aggregates. The rock also contains orange coloured ocellus-like structures filled with Fe-rich chlorite and minor quartz (371/79-1a, Plate V-27, p.135).

5.7.1.4 Trachybasalt

Trachybasalt shows flow texture characterised by parallel orientation of microcrystalline laths of plagioclase (Ghose and Fareeduddin 2011). The trachybasalts contain phenocrysts of plagioclase (555/79-1), autolith (584/79-1) and pelitic xenolith (584/79-3). Interaction of pelitic xenolith with trachybasalt shows development of plagioclase laths and granophyre between alkali feldspar and quartz (584/79-3, Plates V-53, V-54, p.140). The autolith is noted by the absence of chilled margin. Brecciated trachybasalt is traversed by late albite veins (434/79-1). Presence of trachybasalt xenolith in porphyritic basalt indicates that the latter is younger than the trachybasalt (584/79-5). Trachybasalt contains ocellus of green chlorite (580/79-4) and vesicles filled with carbonate and rimmed by chlorite (580/79-30). The rock was affected by late tectonism as evidenced by brecciation (434/79-1). The brecciated fragments are cemented by albite.

5.7.1.5 Crater Facies Lava

Basalts with petrological features usually associated with formation close to a volcanic vent are described in this group. These rocks are characterised by enrichment of volatiles and high silica contents. One sample shows intergranular to intersertal (green glass) texture (567/78-1) with clinopyroxene altered to chlorite and plagioclase laths altered to pure albite. The vesicles are filled with carbonate rimmed by epidote. The groundmass shows new growth of aegirine-augite, titanite and hematite (567/78-11, 567-78-12). Other samples include a plagioclase-phyric basalt containing droplets of basaltic magma in vesicle (334/79-1), silicified albitised basalt (333/79-1) and vesicular basalt with amygdules of carbonate (B69).

5.7.1.6 Spilite

Alteration of basalt through sea water interaction produces an assemblage essentially composed of albite, chlorite, epidote and opaque. Chemically, the altered rock is termed as sodic/albite basalt on account of greater abundance of soda relative to potash. Spilite showing pillow structure with massive core displaying variolitic, sub-ophitic, intersertal and schistose texture and a chilled margin are present in the NHO (Ghose and Singh 1980; Agrawal and Ghose 1986; Singh et al. 2008). It contains albite, green chlorite, relict clinopyroxene and opaque associated with secondary calcite, epidote, serpentine, quartz, rhodochrosite and sphene. Enrichment of large ion lithophile elements (LILE), viz., Rb, Ba and Sr together with Cr and Ni and low Nb/Y ratio of spilite indicates a sub-alkaline basalt affinity (Singh et al. 2008).

5.7.1.7 Hyaloclastite

Hyaloclastites are small angular fragments of basalt resembling tuff, that is formed by the flowing of lava under water. The granulated and shattered mass of glassy lavas show quenched texture characterised by long radiating and acicular needles (27/79-2, quartz tholeiite, Mg# 42, Ghose and Chatterjee 2011) or curved crystals of skeletal plagioclase (37/79-20) and granular/dog-tooth clinopyroxene (37/79-2, Plate Hy-10, Hy-13). Such features are developed on sudden cooling of molten basaltic lava due to a temperature difference of over 1200 °C in contact with cold water. The clinopyroxene occurs as large serrated grains partially altered to chlorite, and as granular grains filling the interstices between the plagioclase laths forming an intergranular texture (27/79-8). Occurrence of tuff in contact with hyaloclastite (37/79-1) indicates explosive volcanic eruption.

Multiple flows can be distinguished on the basis of different orientations of plagioclase crystals in the glassy groundmass containing un-dissolved mesostasis (G/15-1). Consolidation of these resulted in inter-penetration texture between plagioclase and clinopyroxene (G/15-2).

One of the quenched basalts contains clasts of carbonate in vesicular cavities. A clast contains dark brown pleochroic amphibole with a core of clinopyroxene (37/79-9). Post-consolidation deformation of hyaloclastite produced folded carbonate and quartz veins dissected by micro-shear (19/79-1). The carbonates are inter-mixed with chlorite.

5.7.2 Porphyritic Andesite

Compared to the mafic volcanics, intermediate and acid volcanics are rare in the NHO. The porphyritic andesites contain phenocrysts of twinned plagioclase and their groundmass shows an intersertal texture. The plagioclase contains inclusions of devitrified glass. The vesicles are filled with a mixture of sericite and chlorite around a glassy core (62/80-4).

5.8 Volcaniclastics

A wide variety of pyroclastic rocks including agglomerate, bomb, lapilli, breccia, ignimbrite, scoria, pumice, tuff, welded tuff, fine ash and glass occur in the central part of the NHO (Agrawal and Ghose 1986; Ghose and Fareeduddin 2011). The fine tuff and ash fragments are widespread, often interlayered with chert. Presence of granophyre clast in volcanic breccia, and rhyolite ash flow deposit (ignimbrite) containing potash feldspar suggests pronounced fractionation of magma prior to eruption. Hydrous alteration of the primary minerals (e.g. chloritisation) and development of carbonate and quartz veins may be related to the high volatile contents of the magma that caused explosive eruptions with the ejection of lithic fragments, crystals and ash. The pyroclasts are also subjected to deformation resulting in brecciation and ultimately to the formation of pseudo-tachylite veins (L-93-1) similar to those in blueschists. Lithic tuff containing abundant grains of titanite and fragments of trachybasalts may be linked to an alkaline magmatic source. Thus the pyroclastic rocks may have been derived from both sub-alkaline and alkaline basaltic magma sources. Petrographic features of the tuffaceous rocks are presented below. Eroded tuffaceous material subsequently deposited in the Jopi Formation is also discussed in this section to provide a comprehensive account of the processes of formation, weathering and recycling of pyroclastic rocks and ash flow deposits.

5.8.1 Lithic Tuff

The rock fragments of lithic tuff are essentially composed of trachybasalt in association with feldspar and rarely pumiceous glass. The volcanic ejectas were possibly deposited close to vents or craters. The location of lithic volcanic ejectas on either side of the tectonic contacts of the central part of the ophiolite belt, viz., Sataza (western) and Lacham-Nagzu and Lagela-Sutsu (eastern), suggests their origin through deep fissures or fractures in an arc-type oceanic realm prior to emplacement on the continental margin. A fragment of lithic tuff contains abundant titanite (Table 5.2) rimmed by magnetite (439-1, 439-4, Plates Py-13, Py-14, p. 156). Ilmenite occurs as inclusion in titanite. Salient petrographic features of the lithic tuffs are given in Table 5.4.

5.8.2 Vitric Tuff

Glass-rich vitric tuff contains fine crystals of ferromagnesian minerals including a clinopyroxene megacryst. These are interlayered with chert and veined by quartz and carbonate. Pumiceous glass and scoria contains devitrified glass rimmed by chlorite. Coarse ash contains deformed particles and two sets of joints indicate post-consolidation deformation.

5.8.3 Crystal-Vitric Tuff

Tuffs with a high abundance of crystals and absence of lithic fragment are included in this category. The crystals are commonly of white and pink feldspars and opaque embedded in fine ash (439-1, 439-4). Welding is not uncommon at base of the thicker tuff beds due to combined effects of temperature and pressure (17/79-1, Plate Py-9, p.155). The crystals in glassy vitric tuff are either laminated or are surrounded by a rim of glass (17/79) probably reflecting differences in the time taken by the ejecta to settle, their size and weight and thickness of the tuff bed. Cuspate glass shards contain channels or grooves probably related to the escape paths of volatiles upon consolidation (17/80-3). They show post-consolidation micro-folding and faulting (17/79-3, Plate Py-11, p. 155)

5.8.4 Ignimbrite

These are rocks formed by deposition and consolidation of ash. They have been located in a small ephemeral rivulet called Likimero in the volcanic province west of Lacham Lake, near the western tectonic boundary of the NHO. The rock is easily distinguished by the presence of angular and poorly sorted quartz grains of variable dimensions set in a groundmass of chlorite, sericite, brown glass and glass shard (LKM4/89-1). Random orientation of quartz, feldspar and devitrified glass shards in a chlorite groundmass supports solidification of plastic (rheomorphic) flow (LKM3/89-5, Plate Py-22, p. 158). The rock consists of quartz, potash feldspar showing perthitic twinning, opaque, epidote, muscovite, tourmaline (dravite) and plenty of zircons. Rows of opaque and zircon in biotite-rich groundmass define the flow orientation (LKM5/89-2). Zircon also occurs as inclusion in quartz indicating their magmatic origin (LKM3/89-6). Development of kink lamellae of carbonate on opposite sides of quartz grains indicates inhomogeneous distribution of strain in the shear zone (strain shadow, LKM3/89-1).

5.9 Metamorphics

The NHO is associated with very low- to high-grade metamorphic rocks composed of mafic volcanics and pelagic sediments. The fascinating features of mineralogical changes captured in the metabasites, meta-arenites and meta-cherts developed at a converging plate boundary are presented in this section.

5.9.1 Metabasics

5.9.1.1 Very Low-Temperature Metamorphics: Zeolite and Prehnite-Chlorite Facies

Epidote-Chlorite-Natrolite-Albite Association

The earliest stage of metamorphism of basalt is observed in the very low-T zeolite facies rocks that show transformation of plagioclase (albite) to natrolite (K79-6) and epidote (K79-4), and clinopyroxene to chlorite (K79-5) due to reaction with sea water and/or burial metamorphism. The rock retains a relict igneous fabric showing an intergranular texture with chlorite occupying the angular interstices between albite crystals (K79-1). The changes in bulk chemistry are controlled by congruent mineral reactions, in which one phase is transformed into another (Cann 1979). The upper limit of zeolite facies is marked by the disappearance of zeolite and the appearance of albite and chlorite.

Prehnite-Chlorite Schist

The prehnite-chlorite schist is also a very low-T metamorphic rock derived from the mafic volcanic through burial metamorphism. Natrolite is absent in this rock. Sheaf-like aggregates of prehnite with fine chlorite flakes occur as radiating blades (390/70-1, Plates Vlgm-9, 10, p.163). Clinozoisite and opaque occur as accessory phases. This assemblage belongs to prehnite-pumpellyite facies. The mineral paragenesis is formed at temperatures well below 300 °C as determined in an experimental study on the decomposition of calcic plagioclase and diopside in hydrous conditions between 5 and 8 kbar (Heinrichsen and Schurmann 1972).

5.9.1.2 Low-Temperature Metamorphics: Greenschist Facies

Epidote-Albite Actinolite Schist

With rising temperature, prehnite is replaced by epidote and actinolite in greenschist facies metamorphism of metabasic rocks. Pyroxene is also transformed to actinolite in pyroxenite (Act-1). The greenschists show a variety of textures including granoblastic (y38-1), sub-ophitic, schistose and cataclastic. Fractures formed during deformation (three phases) are filled by quartz, albite and ferruginous minerals (43/80-3). The greenschists are characterised by the presence of green coloured minerals, viz., actinolite, chlorite and epidote, together with albite, and subordinate amounts of quartz, rutile, sphene and opaque (23/80-1). Actinolite occurs as prismatic or fibrous and poikiloblasts enclosing plagioclase, sphene, rutile and opaque. Coarse-grained actinolitite (22/80-1) originating from a late intrusion is composed dominantly of actinolite with minor amounts of albite, quartz, epidote and sphene. Recrystallised prisms of such actinolitite or meta-pyroxenite are transformed into fibrous actinolite (Act-1, 140/80-1). Fibrous brown secondary actinolite (Act-2, Plate GS-6– 8, p. 166) is formed on recrystallisation of coarse iron-rich pyroxenes in meta-pyroxenite (140/80-4) between the grain boundaries (140/80-3). Biotite is absent in this assemblage and should not be confused with brown actinolite.

5.9.1.3 High-Pressure Assemblages

The high-pressure rocks of the NHO occur as high angle thrust-bound, allochthonous slivers and lenses within the ophiolite basalts. Eclogite is rare and occurs in the core of some of the lenses that show core-to-margin retrogression with a core of eclogite successively surrounded by garnet-blueschist, glaucophanite and greenschist (Ghose et al. 1984, 2010). The eclogite also occurs as lenses or slivers within ultramafic peridotite (Fig. 3.28). A detailed petrography of the blueschists and eclogite is given below. Salient textural and mineralogical features and mineral associations in selected samples of these rocks confirmed by EPMA are presented in Tables 5.1 and 5.5, respectively.

Glaucophane Schist

Garnet and omphacite are mutually exclusive in the blueschists (glaucophane schists). In the garnet-blueschists, garnet porphyroblasts are highly resorbed and contain chlorite, quartz and phengite veins and inclusions, some of which may pre-date the garnet. However, the high degree of alteration makes it difficult to distinguish the pre-garnet and post-garnet mineral grains within garnet. Smaller garnet grains may represent a second generation of garnet. However, composition of the porphyroblasts and the smaller grains are identical, all containing high-almandine, medium pyrope and grossular, and low but significant spessartine components. Amphibole is of the sodic variety with a magnesio-riebeckite core and a winchite rim. Coarse-grained quartz, phengite and albite + K-feldspar also occur in the matrix. Phengite contains inclusions of apatite and euhedral to subhedral zircon. The omphacite-bearing blueschists are foliated and contain omphacite, glaucophane, epidote, chlorite and phengite porphyroblasts (Plate Gln-4, 9, p.170). Coarse calcite and albite veins are minor components, and rutile and titanite are accessory phases.

A wide variety of glaucophane schists are encountered both in terms of texture and mineral assemblages. They are broadly distinguished into five categories: (1) epidote-albite-chlorite-omphacite glaucophane schist, (2) glaucophane-bearing meta-arenite, (3) epidote-garnet-glaucophane schist, (4) mylonitic glaucophane schists (without micro-shear/pseudotachylite) and (5) ultraclastite glaucophane schists (with micro-shearing and partial melting). Mineralogical compositions of the veins differ in these associations.

The ultra-cataclastite contains quartzofeldspathic, pseudotachylite and carbonate veins; the meta-cherts and mylonites contain mainly quartz veins; non-garnet-bearing glaucophane schists contain quartz, epidote and albite veins and omphacite-bearing glaucophane schist contains phengite vein. The fivefold classification of the glaucophane schists adopted here is based on texture and mineralogy:

(1) Epidote-albite-chlorite-omphacite-glaucophane schist: The rock is composed of omphacite, glaucophane, chlorite, epidote, zoisite, clinozoisite, phengite, potash feldspar and quartz. Minor calcite, albite and K-feldspar are present, and rutile and titanite occur as accessory constituents (212/80, BSE). Omphacite shows transformation to glaucophane and chlorite (212/80-3). K-feldspar borders albite and phengite. Epidote is noted between chlorite and glaucophane. Titanite is present within chlorite. Idioblastic glaucophane Gln-2 across the foliation in quartzofeldspathic gneiss shows variance of orientation with Gln-1 (X-11-1).

(2) Glaucophane-bearing meta-arenites: These rocks show broad warps (M23A-1) or pinching and swelling structure of quartz (N7-1), and comprise epidote, glaucophane and opaque.

(3) Garnet-glaucophane schists: Garnet in these rocks occurs in two modes—one follows the foliation trend (Grt-1, Z74-4) and the other is randomly oriented (Grt-2, 620/79-4, Z74-3). The recrystallised garnet grains (Grt-2) are idioblastic (Z74-3) and display different shapes (620/79-6). The highly resorbed Grt-1 in Z74 is compositionally similar (almandine-spessertine) to that of the idioblastic garnets (Grt-2). The sodic amphibole also shows two different orientations in glaucophane schist (620/79-4). Sodic amphibole in Z74 is zoned, consisting of a magnesio-riebeckite core surrounded by a winchite rim (Plate Gln-25, p.174). The rock contains quartz, phengite and rare zircon. Both apatite and zircon occur as inclusion in phengite (Z74, BSE, Plate Gln-31, 32, p. 176).

(4) Mylonitic glaucophane schist (without pseudotachylite): These metabasic rocks containing glaucophane are involved in shearing to develop pinch-and-swell structures in quartz vein (622/79-1) and refolding of glaucophane (622/79-4, Plate Gln-38, p.178), one-fold axis runs NW–SE and the other is nearly E–W, passing through the knee of the glaucophane (Plate Gln-37, R. S. Sharma, person. commun.). Two generations of foliation are identified in the mylonites, the younger one cross-cuts the shear plane at an acute angle (622/79-2). Prismatic glaucophane is swathed by phengite developing S–C mylonites along shear plane suggesting ductile environment (622/79-6, ML225-2, 78/80-10, Plate Gln-34–36, p.177).

(5) Ultra-cataclastite glaucophane schist (with micro-shear and pseudotachylite veins): Both foliated (76/80-5 and 76/80-3) and massive (128-III-1a,2,4,9 and N146-1,3,6,1a) types of glaucophane schist are included in this category as evidence of high temperature melting (Plate Gln-46, 47, p.180). The former is characterised by microshear defined by actinolite-glaucuphane (76/80-3), resulted in bi-modal glaucophane (a) prismatic following foliation and (b) rhombic with basal cleavages that developed across the foliation (76/80-5). The latter massive type contains brecciated matrix. These dynamically metamorphosed rocks are subjected to extreme pressures along the shear zones that resulted in thrusting and buckling (128-III-2, Plate Gln 49) with generation of pseudotachylite (128-III-1a) and quartzofeldspathic veins (76/80,128-III-2). Small melt generation as a result of eutectic melting at higher temperature is feasible ($\geq$600 °C, at pressure >10 kbar, Tuttle and Bowen 1958; Winkler and Ghose 1973) during deformation and is limited to narrow zones. Carbonates and quartzofeldspathic veins are involved in shear-related S-folds (128-III-1a, Plate Gln-47). Both shearing and buckling may be co-eval (Abhinab Roy, person. commun.). The features like mica-fish (N146-3), glaucophane-fish (128III-2, Plate Gln-49) and veins of pseudotachylite and quartzofeldspathic veins in glaucophane schists, though limited in extent, indicate metamorphism similar to the Barrovian-type in the convergent plate tectonic settings as well. Shearing, brecciation (N146-6) and mortar texture (128III-9, Plate Gln-48) associated with these rocks gives an impression of post-crystalline deformation during exhumation.

Eclogite

Eclogite contains the characteristic garnet + omphacite (with variable jadeite contents) association. It is a plagioclase-free high-pressure rock derived from a basaltic composition at greater depth beyond the stability limit of feldspar. The texture is granoblastic to poikiloblastic with abundant epidote, sodic-calcic amphibole (barroisite), sodic amphibole (glaucophane), and minor garnet, omphacite, phengite, chlorite and quartz (Plate Ec-2, 3, p.183). Interpenetration twins between barroisite and glaucophane have been observed (Plate Ec-5, 9). Accessory rutile is mantled by titanite. Secondary albite and quartz veins cut through the sample. A weak foliation is defined by parallel orientations of amphiboles, phengite and chlorite. Garnet is poikiloblastic and contains inclusions of omphacite, glaucophane, epidote, rutile and quartz. Glaucophane replaces barroisite at rims as a result of retrograde metamorphism. Secondary chlorite veins in garnet and phengite veins in

Table 5.4 Petrographic features of Polymictic tuff breccia in Jopi Formation and Crystal lithic tuff in mafic volcanics of NHO, India

Sample No.	Rock type	General features and texture	Presence of discrete mineral grains	Locality
Jopi formation—post-ophiolite cover sediments				
C21/79	Polymictic tuff breccia	Clasts and fragments of varied rock types and minerals: clasts of serpentinite, pyroxenite, plagioclase phyric basalt, trachyte, rhyolite, anorthosite, serpentinite, radiolarian chert, chert, greywacke, shale, quartzite, phyllite, metabasic and sandstone embellished with plagioclase laths. Brown rhyolite containing chert, lenticular foraminifera and serpentinite. Secondary minerals like sericite in plagioclase, carbonate and chlorite	Presence of discrete grains of both igneous and metamorphic minerals—euhedral crystals of orthopyroxene and plagioclase, prismatic clinopyroxene and glaucophane, anhedral and broken crystals of clinopyroxene and garnet, biotite, quartz, ilmenite rimmed by rutile and chrome-spinel. Orthopyroxene dominates over clinopyroxene in anorthosite. Rare presence of an isolated octahedral opaque mineral with bluish tinge in polarised light and a variegated second order interference colour. Strongly pleochroic green aegirine occurs as discrete grains or forming rim around brown clinopyroxene in trachyte (peralkaline)	Western tectonic contact with Disang, 2 km S12° E of Waziho
Crystal-lithic tuff, Mafic Volcanics				
124/79	Crystal-lithic tuff	Rhyolitic fragment	Phenocryst of pink alkali feldspar, quartz and biotite in glassy groundmass. Granophyric intergrowth between quartz and alkali feldspar	Western tectonic contact, south of Zipu, 5.8 km S32° E of Waziho
439	Crystal-lithic tuff	Clasts of trachy basalt surrounded by aggregates of feldspar laths. Microlites of feldapar are oriented along flow direction. Basalt contains elliptical and cylindrical particles. Contact of basalt and feldspar is stained by brown iron-rich solution	Plagioclase contains inclusions of dark anhedral grains and streaks of opaque along C-axis. Feldspar is rimmed by green chlorite. Skeletal grains of magnetite rimmed with brown chlorite. Sub-rounded and discrete grains or aggregates of brown titanite rimmed by magnetite in ash. Titanite contains inclusions of ilmenite	Eastern tectonic contact with Nimi Formation, 300 m SSE of Ngazu and 400 m away from the tectonic contact
429/79	Crystal-lithic tuff	Rectangular fragment of trachy basalt surrounded by feldspars. Microlites of feldspar in basalt showing flow orientation	Late hydrothermal activity resulted in development of secondary calcite, green chlorite and octahedral spinel	Eastern tectonic contact with Nimi Formation, 0.7 km NW of Ngazu
426/79	Crystal-lithic tuff	Fragments of trachy basalt entrapped in late feldspathic vein	Plagioclase and minor clinopyroxene	Eastern tectonic contact, 1.25 km NW of Ngazu

epidote are common. Evidence of high pressure includes cataclastic zones traversed by multiple veins of brecciated glaucophane and epidote (77/80-5, 6, Plate Ec-14). Quartz as inclusions in garnet (77/80-7, Plate Ec-3) suggests the possibility of existence of coesite formed under ultra-high pressure (UHP) metamorphism that needs to be investigated in the future. However, P–T pseudosection analysis indicates that the eclogites reached maximum pressures of 1.7–2.0 GPa following a clockwise P–T path with prograde metamorphism beginning at ~1.3 GPa/525 °C, peaking at 1.7–2.0 GPa/580-610 °C and retrogressing to ~1.1 GPa/540 °C (Chatterjee and Ghose 2010).

5.9.2 Grunerite-Bearing Metachert

These rocks are composed of quartz and brown chlorite, acicular Mn-rich grunerite amphibole and opaque (855-1, 855-3). The grunerite needles also occur along the quartz grain boundaries. Accessory apatite and rare inclusions of galena and chalcopyrite are present within quartz.

The meta-chert and meta-arenites occur as isolated bodies in contact with the mafic volcanic. They are fine-grained and massive, bluish or buff coloured rocks containing abundant quartz, chlorite and minor blue amphibole and opaque. The bluish amphibole-bearing meta-arenites/greywackes (M23A) are coarser than meta-cherts comprising garnet or clinopyroxene, and phengite in addition to quartz, albite, actinolite and epidote.

5.10 Oceanic Pelagic Sediments

The oceanic sedimentary rocks are represented by clastic (e.g. arenites and shale) and authigenic (e.g. limestone and chert) sediments. They comprise tuffaceous volcano-sedimentary rock, iron-rich sediment, shale, slate, siltstone, arkose, greywacke, bedded limestone and chert. Limestones with clasts of volcanic and *vice versa* are common. Presence of stable accessory minerals like zircon, rutile, tourmaline and muscovite in arenites indicate end products of sedimentation. These stable components are derived through more than one cycle of erosion and deposition.

The oceanic pelagic sediments have undergone at least two phases of deformations with development of joints, micro-folding/faulting/shearing and brecciation. These are manifested in undulose extinction of quartz and mylonitisation in arenites, micro-folding/faulting/shearing in radiolarian cherts/arenites and lamination of graphite flakes in marl (a mixture of clay and carbonates). The rocks are recrystallised into epidote-actinolite-albite schist, phyllite, schist, quartzite, quartzofeldspathic gneiss, crystalline limestone/marble and meta-chert to low temperature greenschist facies. Occurrence of banded chert-magnetite, sphalerite, pyrite and graphite in pelagic sediments give indication of variable supply of oxygen in depositional environment.

5.10.1 Tuffaceous Sediments

The tuffaceous sediments characterise random orientation of amphibole needles. Grain size is rarely coarse. Actinolite typically exhibits small, flat or acicular crystal habit. The mafic tuff (N183-4) is an example of a crenulation cleavage. Deformation during crenulation cleavage formation is possibly in concurrence with a solution-deposition process, in which transport occurred by a combination of localised advection and grain-boundary diffusion. Recrystallisation and hence equilibration of grains in discrete cleavages is recognised by distinct compositional differences between the grains in different domains.

The occurrence of albite and carbonate in fine-grained tuffaceous sediment represents advanced alteration stage (N183-5). The fluid composition is inferred to be H_2O-rich containing $CO_2 + Na^+$ and possibly S. Since the distal alteration halos are dominated by hydrated silicate phases (mainly chlorite), with minor carbonates, fixation of H_2O is indicated. The CO_2 is consumed to form carbonates in the intermediate alteration stage around the actinolite dominated zones. There was possibly a variation in the H_2O:CO_2 ratio of the sulphur-bearing aqueous-carbonic ore fluid, which interacted at varying fluid to rock ratios with progression of the hydrothermal alteration.

5.10.2 Argillaceous Sediments

Foliation is developed in the phyllite due to recrystallisation of clay minerals (567/78-1). Sedimentary features like convolute bedding (567/78-2), gash veins, heterolith and load cast (56-1) are developed as a result of compaction, recrystallisation, strain dilation, density contrast and cleavage formation (Plate OS_Arg 2–4, p. 189). Late crystallisation of actinolite in siliceous argillite (56-1) is caused by change in conditions due to thermal relaxation probably accompanied by uplift and erosion. The carbon phyllite possibly indicates local hypoxic condition of deposition (473/79-1).

5.10.3 Arenaceous Sediment

The deformed sandstone (458/79-1) has a weak S–C fabric developed gritty tectonite with rotational fabrics. The rock may be called coarse-grained polymodal sandstone with well-preserved ‘beard’ texture in the lower part, supporting quick rotation of grains during transport and quick deposition. The feldspathic wacke (446/79-1) shows development of S–C fabric. Rolled over (re-deposited) grains of quartz indicate re-transportation of earlier preserved tractional grains. Presence of an earlier secondary matrix along with the tectonic matrix is an indication of complete absence of thinned skin tectonics and slicing. It rather indicates forced compaction in a cushioning system (vertical compaction). The feldspathic wacke (408/79-1) with concentration of heavies (rutile, sphene, magnetite and apatite) is indicative of a proximity to the source rock and quick disposition. High dispersive pressure and lower primary and secondary matrix indicate slow diagenesis. Bedding planes in banded quartzite (n140-1) show a regularity in distribution, indicating a quiet sedimentation in deep water condition. Resting time for the grains is very high as there is textural maturity. Heavies (sulphides and ilmenite) concentrating within layers suggest a hydrodynamic equilibrium between light and heavies. A ‘drop structure’ like feature (lower central) of a big quartz grain is indicative of re-sedimentation. Upward flow and pseudo-convolutions within the heavies are also visible (upper right).

5.10.4 Carbonates

Fine-to medium-grained carbonates are largely impure siliceous dolomitic limestone. Development of quartz may be

Table 5.5 Petrographic features of blueschists and eclogite in the ophiolite belt of Naga Hills, India

Sample No.	Rock type	Texture and structure	Petrographic features
A. Glaucophane schist (blueschist)			
X-11	Glaucophane schist (meta-arenite)	Gneissic, coarsely foliated	Bands of quartz and feldspar are separated by a mixture of phengite, glaucophane, brown chlorite, epidote and opaque imparting foliation. Glaucophane is pleochroic from brownish yellow to blue. It occurs as prismatic and rhombic idioblast, the latter is developed across foliation
Z-71	Garnet-glaucophane schist	Schistose and cataclastic	Two types of glaucophane—prismatic and sub-hedral, the latter is without cleavage traces. Recrystallised sutured quartz showing polygonal fractures in larger grains. Colourless mica (phengite) occurs both as porphyroblasts or small flakes inducing schistosity. Epidote and opaque occur as accessory minerals
N-7*	Glaucophane-epidote schist	Fine-grained, schistose, minor warps	Epidote forms the most dominant component ~50 % (vol). Other constituents are quartz, albite, glaucophane, green amphibole, chlorite, phengite and opaque. Larger quartz contains streaks of opaque along longitudinal axis, while fine flakes of mica, glaucophane and epidote are randomly oriented. Glaucophane is pleochroic from purple to blue. Feldspars are usually untwinned. A few twinned ones on albite law are present
620/79	Garnet-actinolite-glaucophane schist	Finely foliated, schistose	Quartz dominant (~70 % vol) rock, other minerals comprise glaucophane (25 %), idioblatic granular red garnet (3 %), actinolite, phengite and spinel Randomly oriented garnets are of late origin
622/79	Garnet-glaucophane schist	Mylonitic, presence of mica fish	Refolded glaucophane (lotus-petal appearance) formed due to metamorphic recrystallisation. Two generations of foliation intersecting the shear plane. 'Mica-fish' indicates ductile brittle deformation
N-146	Glaucophane schist	Schistose, brecciated, pseudo-tachylite vein	Development of glaucophane with a core of actinolite/overgrowth. Other components are epidote, zoisite, phengite and caught-up quartz-carbonate fragments. Ultra-metamorphism resulted pseudo-tachylite vein
128(III)	Glaucophane schist	Ultra-cataclastic, mortar texture and development of 'glaucophane-fish' implying ductile brittle environment. Formation of pseudo-tachylite and quartzofeldspathic veins, co-eval thrusting and buckling	Quartzofeldspathic and carbonate veins are involved in shear related S-folds. Formation of pseudo-tachylite veins. Development of 'glaucophane fish' suggests ductile brittle shear along which the ophiolite was emplaced. Fine needles of deep reddish transparent rutile cutting-across glaucophane. Large semi-opaque grains of rutile occur as inclusions in glaucophane. Glaucophane occurs in two modes—prismatic grain with cross-fractures and as recrystallised porphyroblast showing basal cleavage. Greenish diopsidic augite is faintly pleochroic showing moderate extinction angle (15°). Epidote and opaque are common accessories
ML-225*	Glaucophane schist	Mylonitic, mica-fish in shear zone (C-plane) indicates ductile brittle shear	Broad warp in actinolite and glaucophane dotted with epidotes. Actinolite pseudomorphs over glaucophane. Layers of actinolite and rhombic crystals of glaucophane are interwoven with phengite. Quartz contains actinolite inclusion. Groundmass is largely composed of quartz, albite and green chlorite
212/80*	Omphacite-glaucophane schist	Schistose	Essentially composed of albite, green chlorite, actinolite, glaucophane, omphacite and zoisite. Glaucophane cross-cuts large prism of omphacite. Glaucophane in rimmed by calcic-amphibole (actinolite) in close contact with garnet or zoisite. Lenticles of glaucophane and epidote are seen in carbonates. Late vein of phengite cuts across zoisite and carbonate. Pseudomorph of albite (grey drops) is developed in carbonate

(continued)

Table 5.5 (continued)

Sample No.	Rock type	Texture and structure	Petrographic features
43/80	Glaucophane-actinolite-epidote schist	Finely foliated, schistose, presence of micro-shear plane	Relict original lithological band is visible. Two types of schistosity are distinguished, viz., S_1-horizontal and S_2-oblique. Quartz veins and fine granulated rock follow micro-shear plane. Overprinting/pseudomorph of glaucophane is seen on actinolite. Late albite cuts across quartz veins. Epidote, rutile and sericite form accessory minerals
Z-74*	Garnet-glaucophane schist	Medium–grained, boudins of sodic amphibole swathed by phengite	Purple to blue pleochroic sodic amphibole is rimmed by green actinolite. Garnet occurs as inclusion in sodic amphibole. Two types of garnets are discernible (a) small granular type developed along foliation by reaction involving chlorite, phengite and quartz, and (b) late idioblastic or octahedral garnets are found across the foliation. Pseudomorph of sodic amphiboles could be seen in actinolite. Phengite cuts across other constituents. Zoisite is partly rimmed by phengite and green amphibole. Albite, quartz and euhedral rutile are also recorded
855	Grunerite-bearing metachert	Very fine-grained	Fine needles of blue Mn-bearing grunerite cross-cut quartz and chlorite. Formation of blue amphibole is apparently related to streaks of ferruginous material. Recrystallisation of quartz is noted between greenish brown chlorite
ML-69*	Epidote-glaucophane schist	Schistose	Fine-grained glaucophane schist, containing barroisite, chlorite, epidote, phengite and rutile
2/70*	Glaucophane-bearing arenite	Vaguely foliated	Fine fibrous actinolite and granular quartz impart linear fabric
ML-70*	Glaucophane-chlorite-barroisite-epidote rock	Granoblastic	Epidote dominates over other constituents. Glaucophane is pleochroic from pale purple to greenish blue. Chlorite, green amphibole (barroisite), quartz, phengite, ilmenite and opaque form other constituents of the rock
ML-234*	Glaucophane schist	Schistose	A highly carbonated rock containing glaucophane, quartz, phengite, rutile and opaque minerals
76/80*	Actinolite-glaucophane schist (Glaucophanite)	Schistose, micro-shear	Dominantly composed of glaucophane showing faint pleochroism. Green actinolite together with sodic amphibole impart nematoblastic texture. Minor amounts of epidote and opaque form accessories
78/80	Garnet-actinolite-glaucophane schist	Coarsely foliated, sheared and presence of 'mica fish'	Flattened and sutured quartz, colourless octahedral garnet, phengite, epidote and opaques are involved in foliation. Porphyroblasts of glaucophane contain inclusions of epidote and opaque. Prisms of green amphibole show faint pleochroism from colourless to green
139/79*	Garnet-epidote-actinolite-glaucophane rock	Granoblastic	Transformation of green amphibole to glaucophane. Interpenetration twinning between green amphibole (barroisite) and glaucophane. Garnet, albite, phengite, quartz and opaque are other constituents of the rock. Epidote and rutile are present in abundance
B. Eclogite			
77/80*	Barroisite eclogite	Granoblastic and foliated, development of cataclastic zone	Foliated rock with porphyroblasts of omphacite (2 vol%), garnet (8 vol%), barroisite-glaucophane amphibole (47 vol%), epidote (30 vol%) and chlorite (1 vol%). Minor phengite (4 vol%), titanite, rutile, quartz (1 vol%) and albite (in vein) are also present. Poikiloblastic garnet contains inclusions of omphacite, glaucophane, epidote, rutile and quartz. Omphacite contains epidote inclusions. Barroisite is a peak metamorphic mineral and glaucophane replaces barroisite through retrogression. Epidote porphyroblasts have veins of phengite and quartz. A brecciated zone is developed traversed by thin veins of albite. Amphiboles are highly crushed to angular fragments in the cataclastic zone, indicating the rock to have undergone strong deformation. Chlorite with sperulitic texture surrounded by epidote is developed due to late hydrothermal activity

EPMA analyses of minerals are given in Tables 5.1 and 5.2
* Sample analysed by EPMA

the result of extensive development of chemical gradient and fluid migration. Most of the quartz seems to be around the peripheral zones of calcite grains; an indication of early diagenetic dissolution of sponge spicules leading to silica enrichment of interstitial waters from which silica re-precipitated in the overlying horizons. The phengite-magnetite-(188/79-2) and tremolite-magnetite-bearing impure marble may have formed in the mixed layer for mafic volcanic rocks and carbonates at low temperature greenschist facies metamorphism. An admixture of authigenic limestone with the argillceous rocks containing fine flakes of graphite produced sericite-epidote-quartz-graphite-bearing carbonate rocks.

5.10.5 Chert

The cherts are grey, green or red in colour; iron-rich variety is rich in planktonic micro-organisms. They are affected by tectonics with development of crude crenulation (Z55-1), folding (N19-1), series of normal and reverse faults (9/79-1) and brecciation (44/79-3). The cherts are composed of fine cryptocrystalline quartz and chalcedony, very fine needles of grunerite, tremolite, muscovite, sericite, rutile, brown chlorite, albite and opaque. Amphibole grains show tensional patterns (618-1). Biotite is developed late. Stylolite bands (white, sutured) characterise pressure solution (N110-1), formed diagenetically in consolidated rock by differential vertical movement. Similarly, immature meta-arenites are recrystallised under high-pressure into glaucophane-bearing schists or gneisses consisting of quartz, albite, glaucophane, chlorite, opaque, epidote and zoisite.

5.11 Ophiolite-Derived Cover Sediments: The Jopi/Phokphur Formation

The finer sediments derived from ophiolite suite of rocks are classified as polymictic tuff breccia, lithic greywacke, greywacke, feldspathic arenite and arenite. They mostly consist of poorly sorted fragments and clasts of chert (with radiolarian), ultramafic dominated by serpentinite, gabbros, altered volcanic, trachytes, metabasic and meta-arenite containing blue amphibole (glaucophane), greywacke and argillite belonging to ophiolite and Disangs. These are variably cemented in a ferruginous, chlorite or siliceous matrix depending upon the source material. The matrix also contains discrete grains of plagioclase, orthopyroxene, clinopyroxene, glaucophane, garnet, limonite, rutile, quartz, calcite, devitrified glass and opaques. The clast-to-matrix ratio varies from 70:30 to 20:80 (Agrawal 1985). Reworking of some of the clastic rocks is also noted. A wide variation in the assemblages of clasts along the strike has been controlled by underlying ophiolite provenance since the clasts were derived and deposited in a restricted basin with limited transportation. Characteristic features of the polymictic tuff breccia are presented in Table 5.4. The base of sedimentary sequence is feebly metamorphosed and deformed.

5.12 Late Tertiary Granites

The intrusive granitoids range in composition from diorite, granodiorite, granite to quartz porphyry. The rocks usually show hypidiomorphic granular texture, and post-crystalline deformation and brecciation. They contain variable amounts of quartz, alkali feldspar, plagioclase, amphibole, and minor biotite, muscovite, epidote, pyrite, chalcopyrite and magnetite as accessories. The potash feldspar replaces plagioclase or occurs as graphic intergrowth. The plagioclase shows alteration to sassurite. Green amphibole alters to chlorite and contains inclusion of plagioclase.

Base metal sulphide mineralisation of copper and molybdenum is associated with late felsic magmatism which acted as carrier of epigenetic mineralising solution (Ghose et al. 1986). The origin of these rocks has been related to high temperature hydrothermal metamorphism and partial melting of basalt in the contact aureole of a magma chamber beneath an ocean-floor spreading center (Ghose and Chatterjee 2011).

References

Agrawal OP (1985) Geology and geochemistry of the mafic-ultramafic complex of Indo-Burman ranges between Meluri and Awankhoo, Phek district, Nagaland, India. Unpublished Ph.D. Thesis, Patna University, Patna

Agrawal OP, Ghose NC (1986) Geology and stratigraphy of the Naga Hills ophiolite between Meluri and Awankhoo, Phek district, Nagaland, India. In: Ghose NC, Varadarajan S (eds) Ophiolites and Indian plate margin. Sumna Publishers, Patna, pp 163–195

Cann JR (1979) Metamorphism in the ocean crust. In: Talwani M, Harrison CG, Hayes DE (eds) Deep drilling results in the Atlalntic Ocean: ocean crust, Maurice Ewing series 2, American Geophysical Union, Washington, Geodynamic Project. Scientific Report No.48, 230–238

Chatterjee N, Ghose NC (2010) Metamorphic evolution of the Naga Hills eclogite and blueschist, Northeast India: implications for early subduction of the Indian plate under the Burma microplate. J Metamorph Geol 28:209–225

Dick HJB, Bullen T (1984) Chromian spinel as a petrogenetic indicator in abyssal and alpine-type peridotites and spatially associated lavas. Contrib Miner Petrol 86:54–76

Ghose NC, Chatterjee N (2011) Diorite vein in quenched basalt and its implication for the origin of late-granitoid intrusives in Naga Hills Ophiolite, Northeast India. In: Srivastava RK (ed) Dyke swarms: keys for geodynamic interpretation. Springer-Verlag, Berlin, 315–330

Ghose NC, Fareeduddin (2011) Textural fingerprints of magmatic, metamorphic and sedimentary rocks associated with the Naga Hills

Ophiolite, northeast India. In: Ray JS, Sen G, Ghosh B (eds) Topics in igneous petrology: a tribute to professor Mihir K. Bose. Springer, Berlin, 321–351

Ghose NC, Singh RN (1980) Occurrence of blueschist facies in the ophiolite belt of Naga Hills, east of Kiphire, N.E.India. Geol Rdsch 69:41–43

Ghose NC, Agrawal OP, Windley BF (1984) Geochemistry of the blueschist-eclogite association in the ophiolite belt of Nagaland, India. In: Cenozoic crustal evolution of the Indian plate margin. Seminar abstracts, Patna University, Patna, 27–30

Ghose NC, Agrawal OP, Singh RN (1986) Geochemistry of the ophiolite belt of Nagaland, N.E. India. In: Ghose NC, Varadarajan S (eds) Ophiolites and Indian plate margin. Sumna Publishers, Patna, 241–294

Ghose NC, Agrawal OP, Chatterjee N (2010) A geological and mineralogical study of eclogite and glaucophane schists in the Naga Hills Ophiolite, Northeast India. Island Arc 19:336–356

Heinrichsen Th, Schurmann K (1972) Mineral reactions in low-grade metamorphic rocks. In: 24 Intern Geol Cong Section 2, 5–10

Ningthoujam PS, Dubey CS, Guillot S, Fagion A-S, Shukla DP (2012) Origin and serpentinization of ultramafic rocks of Manipur ophiolite complex in the Indo-Myanmar subduction zone, Northeast India. J Asian Earth Sci 50:128–140

Sengupta S, Ray KK, Acharyya SK, de Smeth JB (1990) Nature of ophiolite occurrences along the eastern margin of the Indian plate and their tectonic significance. Geology 18:439–442

Shukla R (1989) Occurrence of rodingite in the ophiolite belt of Manipur. In: Ghose NC (ed) Phanerozoic ophiolites of India. Sumna Publishers, Patna, 189–196

Singh AK (2013) Petrology and geochemistry of abyssal peridotites from the Manipur OphioliteComplex, Indo-Myanmar orogenic belt, Northeast India: implication for meltgeneration in mid-oceanic ridge environment. J Asian Earth Sci 66:258–276

Singh AK, Singh NI, Debala Devi L, Singh RKB (2008) Pillow basalts from the Manipur ophiolite complex, Indo-Myanmar range, Northeast India. J Geol Soc India 72:168–174

Singh AK, Singh N, Ibotombi Singh N, Debala Devi L, Ranjit Th (2010) Mineralogy and geochemistry of ultramafic rocks of northern Manipur ophiolite complex, northeast India. Himalayan Geol. 31:7–18

Singh IB, Debala Devi L, Chani Y (2013) Petrological and geochemical study of serpentinised peridotites from the southern part of Manipur ophiolite compex, Northeast India. J Geol Soc India 82:121–132

Srikanth B, Subba Rao MV, Rao BV, Nirmal Charan S, Balaram V, Ejung OC, (2004) Geochemical signatures in the basaltic rocks of Naga Hills ophiolite belt: implications for petrogenesis and tectonic environment of emplacement. J Appl Geochem 6:177–189

Tuttle OF, Bowen NL (1958) Origin of granite in the light of experimental studies in the system $NaAlSi_3O_8$-$KAlSi_3O_8$-SiO_2-H_2O. Geol Soc Am Mem 74:153

Venkataramana P, Datta AK (1987) Contrasting volcanic suites in Naga Hills and their bearing on the tectonic evolution of the Naga hills ophiolite belt, northeast India. J Geol Soc India 30:33–47

Whitney DL, Evans B (2010) Abbreviations for names of rock-forming minerals. Am Mineral 95:185–187

Winkler HGF, Ghose NC (1973) Further data on eutectics in the system Quartz-Orthoclase-Anorthite-Water. Neues Yearbuch fur Mineralogie Monatsheft 11:481–484

6 Petrogenesis

6.1 Ultramafics

The average composition of peridotite from the ophiolite belt in Manipur shows high MgO (32.0 wt.%), Al_2O_3 (1.28–3.30 wt.%), Ni (1966 ppm), Cr (2005 ppm) and Co (130 ppm) and low TiO_2 (0.12 wt.%), CaO (6.56 wt.%), Na_2O (0.44 wt.%) and K_2O (0.07 wt.%) (Singh et al. 2010; Singh 2013). The average Mg# (molar 100.Mg/[Mg + Fe^{2+}]) varies within a narrow range (90 in dunite and harzburgite, 89 in lherzolite and 83 in wehrlite). The cumulate harzburgite (K-14) exhibits inclusions of pentlandite $\left((Fe,Ni)_9S_8\right)$ and chalcopyrite $(CuFeS_2)$ in olivine and orthopyroxene, and magnetite in clinopyroxene as determined by EPMA.

The ultramafics show a chemical affinity towards the abyssal peridotites and are characterised by forsteritic olivine (Fo_{91-93}), high Al_2O_3 (up to 5.7 wt.%) in clinopyroxene $(En_{48-52}Fs_{4-7}Wo_{44-47})$, enstatite orthopyroxene $(En_{92}Fs_7Wo_1)$ and aluminous spinel (Al_2O_3 51 wt.%) (Table 5.5). The Cr# of chrome spinel is variable (up to 80). Pyroxene thermometry (Lindsley 1983) indicates that the peridotite tectonites (96/80A and ML-77-12) equilibrated at 800–1000 °C at <1.5 GPa pressures. These temperatures are similar to 780–1200°C temperatures estimated from a clinopyroxene composition at 1.9–2.5 GPa pressures and oxygen fugacities between −1.32 and −3.46 in Manipur peridotite (Maibam and Foley 2008). Similar temperatures were also estimated in the Indus-Tsangpo suture ophiolites including metamorphic sole of the Saga ophiolitic mélange (850 °C at 1.2 GPa, Guilmette et al. 2012) and the peridotites associated with the Spontang (1000–1200 °C, Reuber 1986) and the Yungbwa (900–960 °C, Miller et al. 2003) ophiolites. On the other hand, the cumulate ultramafics contain olivine and orthopyroxene with lower forsterite (Fo_{78-85}) and enstatite $(En_{83}Fs_{15}Wo_2)$ contents, and the pyroxenes contain lower alumina (2.2–2.7 wt.%) (Table 5.5). Thus the cumulate ultramafics are equilibrated under a different set of physico-chemical conditions. Estimated temperatures of 650–700 °C from pyroxene thermometry and the presence of hornblende at the pyroxene grain boundaries indicate metamorphic re-equilibration under amphibolite facies conditions.

The peridotites have very low REE concentrations $(\Sigma REE = 0.48-2.14$ ppm). Lherzolites display LREE-depleted patterns $(La_N/Sm_N = 0.14-0.45)$ with a flat to slightly fractionated HREE segments $(Sm_N/Yb_N = 0.30-0.65)$, whereas Cpx-harburgites shows flat to convex-upward LREE patterns $(La_N/Sm_N = 0.13-1.23)$ with more fractionated HREE patterns $(Sm_N/Yb_N = 0.13-0.65)$ than lherzolite (Singh 2013). Abundances of the platinum group elements(PGE) in the peridotite are low (<50 ppb). Mantle-normalised Pd/Ir (1.8–11.9)and Pt/Pt* values (0.2–1.1) show an affinity toward residual mantle (Singh 2013).

6.2 Gabbros

The mafic plutonics and intrusive dolerites of the NHO are sub-alkaline in composition, enriched in iron and show chemical similarity with MORB (Ghose et al. 1986; Venkataramana et al. 1986; Singh et al. 2012a). Calcic plagioclase (An_{93}) and high-Al augite (2.3 wt.% Al_2O_3, $En_{47}Fs_8Wo_{46}$, Table 5.5) likely represent primary products of magmatic crystallisation. The presence of edenite, magnesio-hornblende and magnesio-hastingsite amphiboles (Mg# 76–80) as alteration products of augite indicates an imprint of amphibolite facies metamorphism. The compositions of gabbro occurring in the southern part of the opihiolite belt in Manipur are similar to evolved sub-alkaline basalt with high iron, V and Y contents (Singh et al. 2012). Positive correlations between CaO/Al_2O_3 and Mg#, and Ti and Zr, and negative correlations between SiO_2 and CaO, and FeO and Zr, are consistent with fractionation of both plagioclase and clinopyroxene. Abundances of Th, Nb and LREE in these rocks are higher than in N-MORB.

N. C. Ghose et al., *A Petrographic Atlas of Ophiolite*,
DOI: 10.1007/978-81-322-1569-1_6, © Springer India 2014

High Zr/Nb (4.3–10.8), Y/Nb (1.6–5.0), $(La/Yb)_N$ (4.53–7.34) and $(La/Sm)_N$ (2.58–2.67) ratios indicate derivation from an enriched E-MORB-type mantle source.

6.3 Basalt

A detailed petrographic study of the mafic volcanic rocks of NHO reveals variable texture, mineral compositions and abundances of phenocryst and groundmass. The mafic volcanics are dominated by tholeiitic basalt with minor alkaline trachybasalt and rare andesite. Each variety of basalt is distinctively associated with volcaniclastic rocks. Some tholeiites are characterised by the presence of pyroclastic flow deposit or ignimbrite of rhyolite composition (4/89-1) and tuffs containing clasts of granophyre (124/79-1). Basaltic hyaloclastite with quench-textured breccia (27/79-6) and tuff (37/79-1) originated through phreatomagmatic interaction with water. On the other hand, it appears that low-viscosity magma with low volatile content produced effusive lava flows with limited pyroclastic debris close to the eruptive centres (e.g. Sataza, Zipu and Lacham Lake). Trachybasalt occurs as fragments in a majority of the lithic tuffs. High concentration of titanite in lithic tuff with clasts of trachy basalt relates its derivation from an alkaline magmatic source. Trachybasalt flows are rare. But the high-Ti alkaline trachybasalt (580/79-4), though limited in extent, confirms the co-existence of tholeiitic and alkaline magma types in the ophiolite terrain. Although the basalts are highly altered, primary Fe-rich augite phenocrysts are preserved in both olivine tholeiite 230/79 (6.5 wt.% Al_2O_3, $En_{41}Fs_{14}Wo_{45}$) and clinopyroxene-phyric alkali basalt 585/79 (3.7 wt.% Al_2O_3, $En_{44}Fs_{10}Wo_{46}$) (Table 5.5) that probably crystallised late in the crystallisation sequence.

The alkali basalts with euhedral crystals of clinopyroxene (585/79-2), breccia (434/79-1), amygdules (580/79-3) clasts of carbonates and minor occurrences of titanite, allanite (5.5 wt.% Ce_2O_3) and potassium-rich illite (K_2O 9.9 wt.% in 585/79, Table 5.5) may have erupted in shallow marine conditions as indicated by rock debris around the volcanic vent. Alternatively, it may have erupted above the sea surface and was re-worked by ocean current. Rare porphyritic andesite derived from tholeiitic magma possibly represents the youngest eruptive phase. Spilite formed through interaction of basalt with sea water was followed by low-grade metamorphism and/or metasomatism.

The majority of basalts from the NHO are tholeiitic in composition showing progressive enrichment of total iron and large ion lithophile (LIL) elements (Ghose et al.1986; Venkataramana et al. 1986; Sengupta et al. 1989; Srikanth et al. 2004). They are characterised by a flat and fractionated REE pattern ($(La/Sm)_N = 0.67-2.22$ and HREE ($(Gd/Yb)_N = 0.8-2.9$)), an absence of Eu-anomaly (Eu/Eu* = 0.74–1.09), and negative anomalies of some LIL (Rb, K and Sr) and HFS (Th, Nb and Zr) elements (Sengupta et al. 1989; Srikanth et al. 2004; Rao et al. 2010; Patil et al. 2012). These features indicate negligible plagioclase fractionation and an absence of crustal contamination (Srikanth et al. 2004). An example of a primitive tholeiite is the clinopyroxene/plagioclase-phyric hyaloclastite (27/79-8) collected from the western margin of the NHO. It contains 8.52 wt.% MgO (Mg# 60) and 1.92 wt.% TiO_2, and its trace element concentrations are comparable to N-MORB (Cs 0.10, Rb 2.4, Zr 40, Hf 1.1, Ni 71, La 5.9, Eu 1.8, Yb 4.7, Ta 1.6 and Th 0.5 ppm; XRF/ICP-MS data, Ghose and Chatterjee 2011).

The NHO basalts are similar to both ocean floor basalts and plate margin basalts (Srikanth et al. 2004). These are divided into a low-Ti group and a high-Ti (>2 wt.% TiO_2) group reflecting variable source compositions and degree of partial melting (Sengupta et al. 1989). The Ti/V ratios and REE patterns of the low-Ti group overlap with MORB and island arc basalts. The high-Ti group shows strong REE fractionation and similarities with within-plate basalts erupted at off-axis sea-mounts and ocean islands.

6.4 High-Pressure Metabasic Rocks

The metabasic rocks including glaucophane schists and eclogite are similar in compostion to the low-Ti mafic volcanics discussed above. Therefore, the metabasic rocks are probably the metamorphosed equivalents of the low-Ti mafic volcanic rocks. Their bulk compositions plot in the ocean floor basalt field in a Ti-Zr-Y diagram (Ghose et al. 2010). Although the glaucophane schists and eclogites contain higher HFSE (Ti, Zr, Nb, Y), LREE (La,Ce), Rb and V than the associated greenschists, positive correlations between Ti and Zr among these rocks indicate that they are probably derived from a common basaltic protolith. Low Cr and Ni contents of the eclogites may indicate an evolved nature of the parent magma.

P–T pseudosection analysis indicates that basaltic protolith of the blueschist and eclogites was metamorphosed through a clockwise P–T path (Fig. 6.1, Chatterjee and Ghose 2010). Based on the Grt + Omp + Amp (Brs) + Ep + Chl assemblage and mineral chemical compositions in eclogite, the prograde path may be traced between ~1.3 GPa/525°C and 1.7–2.0 GPa/580–610 °C, the latter representing the peak P–T conditions (Fig. 6.1). The presence of late albite veins and the compositions of the amphibole and epidote rims indicate that the eclogites were retrogressed to ~1.1 GPa/540 °C. A comparison of the

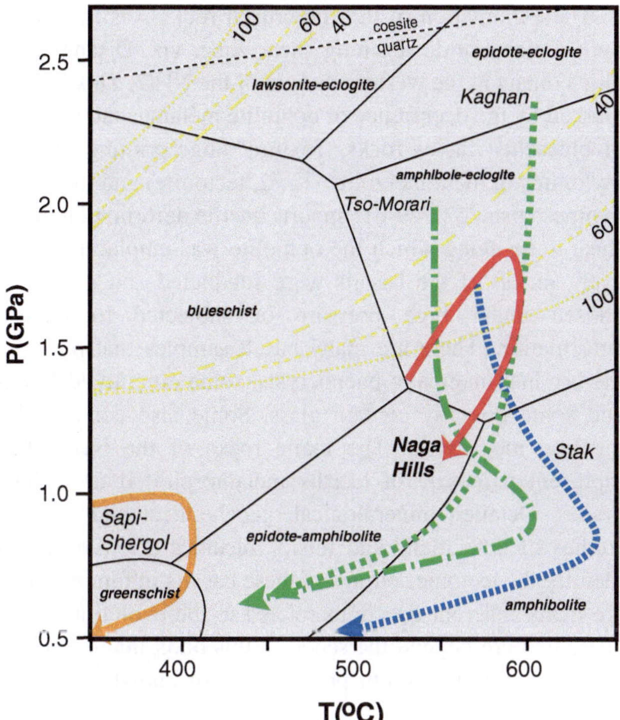

Fig. 6.1 P–T path of metamorphism of the Naga hills eclogite compared with the P–T paths for the Tso-Morari eclogite (de Sigoyer et al. 1997), the Kaghan eclogite (Lombardo et al. 2000) and the Stak eclogite and Sapi-Shergol blueschists (Guillot et al. 2008a) of the northwestern Himalayas (after Chatterjee and Ghose 2010). The petrogenetic grid for metabasaltic bulk-rock compositions is after Okamoto and Maruyama (1999). Also shown are numerically modelled P-T conditions at the *top* (*thick lines, lower curves*) and the *bottom* (*thin lines, upper curves*) of a 5-km-thick subducting crust, calculated by Cagnioncle et al. (2007). The curves correspond with 40, 60 and 100 km/Myr convergence rates for a slab dip of 45°

prograde P–T path with numerical models of plate subduction indicates that the eclogites are probably formed near the top of the subducting slab moving at velocities of ~55–100 mm/yr. These high plate velocities are consistent with the high pre-collision convergence rates between India and Eurasia.

The high-pressure metabasics from the NHO may be compared with the Sapi-Shergol blueschists from the Indus Suture Zone in the northwestern Himalayas. Both are of ocean-floor, tholeiitic affinity and both are emplaced as tectonic slices within Cretaceous-Eocene ophiolite. However, the Sapi-Shergol blueschists were metamorphosed at peak conditions of 0.9–1.0 GPa and 350–420 °C (Mahéo et al. 2006). Therefore, the exhumation of high-pressure rocks related to the subduction of the leading edge of the Indian plate before continental collision was from a greater depth at the eastern margin than at the northern margin. On the other hand, the Eocene eclogites of northwestern Himalayas (Tso Morari and Kaghan) originated from the Indian continental crust as it subducted at the time of the India-Eurasia collision. They were exhumed from much greater depths as indicated by the presence of coesite. Since the subduction dynamics had changed due to the collision, the exhumation mechanism of the Tso Morari and Kaghan eclogites was different compared to the Naga and Sapi-Shergol high-pressure rocks.

6.5 Chromitites

Deformation of chromite grains of NHO reveals that the nodular types are least affected, whereas the massive types preserve brittle as well as plastic processes (Ghosh et al. 2013). This study further suggests that the crystal plastic processes switch over to the brittle regime by mantle dynamics. The content of platinum group of elements (PGE) in peridotite hosting the podiform chromitites is a sensitive indicator of the degree of partial melting of the peridotite and sulphur saturation of basaltic melt. An extensive study of PGE in peridotite and associated chromitite from the southern part of NHO (in Manipur) in recent years (Singh et al. 2012b; Singh 2013) has provided valuable information on the genesis of ophiolite and physical conditions of the mantle. The total PGE concentration in chromitite ranges from <100 to 544 ppb, with greater abundance in high-Cr chromitites (186.2–553.9 ppb) than the high-Al chromitites (81.2–184.5 ppb). The low PGE content of chromitites suggests lack of sulphur saturation of the parent melt in the early stage of crystallisation. The PGE patterns show variable negative Pt-anomalies and enrichment of refractory and compatible IPGE (Os, Ir, Ru) in the high-Cr chromitites compared to the high-Al chromitites (Singh et al. 2012b). The ratio of IPGE to PPGE (Rh, Pt, Pd) is greater in the high-Cr chromitite than the high-Al chromitite. Based upon PGE fractionation patterns of Pd/Ir versus Pt/Pt*, it has been inferred by these authors that the host peridotites are derived by varying degrees of partial melting rather than by magmatic fractionation. The partial melting trend of chromitite from podiform ores shows a subtle difference from that of the Precambrian stratiform bodies. The PGE diversity between the high-Cr and high-Al chromitites may be related to the difference in the composition of magma, degree of melting, magma mixing, sulphur saturation, nucleation, high temperature diffusion, metasomatism and geodynamic setting.

6.6 Evolution of the Naga Hills Ophiolite

The geological, tectonic and petrological features reveal that the NHO was emplaced through a multi-stage process involving the crust and the upper mantle. A first-order description of these stages can be summarised as follows:

(i) Ocean floor was created with the eruption of tholeiitic basalt at a spreading centre in the Neotethys Ocean. The tholeiitic basalt originated through melt extraction from depleted mantle peridotite that generates MORB.

(ii) Other basalts such as high-Ti tholeiites and minor alkaline basalts probably erupted on the ocean floor from off-axis sea-mounts and oceanic islands. Shallow submarine/sub-aerial hyaloclastite and crater facies amygdaloidal lava may have originated on the sea-mounts/oceanic islands. The ocean floor was part of the leading edge of the Indian plate that subducted under the Eurasian plate (or Myanmar microplate).

(iii) An island arc may have developed on the oceanic part of the Myanmar microplate. Rare andesite, felsic volcanics, volcaniclastics and sub-aqueous ash flow (ignimbrite) probably originated in the island arc. Felsic volcanism is common in continental margin arcs, and submarine arcs are composed of pillow lava, voluminous hyaloclastite, tuff and breccia (Condie 1989).

(iv) Subduction of the oceanic plate resulted in the metamorphism of the ocean-floor basalts under high pressure and the formation of eclogite containing garnet, omphacite, barroisite, epidote and chlorite.

(v) Initial exhumation of the eclogite was marked by the disappearance of porphyroblastic garnet (garnet-1), replacement of barroisite by glaucophane, and transformation of eclogite to glaucophane schist.

(vi) Exhumation to lower crust was marked by the disappearance of omphacite, appearance of plagioclase and the first phase of deformation (phase 1).

(vii) Intense deformation (phase 2), shearing, cataclasis, mortar texture formation, brecciation, partial melting (formation of pseudotachylite and quartzofeldspathic veins) and appearance of second-generation idioblastic garnet-2 (620/79-4, 620/79-5) and broad platy glaucophane-2 (76/80-5) occurred on further uplift to upper crust.

(viii) Final uplift causing transformation of glaucophane schist to greenschist with increased activity of water.

(ix) Erosional debris from the ophiolite sequence, chert and marine sediments were deposited on top of the ophiolite that formed a cover sequence of sedimentary rocks (Jopi Formation).

The exhumation of the sub-crustal rocks was facilitated by a major mantle reaching shear zone, viz. Disang-ophiolite contact at the western margin of the NHO. This zone is marked by the occurrence of ophiolite mélange and an array of blueschist facies rocks. Textural fingerprinting of S–C mylonites in meta-lherzolite (K-72, tectonite) and mica-fish in blueschists (622/79-6) supports ductile deformation in the shear zone along which the ophiolite was emplaced. However, not all of the basalts were subducted and metamorphosed under high pressure or subjected to intense deformation. There are many basalt samples that preserve the original magmatic phenocrysts, some are amygdaloidal and some possibly contain glass. Some also contain plagioclase megacrysts. The mafic rocks of the NHO thus represent a mosaic of basalts metamorphosed to various grades. Detailed mineralogical, geochemical and isotopic studies of both mafic and felsic volcanics are essential to identify the tectonic settings of these basalts in future. There are many inherent questions related to the evolution of the NHO that are beyond the scope of this book that cannot be easily answered with our present state of knowledge.

References

Cagnioncle AM, Parmentier EM, Elkins-Tanton LT (2007) Effect of solid flow above a subducting slab on water distribution and melting at convergent plate boundaries. J Geophy Res 112, B09402

Chatterjee N, Ghose NC (2010) Metamorphic evolution of the Naga Hills eclogite and blueschist, Northeast India: implications for early subduction of the Indian plate under the Burma microplate. J Meta Geol 28:209–225

Condie KC (1989) Plate tectonics and crustal evolution. 3rd, edition, Pergamon Press, Oxford, 476

de Sigoyer J, Guillot S, Lardeaux JM, Mascle G (1997). Glaucophane-bearing eclogites in the Tso Morari dome (eastern Ladakh, NW Himalaya). Europe J Min 9:1073–1083

Ghose NC, Shrivastava MP (1986) Podiform chromite of Naga Hills ophiolite, N.E. India. In: Petrascheck W et al. (eds.) Chromites. Theophrastus Publication, Athens, 263–284

Ghose NC, Chatterjee N (2011) Diorite vein in quenched basalt and its implication for the origin of late-granitoid intrusives in Naga Hills ophiolite, Northeast India. In: Srivastava RK (ed) Dyke swarms: keys for geodynamic interpretation. Springer, Heidelberg, 315–330

Ghose NC, Agrawal OP, Chatterjee N (2010) A geological and mineralogical study of eclogite and glaucophane schists in the Naga Hills ophiolite, Northeast India. Island Arc 19:336–356

Ghosh B, Ray J, Morishita T (2013) Grain-scale plastic deformation of chromite from podiform chromitite of the Naga-Manipur ophiolite belt, India: Implication to mantle dynamics. Ore Geology Review, doi:10.1016/j.oregeorev.2013.09.001

Ghosh B, Ray, J. Morishita, T (2013) Grain-scale plastic deformation of chromite from podiform chromitite of the Naga-Manipur ophiolite belt, India: Implication to mantle dynamics. Ore Geology Review, doi:10.1016/j.oregeorev.2013.09.001

Guillot S, Maheo G, de Sigoyer J, Hattori KH, Pecher A (2008) Tethyan and Indian subduction viewed from the Himalayan high- to ultra-high pressure metamorphic rocks. Tectonophysics 451:225–241

Guilmette C, Hébert R, Dostal J, Indares A, Ullrich T, Bédard E, Wang C (2012) Discovery of a dismembered metamorphic sole in the saga ophiolitic mélange, South Tibet: assessing an Early Cretaceous disruption of the Neo-Tethyan supra-subduction zone and consequences on basin closing. Gondwana Res 22(2):398–414

Lindsley DH (1983) Pyroxene thermometry. Am Mineral 68:477–493

Lombardo B, Rolfo F, Compagnoni R (2000) Glaucophane and barroisite eclogites from the Upper Kaghan nappe- implications for the metamorphic historyof the NW Himalaya. Geol Soc, London, Special Publications 170:411–430

Mahéo G, Fayoux C, Guillot S, Garzanti E, Capiez P, Mascle G (2006) Geochemistry of ophiolitic rocks and blueschists from the Sapi-Shergol mélange (Ladakh, NW Himalaya, India): implication for the timing of the closure of the Neo-Tethys ocean. J Asian Earth Sci, 26:695–707

Maibam B, Foley S (2008) Abyssal peridotites from the ophiolite belt of northeastern India: implications for the equilibrium conditions and tectonic setting. Seminar on Indo-Myanmar ranges in the tectonic framework of Himalaya and southeast Asia, Imphal, 84–85

Miller C, Thöni M, Frank W, Schuster R, Melcher F, Meisel T, Zanetti A (2003) Geochemistry and tectonomagmatic affinity of the Yungbwa ophiolite, SW Tibet. Lithos 66:155–172

Okamoto K, Maruyama S (1999) The high-pressure synthesis of lawsonite in the MORB + H_2O system. Amer Min 84:362–373

Patil SK, Thong GT, Watitemsu T, Rao BV (2012) Geochemistry and paleomagnetism of the basalt of the ophiolite suite in parts of Phek district, Nagaland. National symposium on recent advances in applied geochemistry: current status and future trends, Indian Soc Appl Geochemists, 20–21

Rao BV, Ezung C, Nayak R (2010) Geology, genesis and tectonic setting of volcanic rocks from Naga ophiolite belt. Mem Geol Soc India 75:317–328

Reuber I (1986) Geometry of accretion and oceanic thrusting of Spontang ophiolite, Ladakh-Himalaya. Nature 321:592–596

Sengupta S, Acharyya SK, Van Den Hul HJ, Chattopadhyay B (1989) Geochemistry of volcanic rocks from Naga Hills ophiolites, north east India and their inferred tectonic setting. J Geol Soc 146:491–498

Singh AK (2013) Petrology and geochemistry of abyssal peridotites from the Manipur ophiolite complex, Indo-Myanmar orogenic belt, northeast India: implication for melt generation in mid-oceanic ridge environment. J Asian Earth Sci 66:258–276

Singh AK, Singh N, Ibotombi Singh N, Debala Devi L, Ranjit Th (2010) Mineralogy and geochemistry of ultramafic rocks of northern Manipur ophiolite complex, northeast India. Himalayan Geol 31:7–18

Singh AK, Singh NB, Debala Devi L, Singh RKB (2012a) Geochemistry of mid-ocean ridge mafic intrusives from the Manipur ophiolite complex, Indo-Myanmar orogenic belt, N.E.India. J Geol Soc India 80:231–240

Singh AK, Devala Debi L, Ibotombi Singh N, Subramanyam KSV (2012b) Platinum group elements and gold distributions in the peridotites and associated podiform chromitites of the Manipur ophiolite complex, Indo-Myanmar orogenic belt, northeast India. Chemie der Erde, http://dx.doi.org/10.1016/j.chemer.2012.07.004

Srikanth B, Subba Rao MV, Rao BV, Nirmal Charan S, Balaram V, Ejung OC (2004) Geochemical signatures in the basaltic rocks of Naga Hills ophiolite belt: implications for petrogenesis and tectonic environment of emplacement. J Appl Geochem 6:177–189

Venkataramana P, Datta A.K, Acharyya, SK (1986) Petrography and petrochemistry. In: Mitra ND et al. (eds.) Geology of Nagaland ophiolite, Mem Geol Surv India 119:33–63

Part II

Photomicrographs and Backscattered Electron Images

7 Peridotite Tectonite

7.1 Dunite

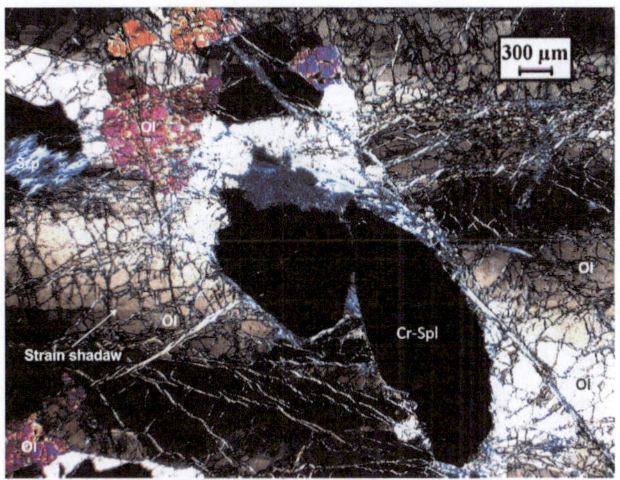

Plate PT-1 Highly fractured olivine showing strain shadow adjacent to Cr-spinel grains in dunite between crossed polars. The fractures are filled with serpentine (384/79-1)

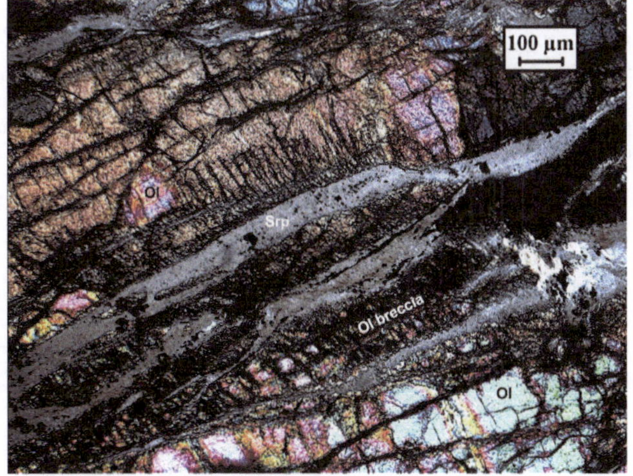

Plate PT-2 Olivine porphyroclasts traversed by serpentine veinlets in dunite between crossed polars. Brecciated granular olivine neoblasts are trapped in serpentine (100/79-3)

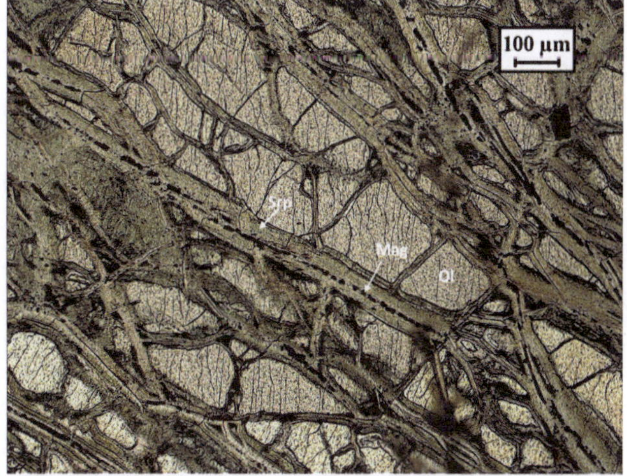

Plate PT-3 Olivine porphyroclast with cross fractures traversed by serpentine veins in dunite. Trails of magnetite occur within the serpentine veins (105/80-2)

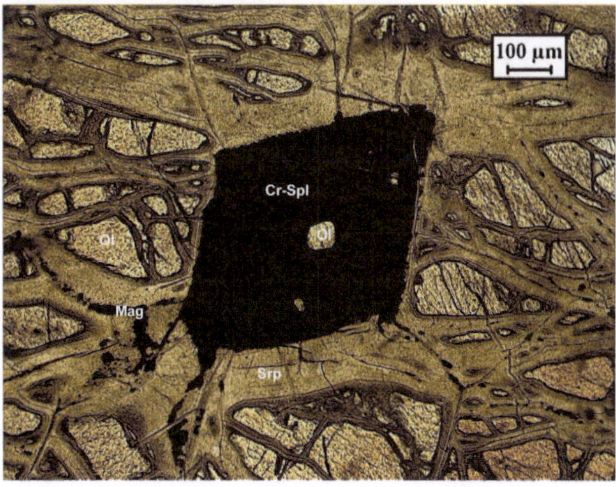

Plate PT-4 Magnified view of rhombic Cr-Spinel containing inclusion of olivine in dunite (197/80-2)

N. C. Ghose et al., *A Petrographic Atlas of Ophiolite*,
DOI: 10.1007/978-81-322-1569-1_7, © Springer India 2014

Plate PT-5 Fractured crystals of olivine in dunite traversed by serpentine veins cross-cutting each other (107/80-2)

Plate PT-7 Serpentinised dunite tectonite with altered olivine porphyroclasts and strongly resorbed chrome spinel in a serpentine matrix. The olivine porpyroclasts are cut by kinked serpentine veins indicating shear deformation. Spinel contains olivine inclusions (108/80-1)

Plate PT-6 Preferentially oriented serpentine veins cutting across olivine and opaque in dunite (105/80-1)

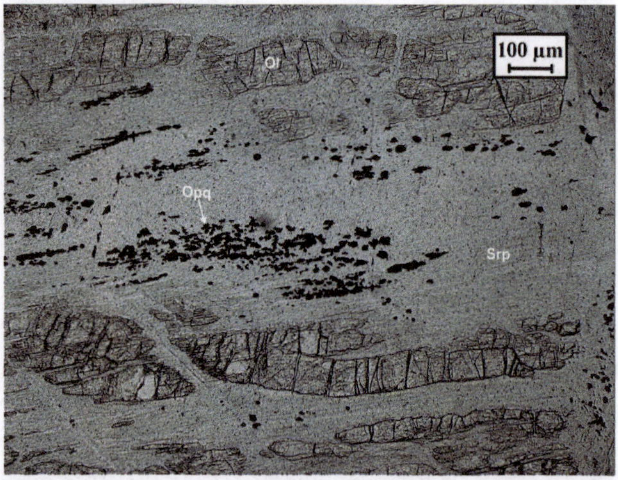

Plate PT-8 Magnified view of sheared dunite in PT-7 elongated remnant olivine with cross fractures filled with serpentine and fine-grained magnetite. The opaque stringers are oriented across the olivine shear planes, indicating two stages of deformation and recrystallisation (108/80-4)

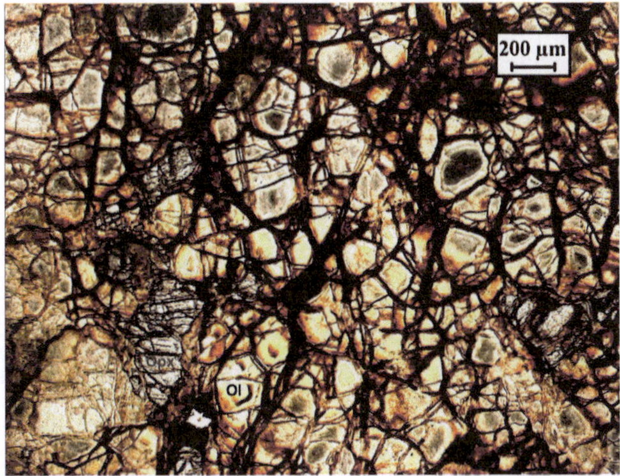

Plate PT-9 Equigranular dunite containing fractured grains of olivine, orthopyroxene and iron oxide. Some olivine grains have smoothly curved or straight boundaries with a core containing glassy material (R16-1)

Plate PT-11 Equigranular dunite of PT-9 and PT-10 between crossed polars showing isotropic amorphous cores of olivine (R16-3)

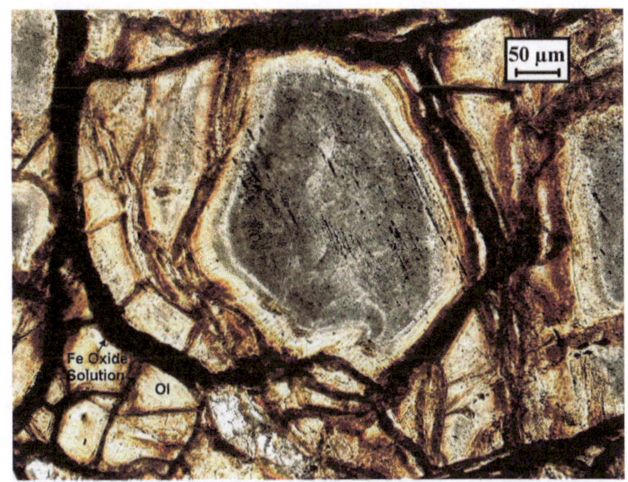

Plate PT-10 Magnified view of remnant olivine in dunite of PT-9 rimmed by layers of iron oxide. The altered core of olivine is composed of amorphous grey material diagonally cross-cut by fine streaks of opaque minerals (R16-2)

7.2 Harzburgite

Plate PT-12 Olivine in harzburgite showing strain shadow between crossed polars. Unaltered olivine, non-pleochroic orthopyroxene and disseminated grains of chrome spinel are the essential components of the rock (380/79-1)

7.3 Lherzolite

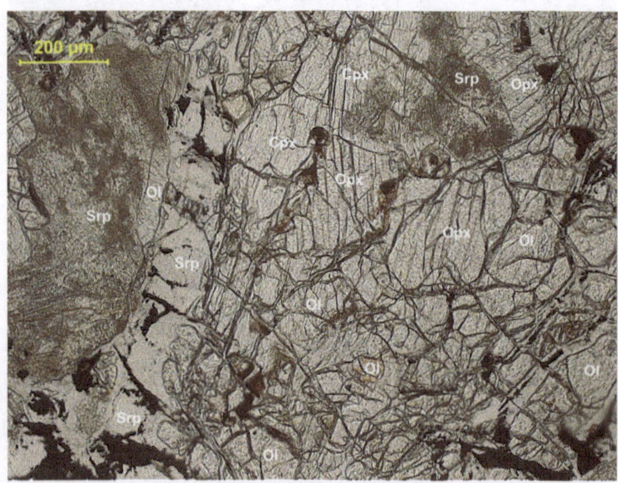

Plate PT-13 Partially serpentinised lherzolite with fractured grains of olivine, clinopyroxene and orthopyroxene (96/80-2)

Plate PT-14 Partially serpentinised lherzolite of PT-13 between crossed polars showing olivine, clinopyroxene and orthopyroxene grains in equilibrium (96/80-3)

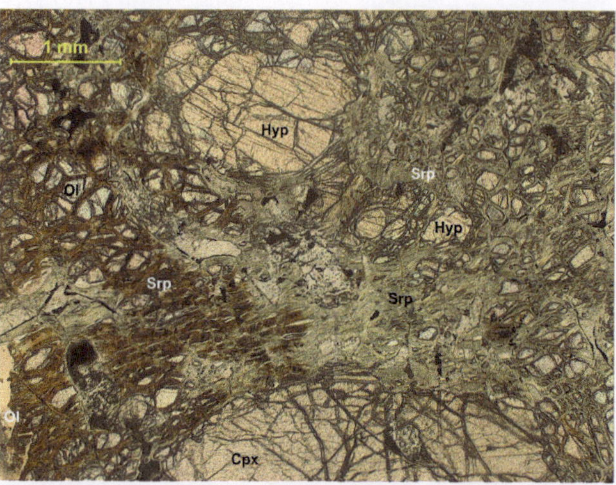

Plate PT-15 Lherzolite containing porphyroclasts of pleochroic orthopyroxene (Hyp), clinopyroxene, relict olivine and resorbed chrome spinel in a serpentinised matrix (ML77/30-3)

7.4 Wehrlite

Plate PT-16 Fragments of resorbed clinopyroxene in serpentinised wehrlite close to a tectonic contact. In hand specimen, the rock shows some deformation and minor presence of orthopyroxene and opaque (606-1)

8 Spinel Peridotite

8.1 Spinel Harzburgite

Plate SP-1 Spinel harzburgite consisting of serpentinised olivine, non-pleochroic orthopyroxene and partially resorbed Cr-spinel with inclusions of serpentine (M77/20-2)

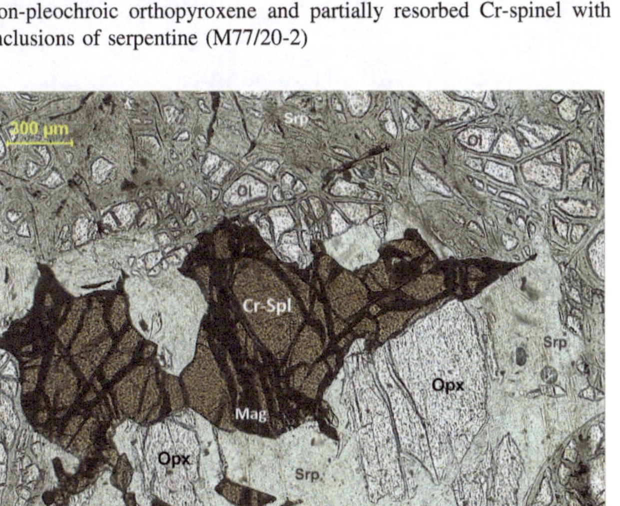

Plate SP-2 Cr-spinel with magnetite veins surrounded by serpentinised olivine and non-pleochroic orthopyroxene in spinel harzburgite (ML77/12-1)

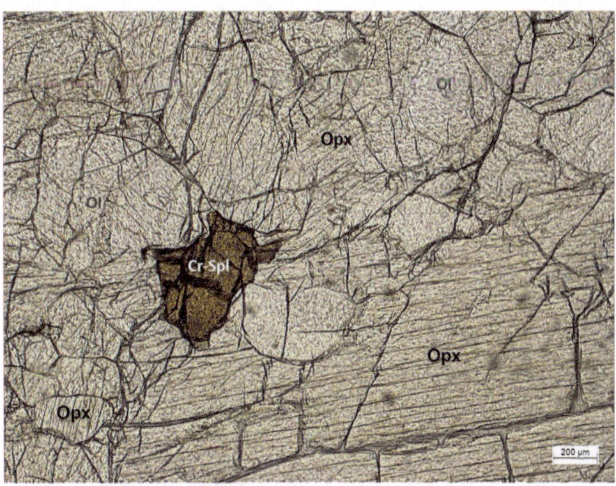

Plate SP-3 Xenomorphic Cr-spinel in relatively unaltered and non-foliated spinel harzburgite containing long prisms of orthopyroxene and olivine (M-1)

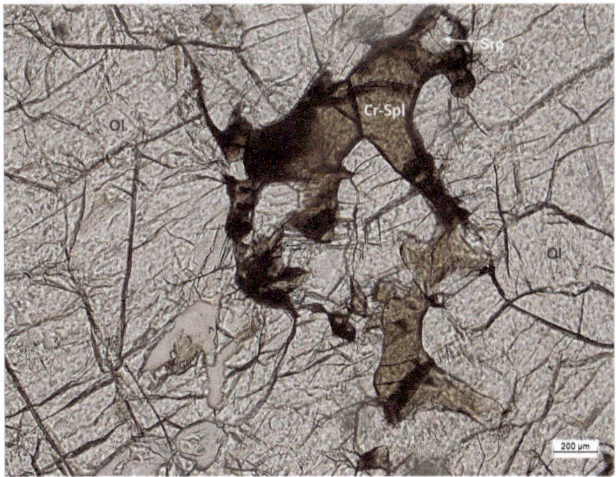

Plate SP-4 Skeletal Cr-spinel showing reaction texture with an olivine dominated matrix containing minor serpentine in harzburgite (M-3)

N. C. Ghose et al., *A Petrographic Atlas of Ophiolite*,
DOI: 10.1007/978-81-322-1569-1_8, © Springer India 2014

Plate SP-5 Coarse, protogranular harzburgite containing olivine remnants traversed by serpentine, large crystals of non-pleochroic orthopyroxene and skeletal Cr-spinel. Both serpentine and orthopyroxene are trapped within spinel on *right*. The spinel band on *left* is flanked by orthopyroene on either side (R-16-5)

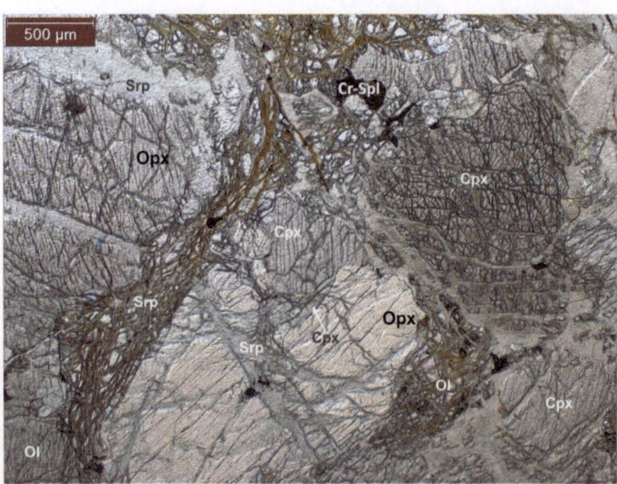

Plate SP-7 Coarse orthopyroxene in spinel harzburgite containing serpentinised olivine and xenomorphic clinopyroxene and Cr-spinel. Protoclasts of highly fractured clinopyroxene and orthopyroxene are traversed by secondary veinlets of serpentine (R-16-3)

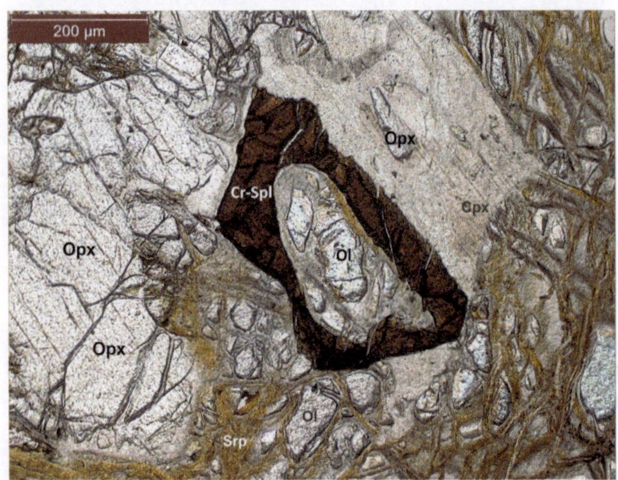

Plate SP-6 Skeletal spinel containing a large inclusion of serpentinised olivine. The prismatic grain of *greyish* clinopyroxene on *right* contains inclusions of orthopyroxene (R-16-1)

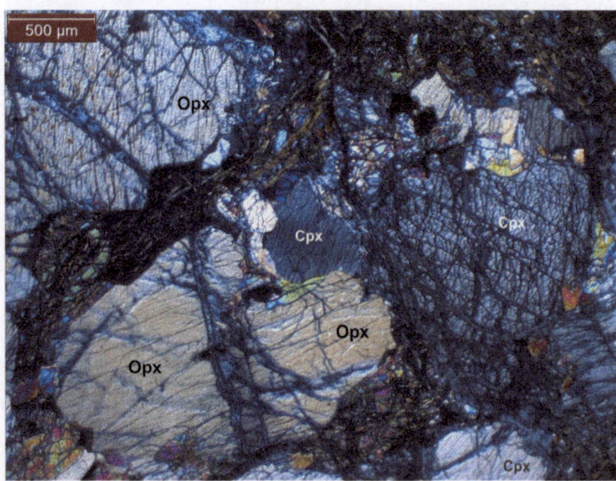

Plate SP-8 Spinel harzburgite of SP-7 between crossed polars (R-16-4)

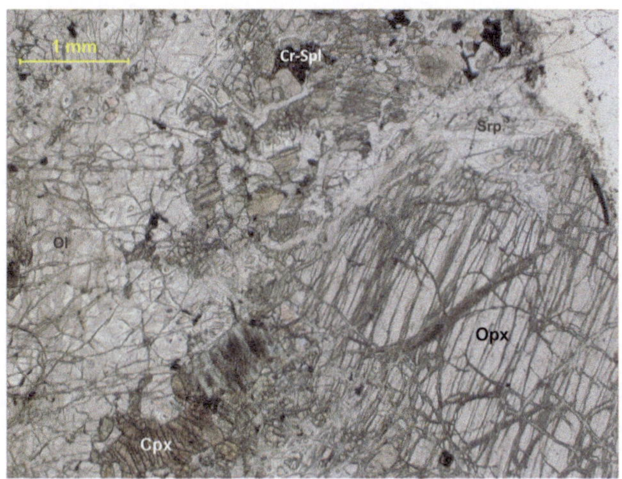

Plate SP-9 Spinel harzburgite consisting of coarse orthopyroxene, olivine, *brown* clinopyroxene and spinel. The rock is intricately veined by serpentine (n179-1)

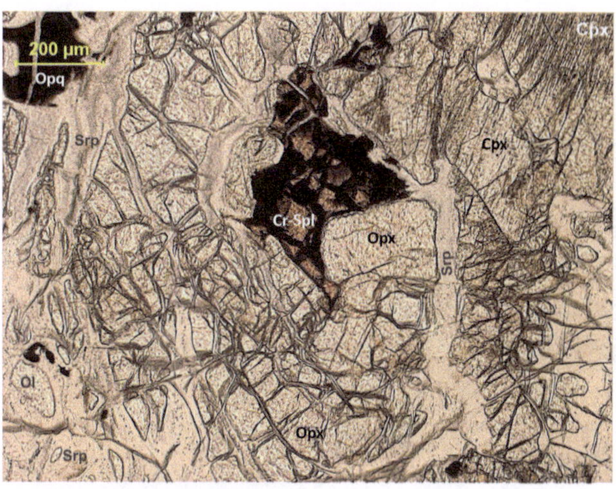

Plate SP-11 Spinel harzburgite consisting of serpentinised olivine, orthopyroxene, clinopyroxene and spinel (n179-2)

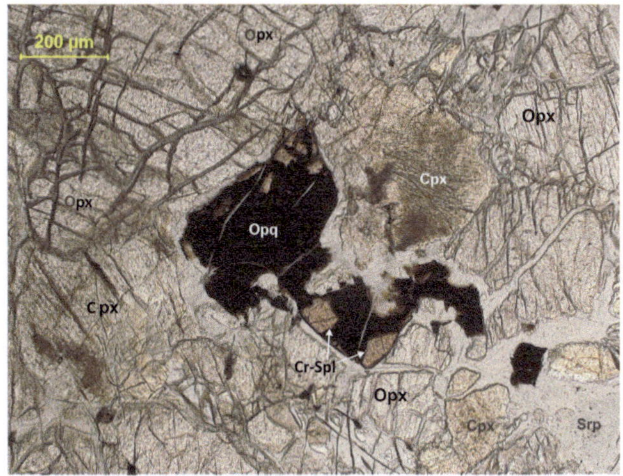

Plate SP-10 Magnified view of spinel in harzburgite of SP-9 (n179-4)

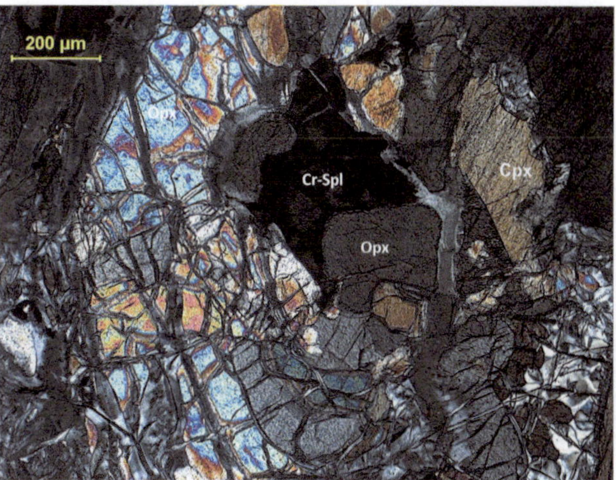

Plate SP-12 Spinel harzburgite of SP-11 between crossed polars (n179-3)

8.2 Spinel Lherzolite

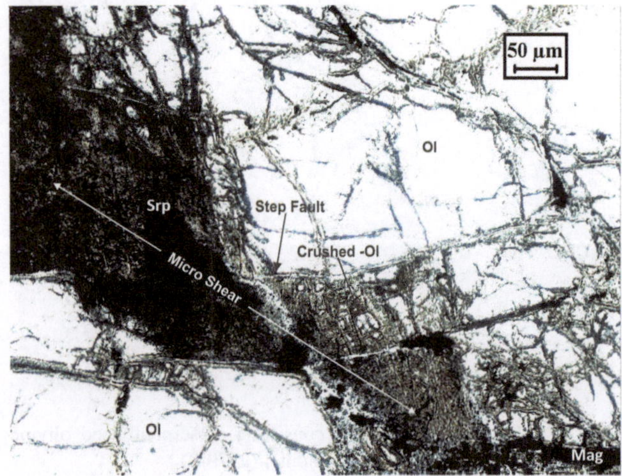

Plate SP-13 S-C mylonite in spinel lherzolite showing microshear dissected by step faulting. Pulverised grains of olivine and displaced magnetite vein along the shear plane are noticeable (K72)

Plate SP-15 Micro-shear in spinel lherzolite containing pulverised and brecciated grains of clinopyroxene and serpentine (96/80b)

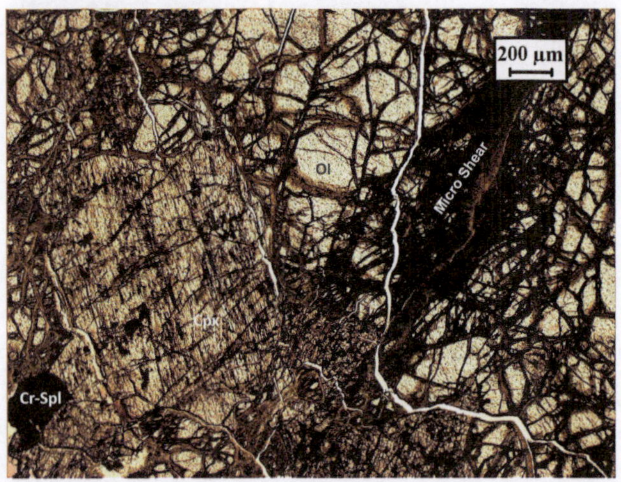

Plate SP-14 Spinel lherzolite with micro-shear containing olivine and clinopyroxene with streaks of opaque along the cleavage (K72-8)

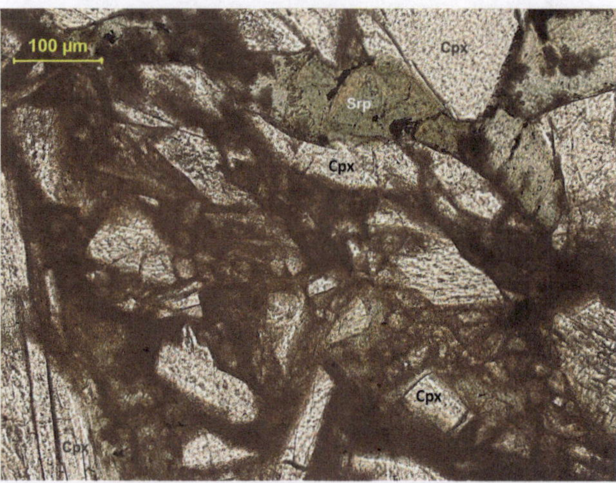

Plate SP-16 Brecciated non-pleochroic clinopyroxene and serpentine at the terminal part of micro-shear of spinel lherzolite in SP-15 showing protoclastic texture (96/80b-3)

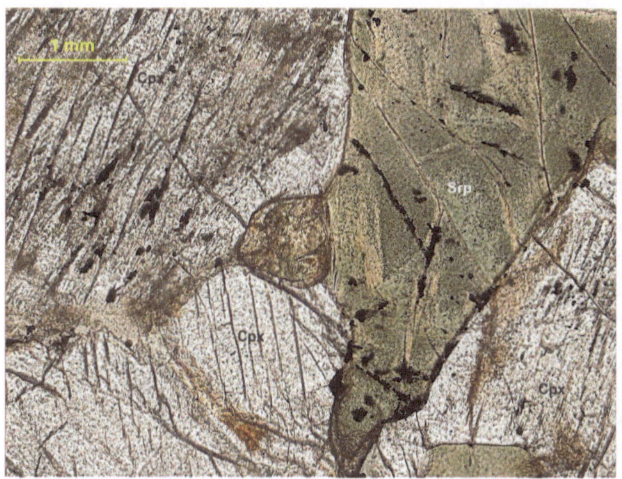

Plate SP-17 Spinel lherzolite of SP-15 showing triple-point contacts of clinopyroxene and *green* serpentine (96/80b-2)

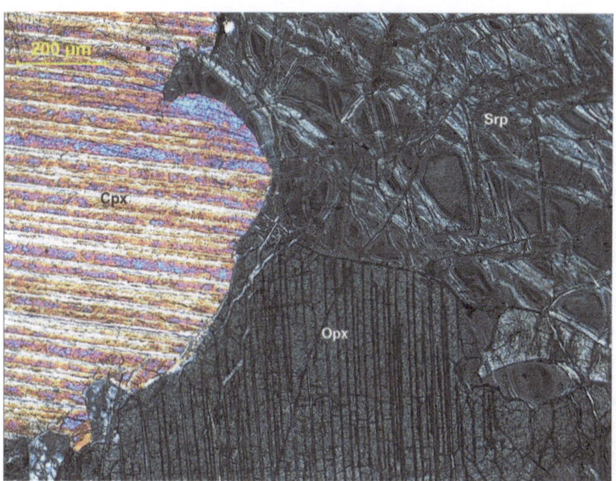

Plate SP-19 Spinel lherzolite between crossed-polars showing triple-junction contact of clinopyroxene, non-pleochroic orthopyroxene and serpentine (after olivine). Note the lobate shape of the clinopyroxene grain and embayed margin of the serpentinised olivine indicating reaction (ML77-11-2)

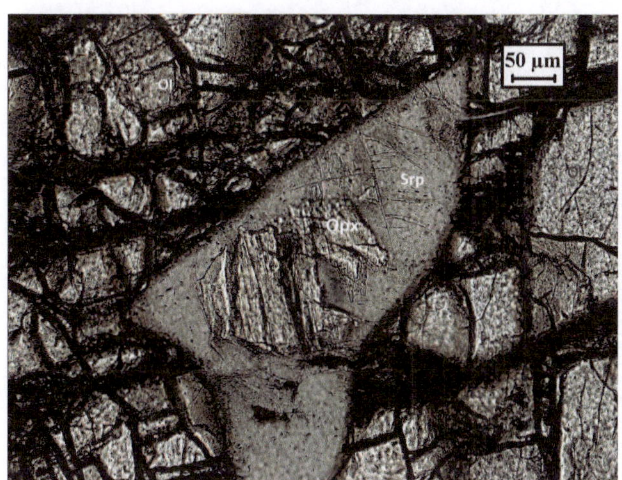

Plate SP-18 High magnification image of non-pleochroic resorbed orthopyroxene surrounded by olivine in spinel lherzolite (K72)

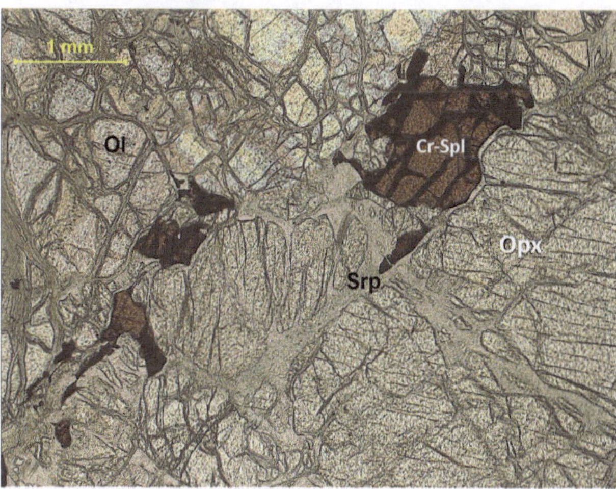

Plate SP-20 Spinel lherzolite of SP-19 showing spinel in a serpentinised matrix of olivine and orthopyroxene (ML77-11-1)

Plate SP-21 Skeletal spinel in spinel lherzolite containing clinopyroxene and serpentine with relict olivine (M77-8-2)

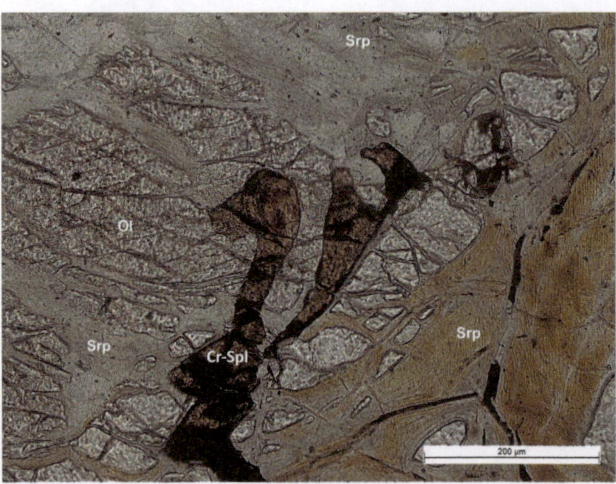

Plate SP-23 Intergrowth of spinel and olivine traversed by serpentine veins in spinel lherzolite (M77-19-3)

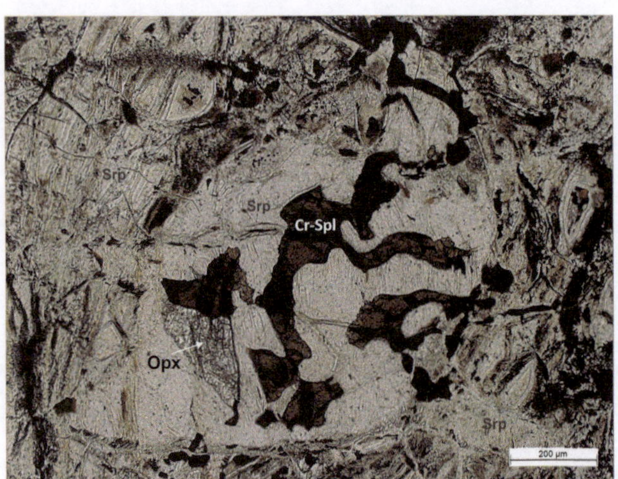

Plate SP-22 Vermicular spinel and non-pleochroic orthopyroxene in a serpentinised matrix in spinel lherzolite (M77-8-1)

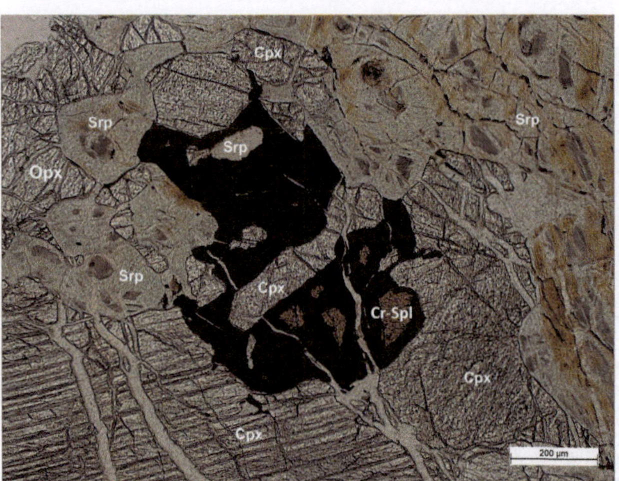

Plate SP-24 Spinel lherzolite consisting of olivine remnants in serpentine, clinopyroxene, non-pleochroic orthopyroxene and spinel (M77-19-1)

8.3 Serpentinised Spinel Peridotite

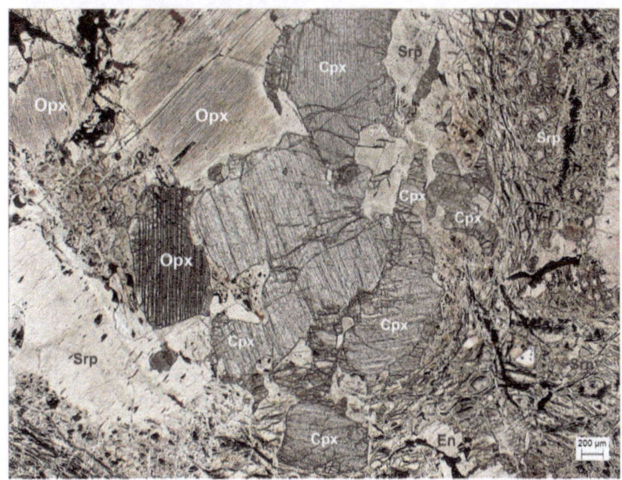

Plate SP-25 A cluster of non-pleochroic ortho- and clino-pyroxenes encircled by serpentine in spinel lherzolite (77-8-3)

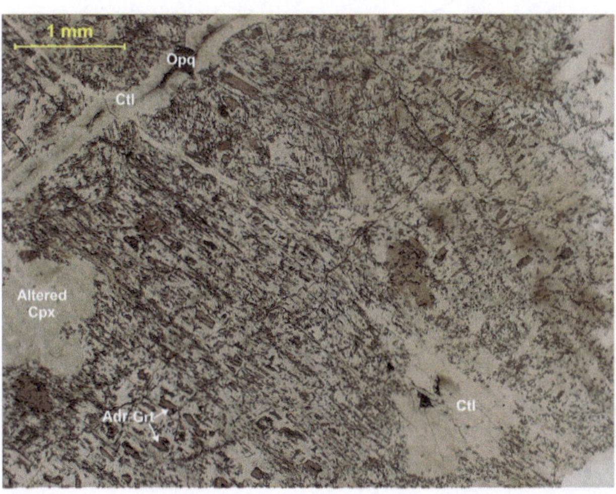

Plate SP-27 Brecciated blocks of serpentinised peridotite containing clinopyroxene, serpentine (chrysotile) and small subhedral andradite garnet of metasomatic origin. The rock shows strong foliation (202-80-2)

Plate SP-26 Spinel lherzolite of SP-25 between crossed polars. Clinopyroxene shows strong birefringence (77-8-4)

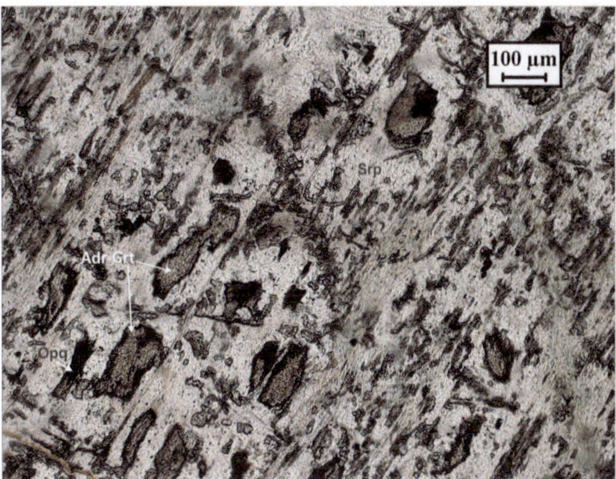

Plate SP-28 High magnification view of serpentinised spinel peridotite in SP-27 showing linear alignment of metasomatic andradite garnet and opaque in serpentine along the foliation plane (202-80-3)

Plate SP-29 Carbonate-rich spinel peridotite, consisting of serpentine, carbonate, clinopyroxene and spinel. The rock is not foliated (77-49-1)

Plate SP-30 Carbonate-rich spinel peridotite of SP-29 containing a xenolithic fragment with chilled margin in a groundmass dominated by carbonate (77-49-2)

9 Pyroxenite

9.1 Olivine Websterite

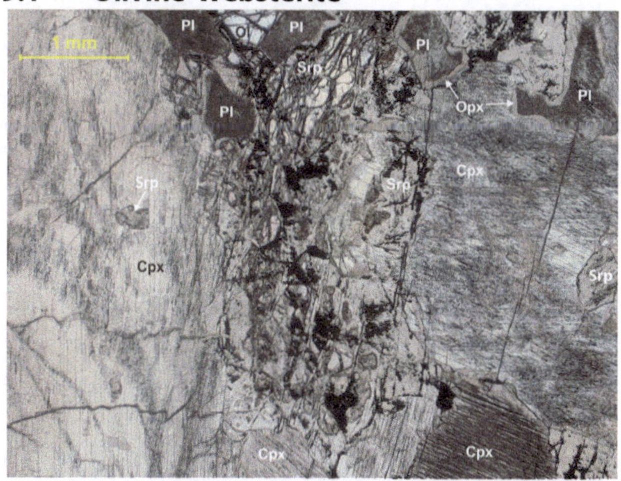

Plate Px-1 Coarse-grained olivine websterite consisting of serpentine with remnant olivine, clinopyroxene, opaque and plagioclase–orthopyroxene rim (refer to Px-5). Clinopyroxene contains inclusions of serpentinised olivine (*on left*) and wavy exsolution lamellae of orthopyroxene (*grey*) across cleavage on *right* (160/80-4)

Plate Px-3 Olivine websterite with bimodal grain size of pyroxenes showing exsolution lamellae of orthopyroxene across cleavage planes (160/80-5)

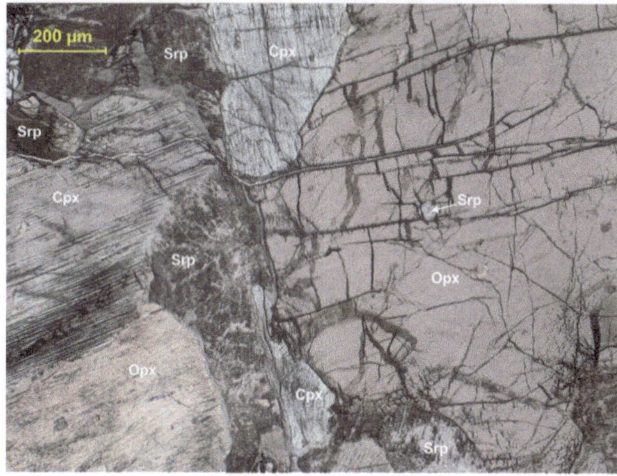

Plate Px-2 Magnified view of olivine websterite of 160/80-4 containing serpentine and coarse ortho- and clino-pyroxenes. The orthopyroxene is rimmed by serpentine and contains inclusions of serpentinised olivine (160/80-3)

Plate Px-4 Olivine websterite of 160/80-5 viewed between crossed polars. The fractures in clinopyroxene are filled with orthopyroxene (160/80-6)

N. C. Ghose et al., *A Petrographic Atlas of Ophiolite*,
DOI: 10.1007/978-81-322-1569-1_9, © Springer India 2014

Plate Px-5 Magnified view of the reaction between plagioclase and orthopyroxene in olivine websterite in contact with clinopyroxene, serpentine and secondary opaque. The clinopyroxene contains orthopyroxene lamellae (160/80-1)

Plate Px-6 Olivine websterite of 160/80-1 viewed between crossed polars (160/80-2)

9.2 Hornblende Websterite

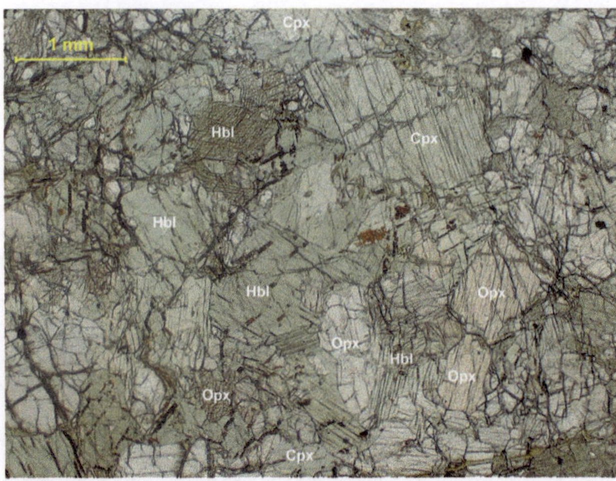

Plate Px-7 Granoblastic texture in hornblende websterite containing pleochroic orthopyroxene, clinopyroxene, hornblende and minor opaque. Fe–Ti oxides occur along the pyroxene-hornblende grain contacts (ML55-1)

Plate Px-8 Hornblende websterite of ML55-1 viewed between crossed polars. Clinopyroxene neoblasts occur between the interstices of larger grains (ML55-2)

9.3 Olivine Clinopyroxenite

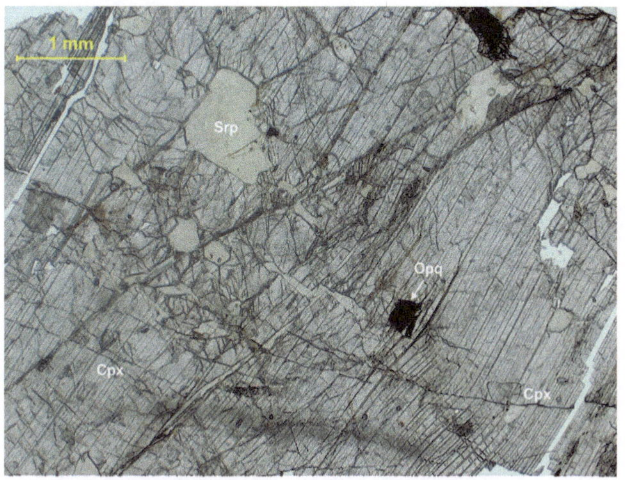

Plate Px-9 Poikilitic clinopyroxene shows deformed cleavage and inclusions of opaque and serpentinised olivine in olivine clinopyroxenite. The white patches represent cracks in slide (655-1)

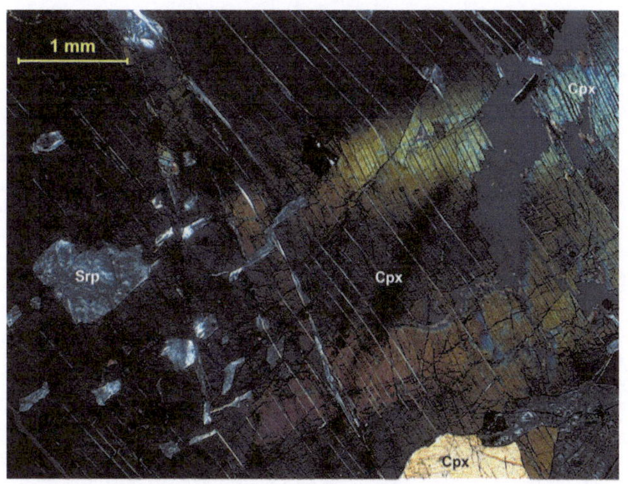

Plate Px-10 Olivine clinopyroxenite of 655-1 viewed between crossed polars showing kinked cleavage and strain shadow in clinopyroxene. The rock occurs close to a tectonic zone. Serpentine is developed across the cleavage planes (655-2)

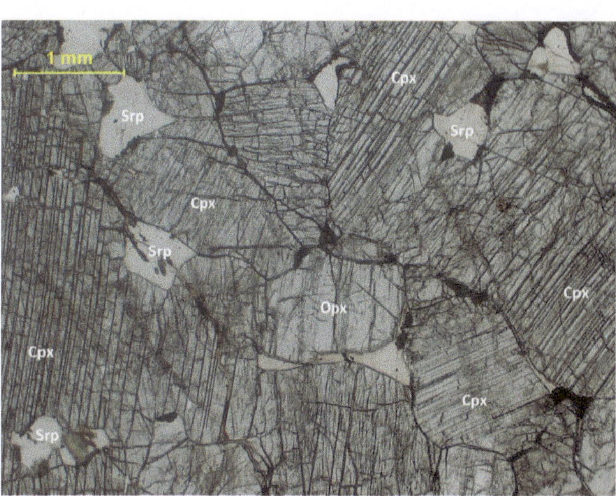

Plate Px-11 Intercumulus grains of serpentine and Fe–Ti oxides between grain boundaries of pyroxene in olivine clinopyroxenite without any deformation (10-1)

9.4 Clinopyroxenite

Plate Px-12 Polysynthetic twinning in clinopyroxene between crossed polars in clinopyroxenite (194/84)

10 Serpentinite

Plate S-1 Fibrous ribbons of serpentine in serpentinite between crossed polars (111/80)

Plate S-2 Reticulate texture in serpentinite between crossed polars (Z-49-4)

Plate S-3 Serpentinite containing chrysotile veins and relict pyroxene between crossed polars (144/80)

Plate S-4 Ribbon texture in serpentinite between crossed polars (192/80)

N. C. Ghose et al., *A Petrographic Atlas of Ophiolite*,
DOI: 10.1007/978-81-322-1569-1_10, © Springer India 2014

Plate S-5 Folded serpentinite showing concentration of opaque and talc fibres along fold closure (S-9-1)

Plate S-7 Serpentinite breccia showing a clinopyroxene fragment traversed by secondary serpentine (N28-3)

Plate S-6 Folded and faulted serpentine band in serpentinite. Syntectonic opaques developed after serpentinisation (153/79-1)

Plate S-8 Brecciated serpentinite (116/80)

11 Rodingite

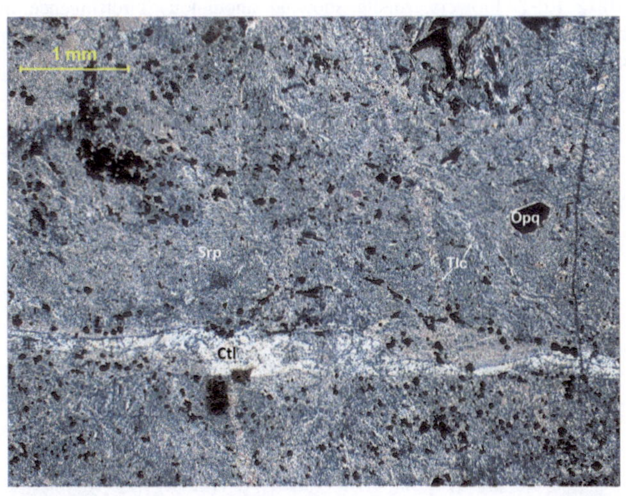

Plate Rd-1 Rodingite between crossed polars showing an opaque-talc-serpentine intergrowth traversed by veins of chrysotile and talc (70/79-2)

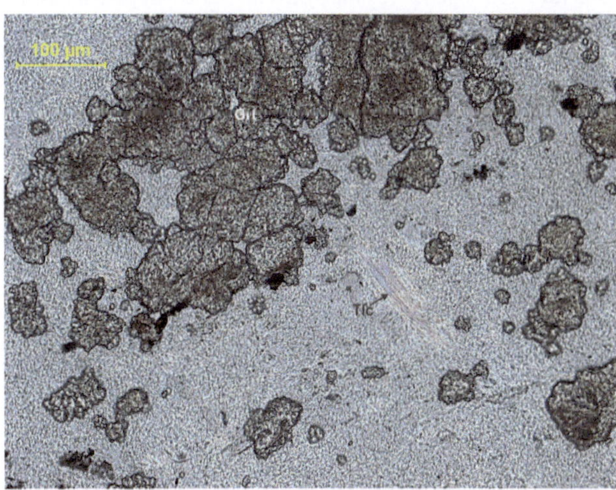

Plate Rd-3 Rodingite containing granular xenoblastic garnet and opaque in a matrix of serpentine and minor talc (flake near centre) (70/79-2)

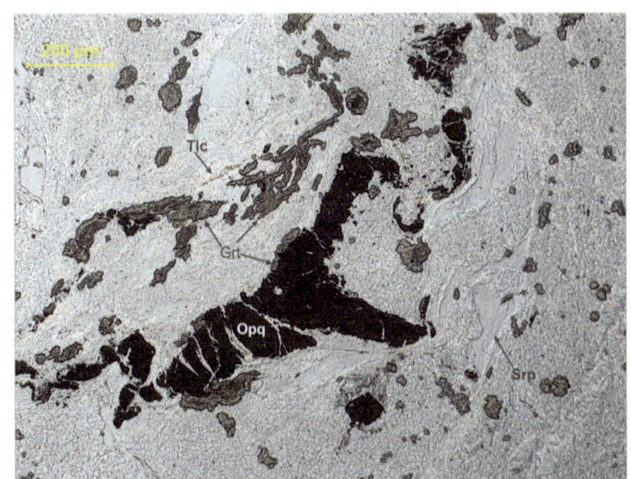

Plate Rd-2 Late development of garnet around skeletal opaque in talc-serpentine rodingite (70/79-5)

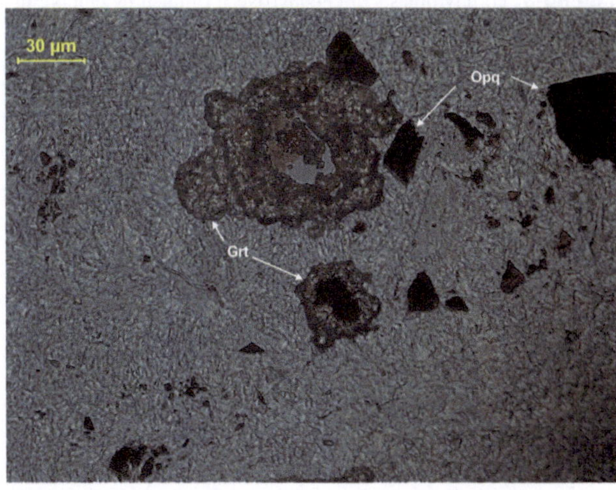

Plate Rd-4 Magnified view of garnet xenoblasts in rodingite of 70/79-2 showing atoll-shaped garnet grains with opaque-filled cores (70/79-4)

N. C. Ghose et al., *A Petrographic Atlas of Ophiolite*,
DOI: 10.1007/978-81-322-1569-1_11, © Springer India 2014

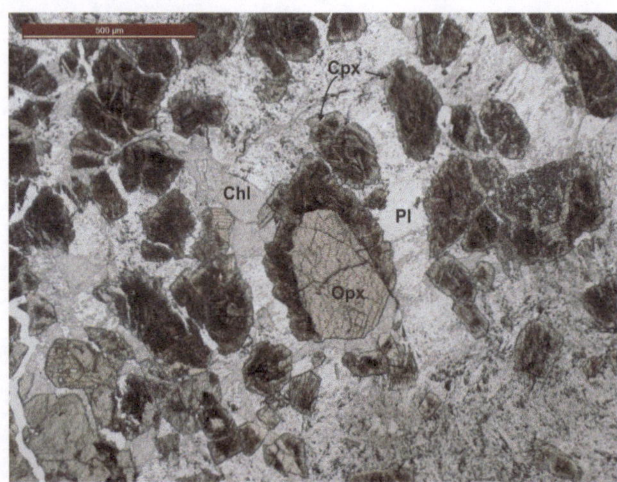

Plate Rd-5 Xenoblasts of orthopyroxene and clinopyroxene associated with plagioclase in rodingite of gabbroic parent rock. Clinopyroxene forms a rim around idioblastic orthopyroxene. Secondary chlorite occurs in the matrix (122-80-1)

Plate Rd-7 Flowery zoisite showing anomalous birefringence in rodingite. Idioblastic grains of clinopyroxene and plagioclase under crossed polars are randomly oriented (122-80-9b)

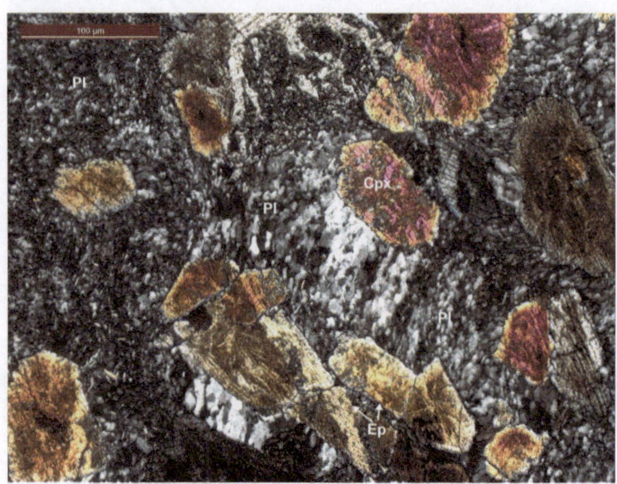

Plate Rd-6 Porphyroblasts of clinopyroxene rimmed by epidote in a mosaic of plagioclase in rodingite between crossed polar (22-80-7)

Plate Rd-8 Metasomatic replacement of clinopyroxene by plagioclase in rodingite between crossed polars. The pyroxene is rimmed by epidote and cut across by a vein of plagioclase (122/79-10)

12 Chromitite

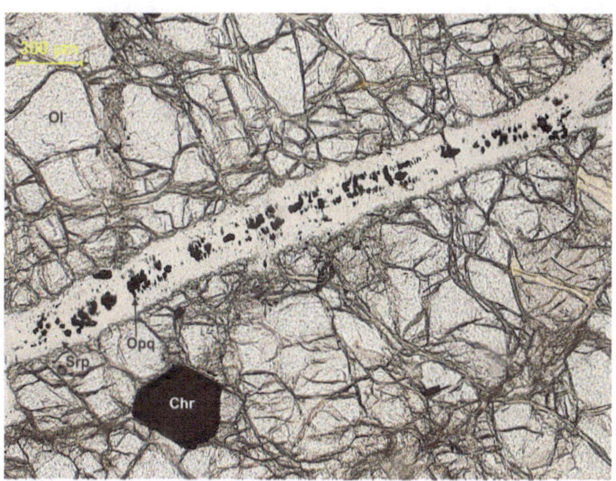

Plate Chr-1 Brecciated olivine with euhedral chromite is traversed by a vein of serpentine containing granular magnetite in harzburgite tectonite (380/79-3)

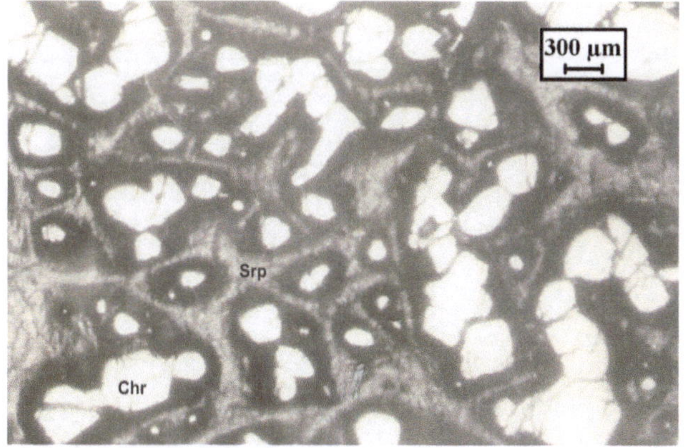

Plate Chr-2 Nucleation of chromite in serpentinised olivine grains under reflected light in disseminated ore (z-76-160)

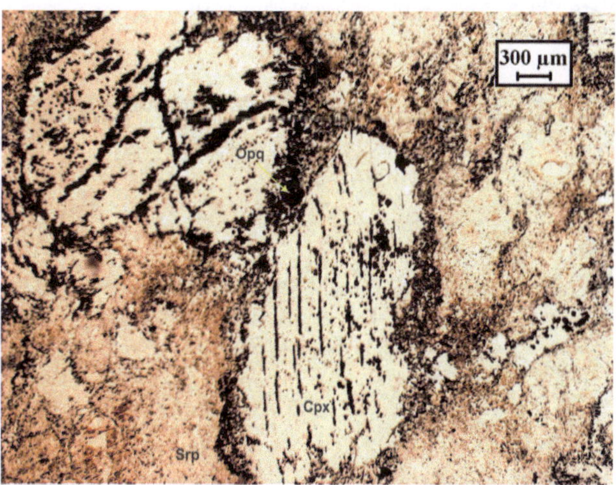

Plate Chr-3 Opaque lamellae in clinopyroxene possibly indicating Cr exsolution in response to cooling in serpentinised peridotite (Z-76-14)

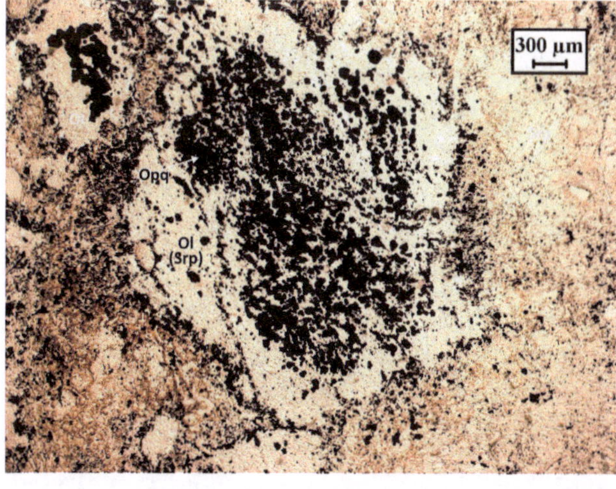

Plate Chr-4 Altered olivine with serpentine and granular magnetite in serpentinised peridotite (Z-76-15)

N. C. Ghose et al., *A Petrographic Atlas of Ophiolite*,
DOI: 10.1007/978-81-322-1569-1_12, © Springer India 2014

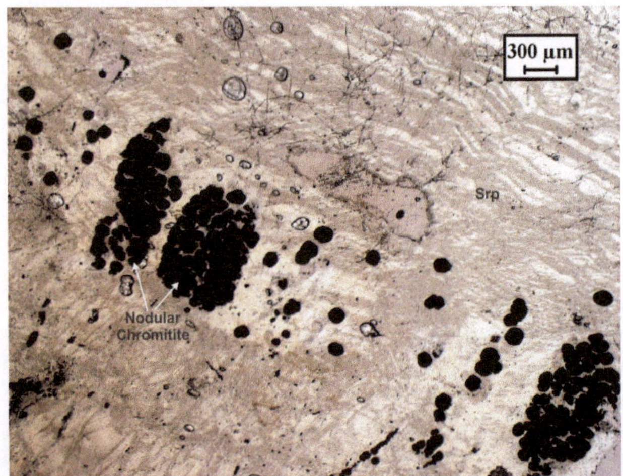

Plate Chr-5 Nodules of octahedral chrome spinel across serpentine fibres in serpentinised peridotite (Z-76-11)

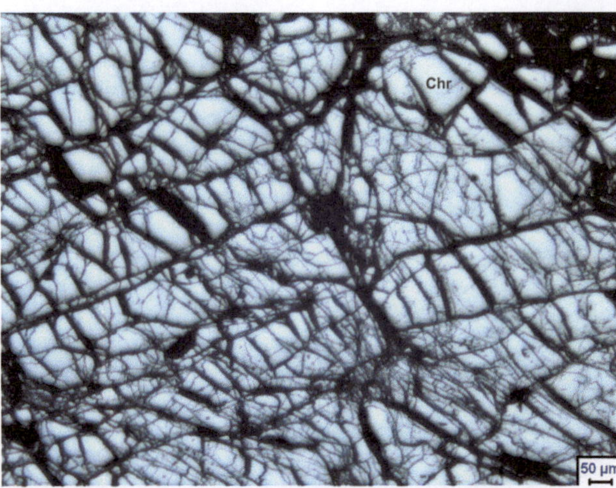

Plate Chr-8 Highly fractured chromite in massive chromitite under reflected light (m-140-01)

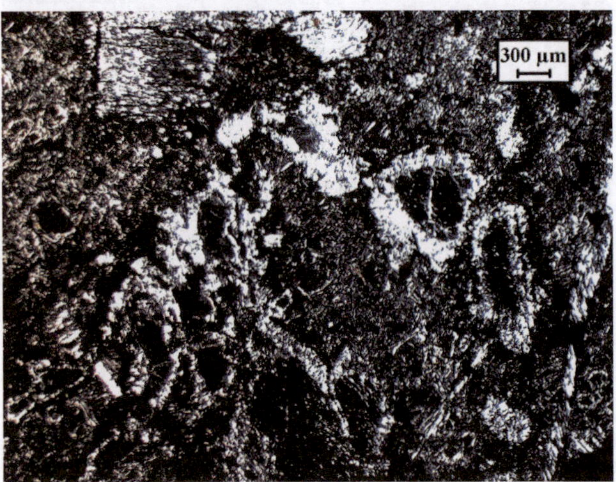

Plate Chr-6 Nodular chromite grains surrounded by serpentine between crossed polars (z-76-16)

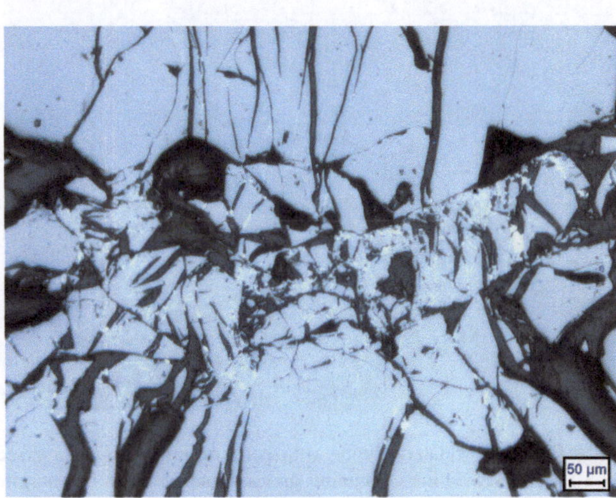

Plate Chr-9 Micro-brecciated zone of chromite within massive chromitite under reflected light (m-34-05)

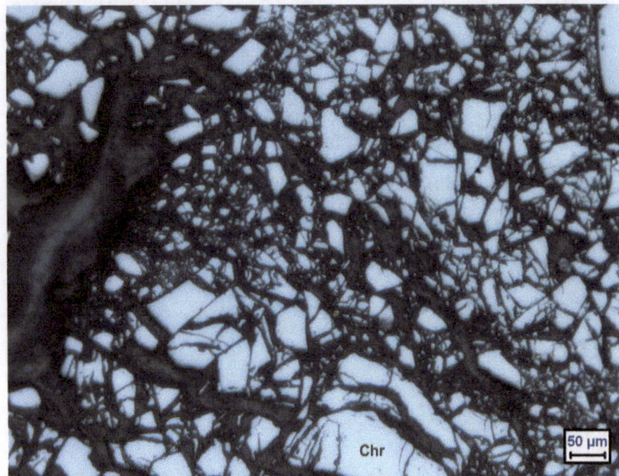

Plate Chr-7 Brecciated chromite in massive chromitite under reflected light (m-140-02)

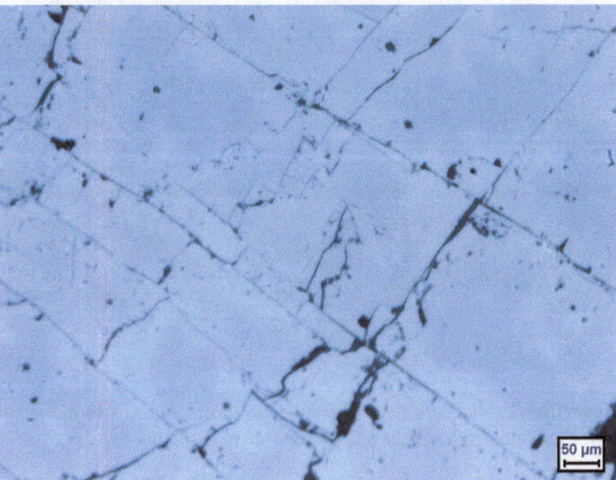

Plate Chr-10 Large chromite grain with cleavage in massive chromitite under reflected right (m-45-02)

Plate Chr-11 Inclusion of euhedral clinopyroxene in coarse-grained massive chromitite under reflected light (m-150-02)

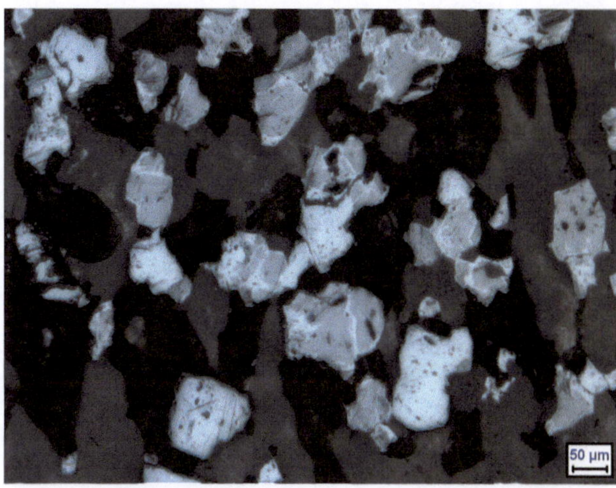

Plate Chr-13 *Rounded silicate* and pentlandite with possible platinum group elements (PGE) in nodular chromitite under reflected light (n-24-01)

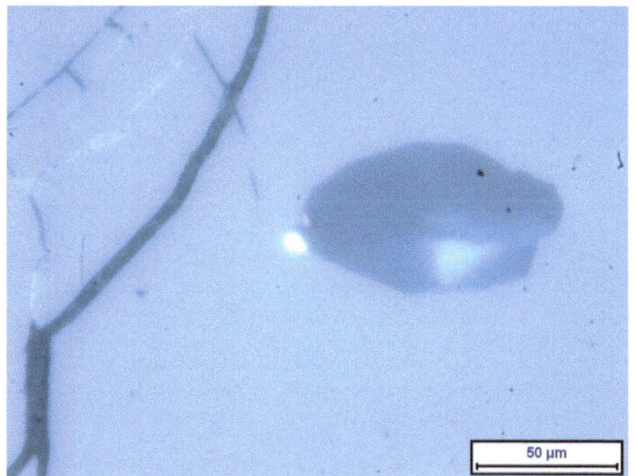

Plate Chr-12 Massive chromitite under reflected light containing silicate (*grey*) and sulphide (*bright*) inclusions (m-34-02)

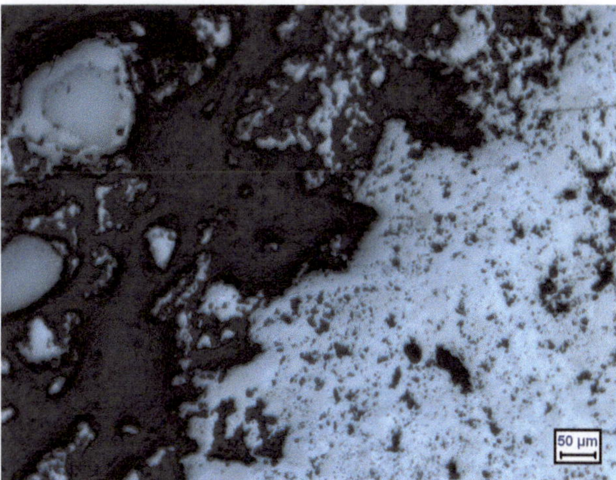

Plate Chr-14 Intergrowth of silicate and sulphide minerals in chromitite (M-10-03)

13 Peridotite Cumulate

13.1 Dunite

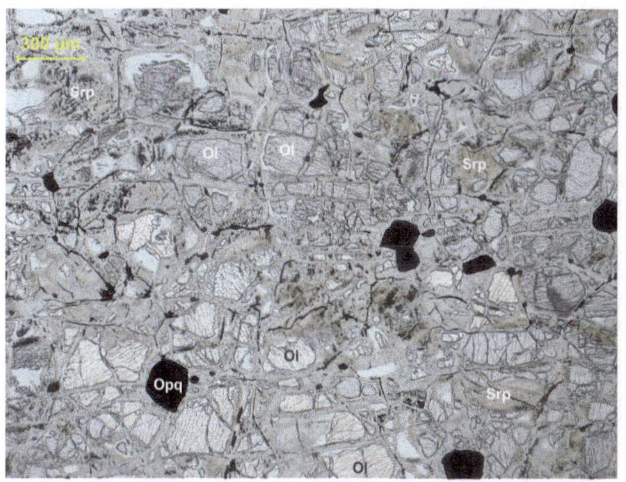

Plate Pc-1 Small grains of olivine remnants and opaque in serpentinised dunite. The olivine grains show uniform extinction indicating absence of strain (126/80-2)

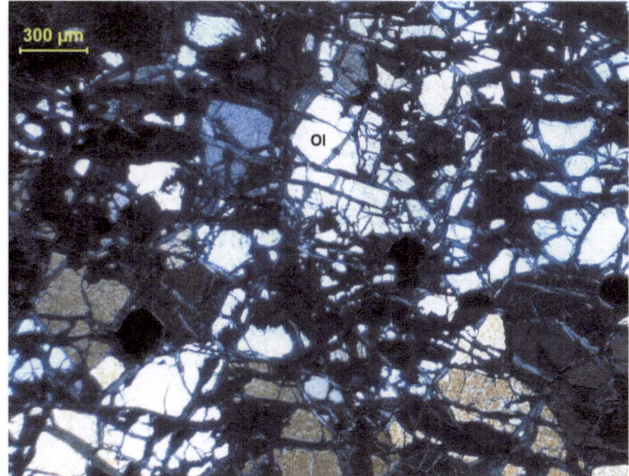

Plate Pc-2 Dunite of Pc-1 between crossed polars (126/80-3)

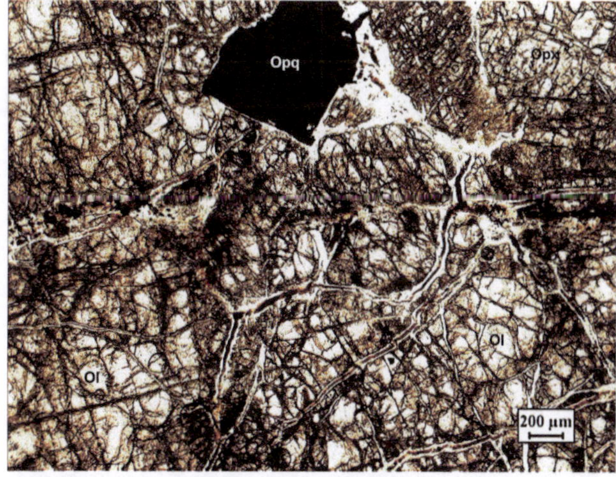

Plate Pc-3 Granular dunite containing olivine, minor non-pleochroic orthopyroxene and opaque. Late hydrothermal alteration developed serpentine along irregular fractures (383/79)

N. C. Ghose et al., *A Petrographic Atlas of Ophiolite*,
DOI: 10.1007/978-81-322-1569-1_13, © Springer India 2014

13.2 Harzburgite

Plate Pc-4 Serpentinised harzburgite containing relict olivine, non-pleochroic orthopyroxene and chrome spinel (M77/10-1)

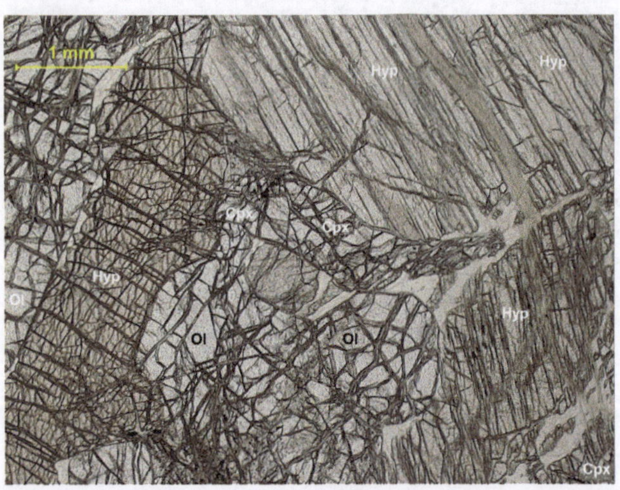

Plate Pc-5 Coarse-grained undeformed harzburgite containing hypersthene, olivine and minor clinopyroxene with fractures filled by serpentine. The sample also contains traces of chrome spinel and Fe-oxides (9-1)

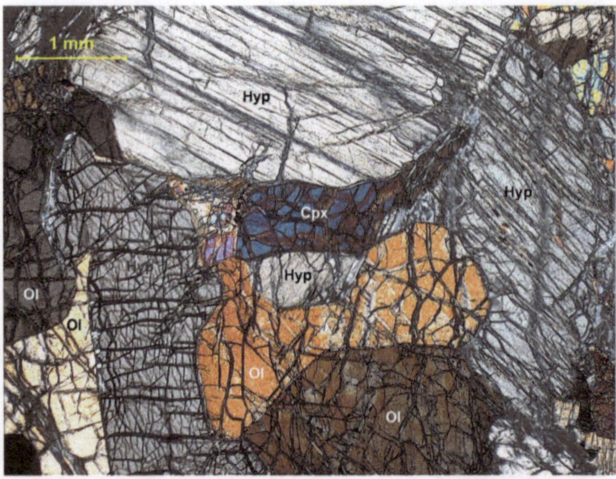

Plate Pc-6 Coarse-grained harzburgite of Pc-5 between crossed polars (9-3)

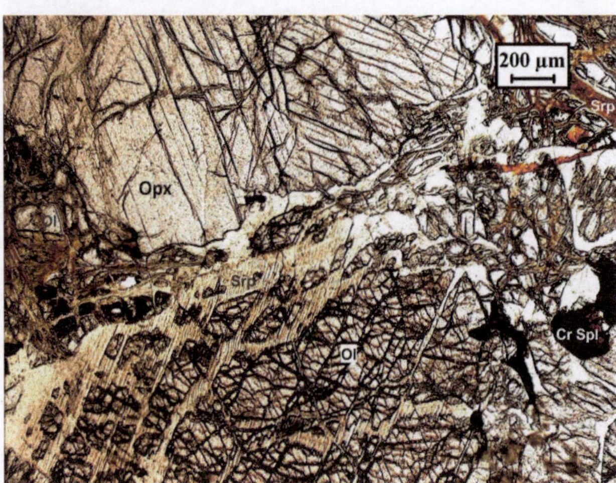

Plate Pc-7 Coarse-grained harzburgite containing remnant olivine in serpentine, non-pleochroic orthopyroxene and chrome spinel (R-16)

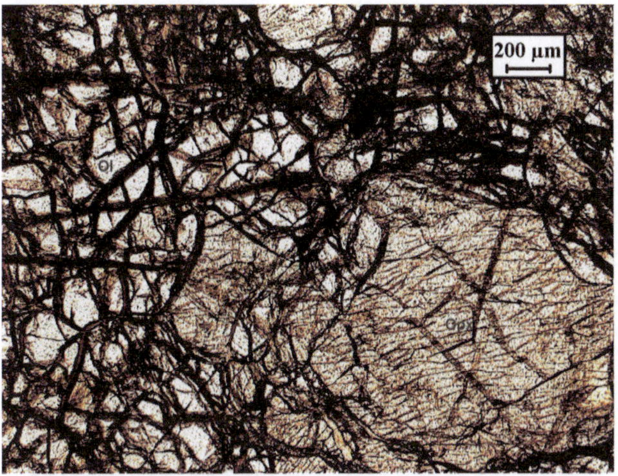

Plate Pc-8 Protogranular harzburgite containing coarse non-pleochroic orthopyroxene surrounded by highly fractured olivine filled with Fe-oxides (K14-1)

Plate Pc-9 Harzburgite containing xenomorphic granular olivine, prisms of non-pleochroic orthopyroxene and both primary and secondary opaques. The secondary opaques are developed at grain boundaries and in fractures during serpentinisation (2-1)

Plate Pc-10 Harzburgite of Pc-9 between crossed polars (2-2)

13.3 Lherzolite

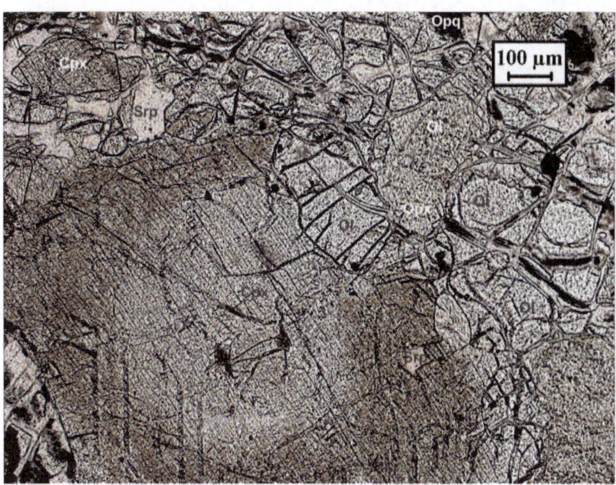

Plate Pc-11 Lherzolite with zoned poikilitic clinopyroxene containing inclusions of serpentine (after olivine), opaque and orthopyroxene exsolution lamellae diagonally across the cleavage (131/80-2)

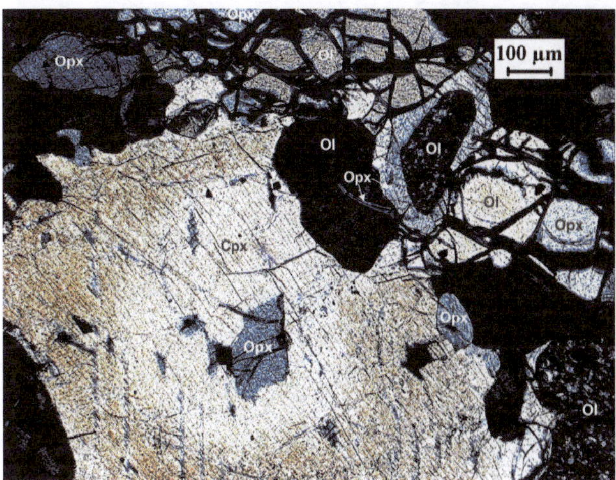

Plate Pc-12 Zoned poikilitic clinopyroxene in lherzolite of Pc-11 between crossed polars containing inclusions of orthopyroxene (*blue*) and olivine (*dark grey*) in the core. Orthopyroxene also occurs as exsolution lamellae within clinopyroxene and small grains with olivine inclusion in the adjoining region (131/80-1)

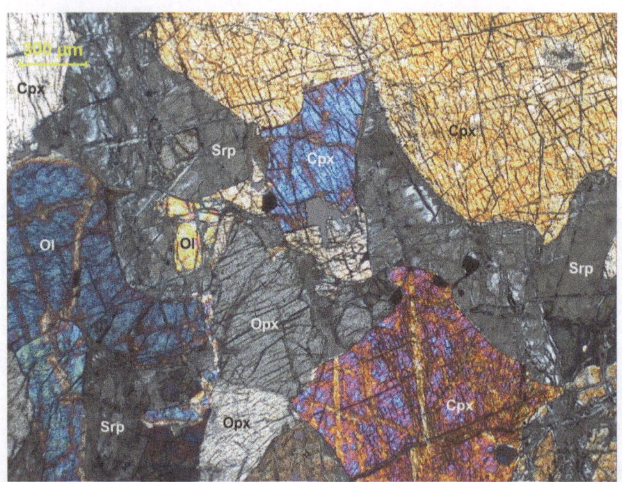

Plate Pc-13 Lherzolite between crossed polars showing serpentinised olivine, clinopyroxene, orthopyroxene and opaques (5-2)

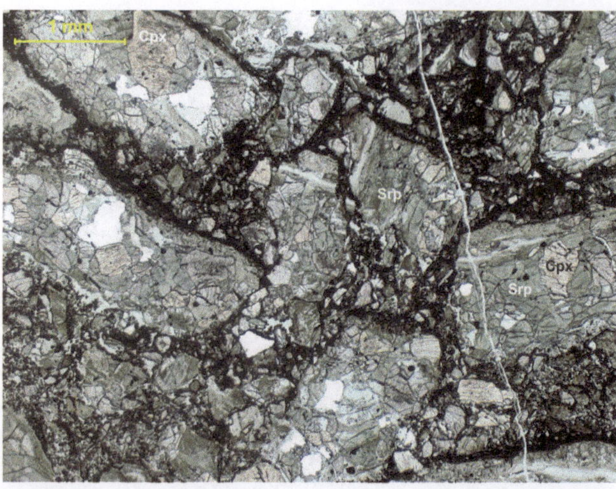

Plate Pc-15 Brecciated wehrlite containing serpentine, clinopyroxene and minor opaque traversed by a vein of carbonate (174/80-1)

13.4 Wehrlite

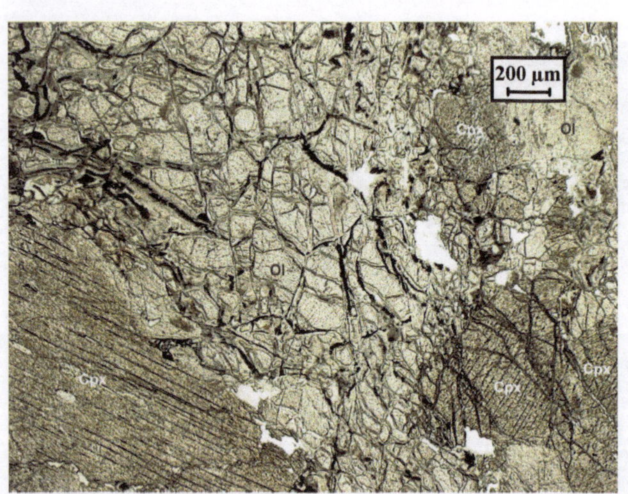

Plate Pc-14 Coarse-grained protogranular wehrlite containing clinopyroxene and highly fractured olivine filled with secondary opaque and serpentine (129/80-1)

Plate Pc-16 Brecciated wehrlite of Pc-15 between crossed polars containing serpentine and resorbed clinopyroxene (174/80-2)

Gabbro

14

14.1 Olivine Gabbro

Plate G-1 Olivine gabbro between crossed polars showing a large grain of plagioclase surrounded by olivine and clinopyroxene in equilibrium. The plagioclase is saussuritised and olivine is partly serpentinised. The plagioclase contains an olivine inclusion (158/80-2)

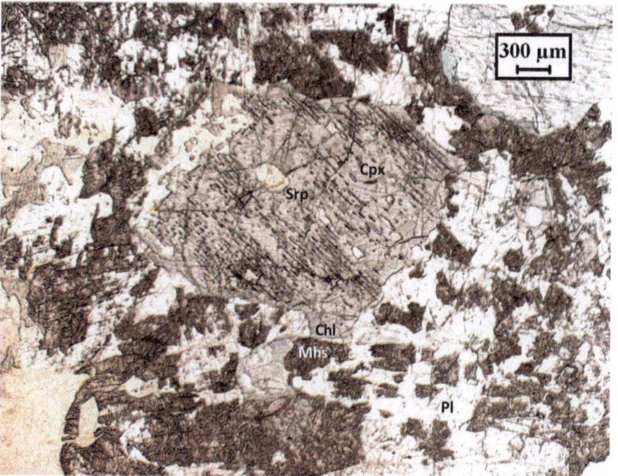

Plate G-2 Porphyritic olivine gabbro showing a *pale brown* clinopyroxene phenocryst surrounded by *dark brown* magnesio-hastingsite and calcic plagioclase. The poikilitic clinopyroxene contains inclusions of serpentinised olivine and exsolution lamellae of orthopyroxene diagonally across the cleavage. The pyroxenes are partially altered to chlorite near their rims (155-80-1)

N. C. Ghose et al., *A Petrographic Atlas of Ophiolite*,
DOI: 10.1007/978-81-322-1569-1_14, © Springer India 2014

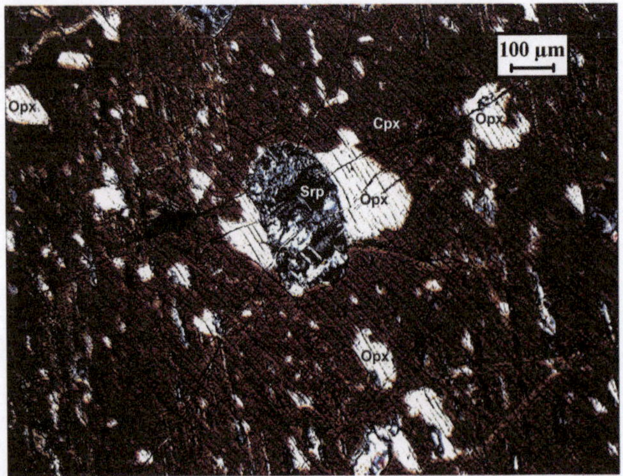

Plate G-3 Magnified view of poikilitic clinopyroxene in olivine gabbro 155/80-1 between crossed polars showing inclusions of serpentine and orthpyroxene, and orthopyroxene exsolution lamellae (155-80-3)

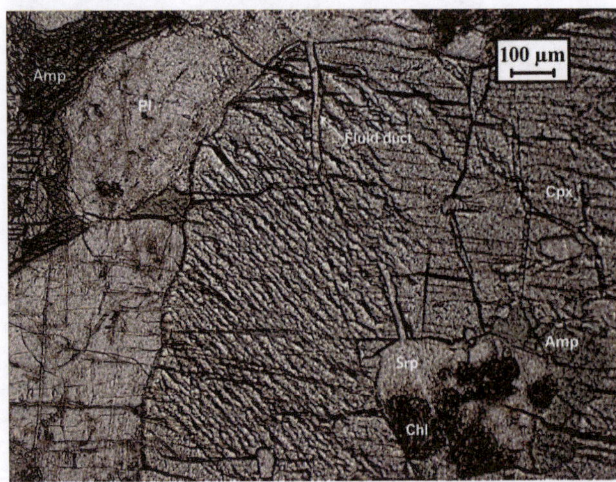

Plate G-4 Magnified view of poikilitic clinopyroxene surrounded by plagioclase and amphibole in olivine gabbro 155/80-1. The clinopyroxene contains inclusion of serpentine + brown chlorite connected to the outside of the grain through a vein that evidently enables the inside of the grain to be altered. *Green* amphibole occurs at the contact between the inclusion and the host clinopyroxene (155-80-4)

14.2 Olivine-Hornblende Gabbro

Plate G-5 Olivine-hornblende gabbro showing granoblastic texture and containing hornblende, clinopyroxene, plagioclase, serpentine and minor chrome spinel and magnetite (ML129-2)

Plate G-6 Olivine-hornblende gabbro of ML129-2 between crossed polars showing recrystallisation of amphibole around serpentine and pyroxene. Iron released in the transformation resulted in formation of Fe-oxide (ML129-3)

14.3 Norite

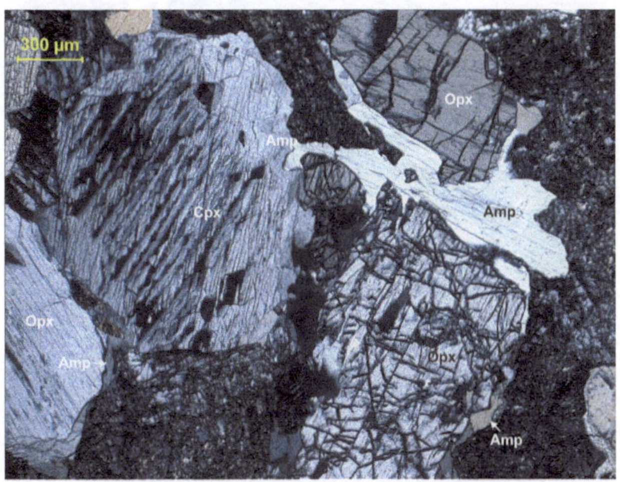

Plate G-7 Norite containing pleochroic orthopyroxene (*pale green* to *pink*), clinopyroxene, amphibole and plagioclase. Clinopyroxene shows coarse exsolution lamellae of magnetite across the cleavage. Amphiboles are developed at the pyroxene grain boundaries (164/80-2)

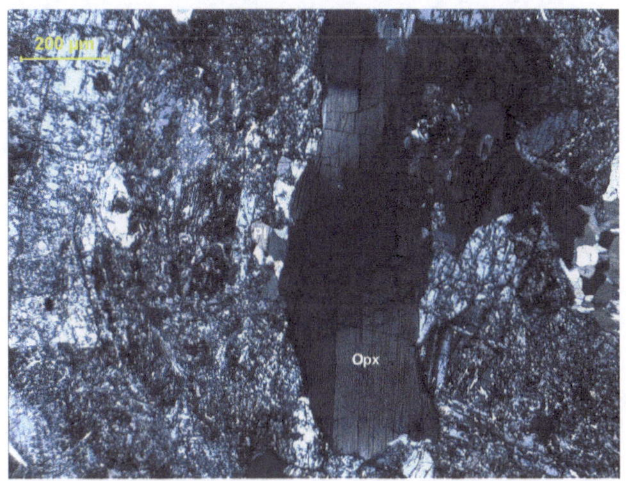

Plate G-8 Orthopyroxene displaying strain shadow in norite of 164/80-2. The rock shows brecciation. Plagioclase shows strain effects and saussuritisation (164/80-4)

14.4 Gabbronorite

Plate G-9 Gabbronorite with granoblastic texture containing pleochroic orthopyroxene, clinopyroxene, plagioclase and abundant opaque distributed uniformly. The orthopyroxene at the centre contains olivine (*grey*) and opaque inclusions (161/80-1)

Plate G-10 Gabbronorite of 161/80-1 between crossed polars (161/80-2)

Plate G-11 Transformation of pleochroic orthopyroxene to *greenish* amphibole in gabbronorite. Chlorite occurs between the amphibole rim of the orthopyroxene and plagioclase (161/80-4)

Plate G-13 A cluster of plagioclase grains surrounded by ortho- and clino-pyroxenes in gabbronorite. The cluster is traversed by fractures or micro-shear causing brecciation (161/80-9)

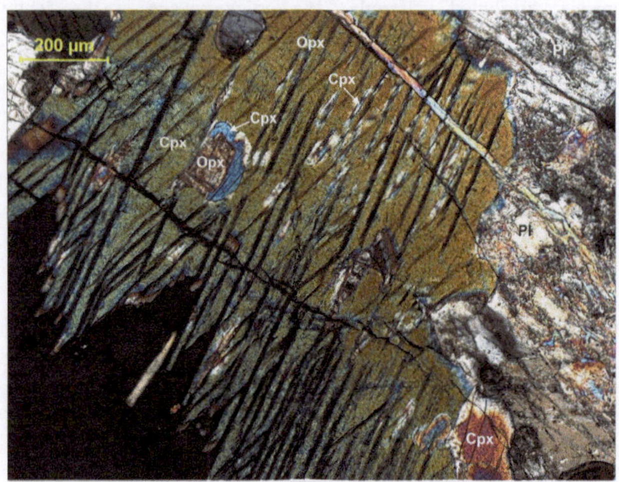

Plate G-12 Magnified view of the gabbronorite 161/80-4 between crossed polars. The orthopyroxene core shows exsolution lamellae of inverted clinopyroxene across the cleavage (161/80-5)

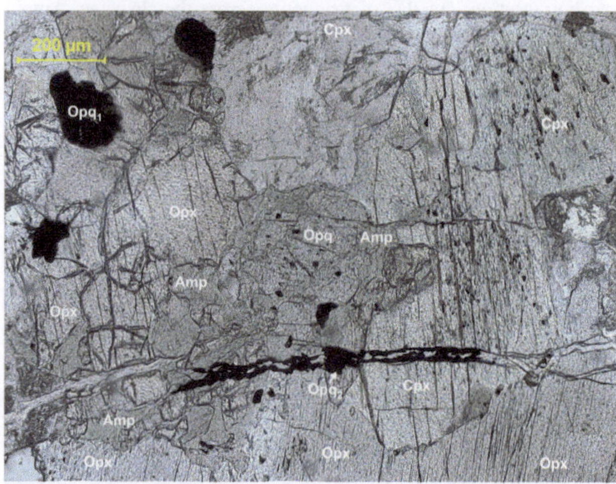

Plate G-14 Polyphase mineralisation of Fe–Ti oxides in gabbronorite 161/80-9 containing both primary disseminated/granular opaque (Opq_1) and secondary opaque (Opq_2) in vein (161/80-10)

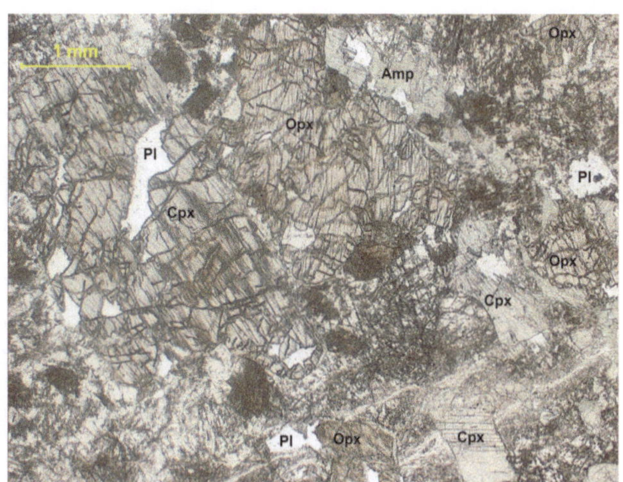

Plate G-15 Medium-grained gabbronorite consisting of clinopyroxene, pleochroic orthopyroxene, *green* amphibole and altered calcic plagioclase (163/80-1)

Plate G-17 Gabbronorite containing pleochroic *green* orthopyroxene, clinopyroxene, plagioclase and opaque. The pyroxenes are marginally altered to amphibole. Prismatic clinopyroxene contains exsolution lamellae of orthopyroxene (169/80-1)

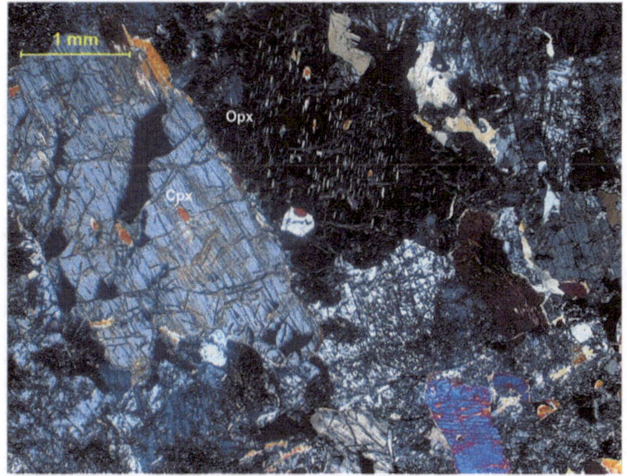

Plate G-16 Gabbronorite 163/80-1 between crossed polars showing dark prismatic orthopyroxene with exsolution lamellae of inverted clinopyroxene and inclusions of plagioclase and smaller pyroxene (163/80-2)

Plate G-18 Gabbronorite 169/80-1 between crossed polars showing exsolution lamellae in clinopyroxene (169/80-2)

14.5 Hornblende Gabbro

Plate G-19 Hornblende gabbro consisting of clinopyroxene, plagioclase, minor pleochroic orthopyroxene and opaque (149-1)

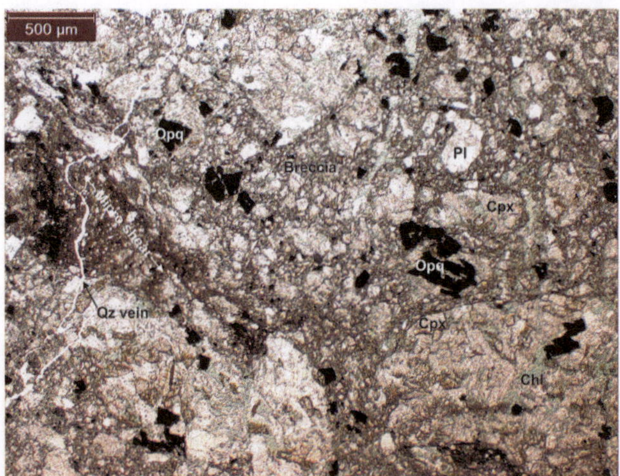

Plate G-21 Highly brecciated gabbro showing angular fragments and blocks of plagioclase, chloritised pyroxene and opaque. Secondary quartz veins cut across the rock. The opaque contains inclusion of chlorite (297/79-1)

14.6 Brecciated Gabbro

Plate G-20 Brecciated gabbro with angular fragments of plagioclase and deformed clinopyroxene (708-1)

Plate G-22 Brecciated gabbro 297/79-1 between crossed polars showing highly crushed and granulated plagioclase (*grey*) and clinopyroxene (*orange* and *yellow*), quartz and chlorite (297/79-2)

15 Plagiogranite

Plate PG-1 Hypidiomorphic granular texture in tonalite between crossed polars consisting of twinned plagioclase, quartz, amphibole, chlorite and epidote. Chlorite is more abundant than amphibole (M-1-3)

Plate PG-3 Tonalite showing vague foliation and containing plagioclase, quartz, *green amphibole* and opaque. Opaques usually occur at the amphibole grain boundaries (310/79-1)

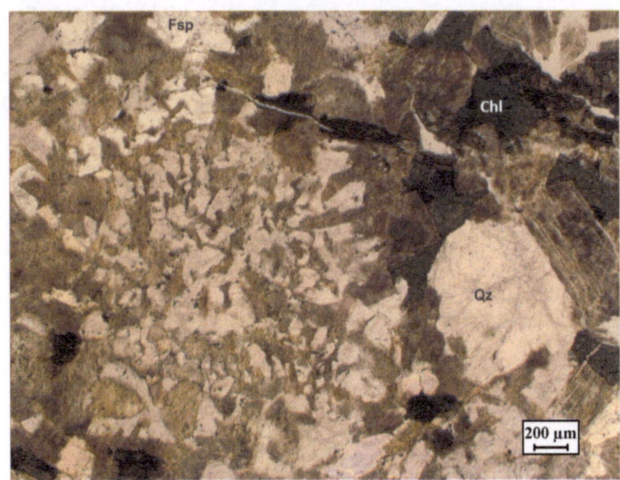

Plate PG-2 Tonalite M-1-3 showing myrmekitic intergrowth of oligoclase and vermicular quartz replacing potassium feldspar. Amphiboles are altered to *green chlorite* (M-1-2)

Plate PG-4 Recrystallised tonalite between crossed polars showing hypidiomorphic granular texture. The rock contains serrated quartz, rectangular laths of plagioclase, amphibole, flakes of biotite and minor epidote. Plagioclase shows normal zoning and contains inclusion of mafic minerals (369-79-2)

N. C. Ghose et al., *A Petrographic Atlas of Ophiolite*,
DOI: 10.1007/978-81-322-1569-1_15, © Springer India 2014

Plate PG-5 Panidiomorphic granular texture in tonalite containing plagioclase, quartz and amphibole partially altered to chlorite and opaque. Amphibole occupies the core of plagioclase in the plagioclase +amphibole intergrowth (16r/80-1)

Plate PG-7 Quartz diorite between crossed polars showing interstellar twinned plagioclase with interstitial quartz, hornblende and opaque between plagioclase laths (M-2-3)

Plate PG-6 Tonalite 16r/80-1 containing partially altered amphibole and plagioclase, and minor opaques, epidote and apatite. Amphibole is partially altered to chlorite, and a late quartz vein cuts through the rock (16r/80-2)

Plate PG-8 Diorite between crossed polars showing hypidiomorphic granular texture consisting of altered plagioclase, green amphibole and accessory epidote (K32-1)

Plate PG-9 Diorite between crossed polars dominantly composed of plagioclase and minor quartz. Post-emplacement hydrothermal activity resulted in the growth of mica, epidote and carbonate (294/79-2)

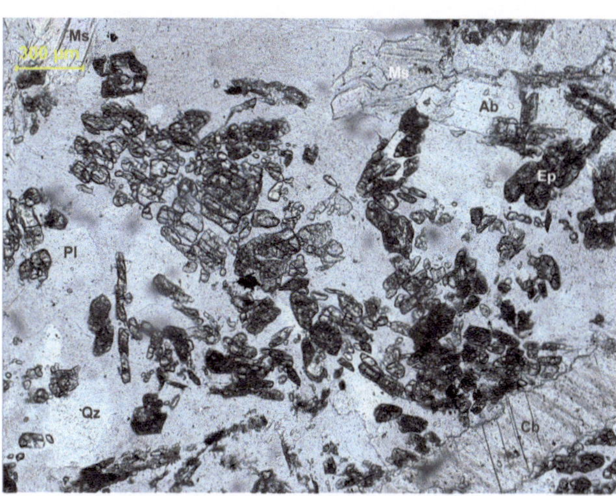

Plate PG-10 Diorite 294/79-2 with secondary epidote, muscovite, albite, carbonate and minor opaque (294/79-3)

16 Anorthosite

Plate An-1 Hypidiomorphic granular anorthosite between crossed polars containing unaltered twinned calcic plagioclase, chloritised hornblende, epidote, and minor sphene, opaque and carbonates (822-1)

Plate An-3 Coarse anorthosite between crossed polars showing panidiomorphic-granular texture and consisting of calcic plagioclase (~85 vol %) and intercumulus grains of prismatic amphibole and opaque (T-74-1)

Plate An-2 Anorthosite 822-1 containing two types of Fe–Ti oxides: massive-type (Opq_1) filled with intercumulus chloritised amphibole and plagioclase, and skeletal octahedral grain (Opq_2) laced with fine carbonate veins (822-2)

Plate An-4 Poikilitic *yellow–brown* amphibole (Amp_1) in anorthosite T-74-1 containing inclusion of *brown* hexagonal amphibole (Amp_2) and *grey* chlorite (Chl) showing replacement features (T-74-3)

N. C. Ghose et al., *A Petrographic Atlas of Ophiolite*,
DOI: 10.1007/978-81-322-1569-1_16, © Springer India 2014

Plate An-5 High magnification view of poikilitic amphibole Amp_1(magnesio-hornblende) in anorthosite T-74-3 between crossed polars showing alteration to chlorite (T-74-4)

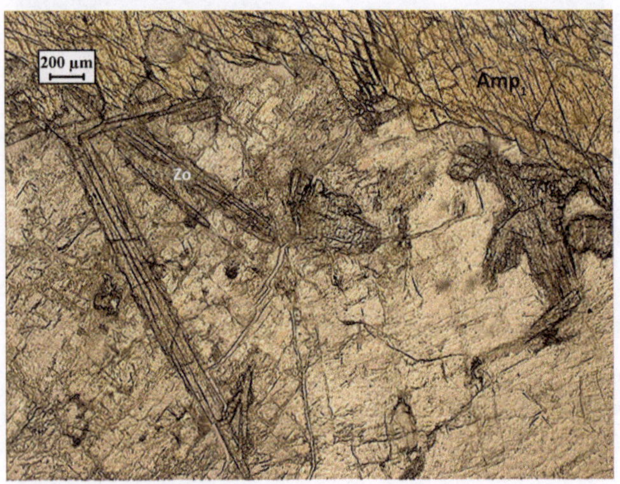

Plate An-6 High magnification view of prismatic zoisite (Zo) in anorthosite T-74 in association with magnesio-hornblende (Amp_1) (T-74-5)

Plate An-7 High magnification view of rhombic and prismatic zoisite in anorthosite T-74 between crossed polars (T-74-8)

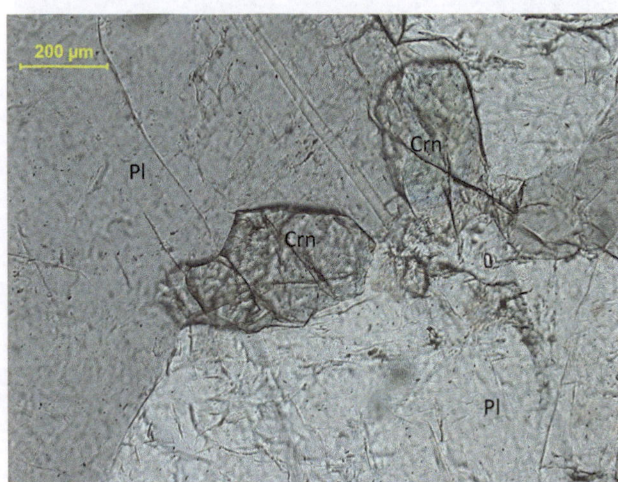

Plate An-8 Anorthosite T-74 showing growth of corundum between large plagioclase grains (T-74-7)

17 Dolerite

Plate D-1 Medium-grained dolerite under crossed polars showing ophitic to sub-ophiotic texture. Plagioclase shows twinning, clinopyroxenes are birefringent, and accessory opaques occur as inclusions in clinopyroxene and plagioclase (Y38)

Plate D-2 Medium-grained diabase showing relict ophitic texture with highly altered plagioclase and clinopyroxene. The equant anhedral clinopyroxene grains are altered to a mixture of fine needles of tremolite, epidote and opaque minerals. The outlines of plagioclase are not visible and altered to albite. The groundmass is composed of secondary minerals that include chlorite (123-80-1)

N. C. Ghose et al., *A Petrographic Atlas of Ophiolite*,
DOI: 10.1007/978-81-322-1569-1_17, © Springer India 2014

18 Mafic Volcanics

18.1 Olivine Basalt

Plate V-1 Pseudomorph of chlorite + epidote after olivine phenocryst in porphyritic olivine basalt. The fine-grained groundmass shows an intersertal to intergranular texture (230/79-2)

Plate V-3 Phenocryst of albite in basalt showing longitudinal cracks filled with secondary chlorite. The groundmass contains acicular clinopyroxene and green chlorite rimmed by epidote (230/79-4)

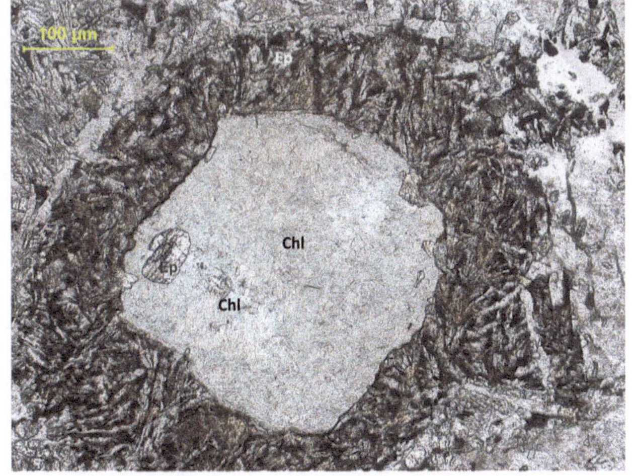

Plate V-2 Basalt 230/79-2 showing a kernel of highly altered olivine to chlorite with epidote inclusion, and a mantle of tiny prisms of epidote (230/79-1)

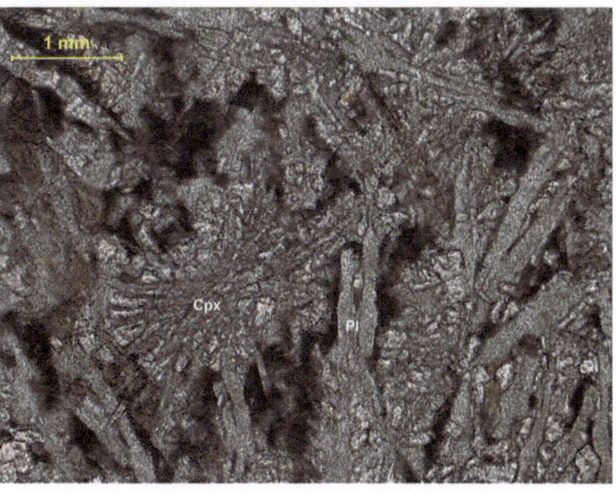

Plate V-4 Basalt 230/79-2 showing plagioclase laths and acicular sheafs of clinopyroxene. Small anhedral grains are epidote (230/79-8)

N. C. Ghose et al., *A Petrographic Atlas of Ophiolite*,
DOI: 10.1007/978-81-322-1569-1_18, © Springer India 2014

Plate V-5 Basalt (V-1) showing a large plagioclase phenocryst surrounded by a fine-grained groundmass with an intergranular to intersertal texture. Longitudinally aligned glass in plagioclase is developed along cleavage planes (230/79-9)

Plate V-6 Basalt (V-1) between crossed polars showing sectorial twinning in chlorite at the centre surrounded by granular epidote and laths of plagioclase (230/79-6)

18.2 Plagioclase-Phyric Basalt

Plate V-7 Plagioclase-phyric basalt showing intersertal texture. Rounded resorbed plagioclase appears to have formed earlier than the elongated phenocrysts. The groundmass is composed of plagioclase microlite and glass (W11-1)

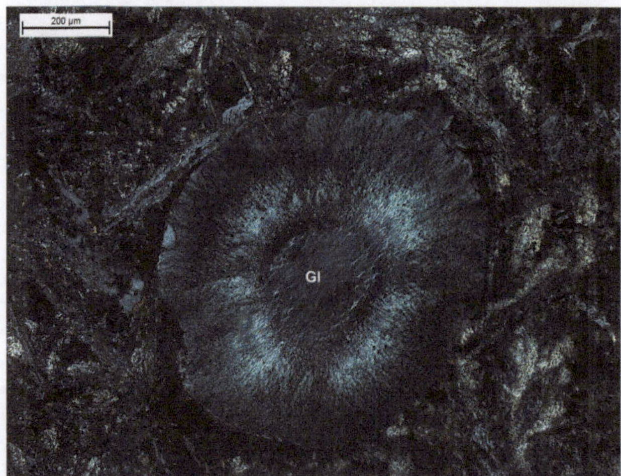

Plate V-8 Plagioclase-phyric basalt (V-7) between cross-polars showing spherule of glass with radiating fibres. The zoned spherulitic structure is developed at very high temperature (W11-4)

Plate V-9 Plagioclase-phyric basalt showing intergranular to intersertal texture. It consists of skeletal plagioclase and pyroxene altering to chlorite and iron oxides (230/70-1)

Plate V-10 Porphyritic basalt with a groundmass of feldspar microlites and interstitial pyroxene altered to chlorite. The pyroxene phenocrysts are pseudomorphed by either chlorite or a mixture of chlorite and carbonate (410/79-1)

18.3 Pyroxene-Phyric Basalt

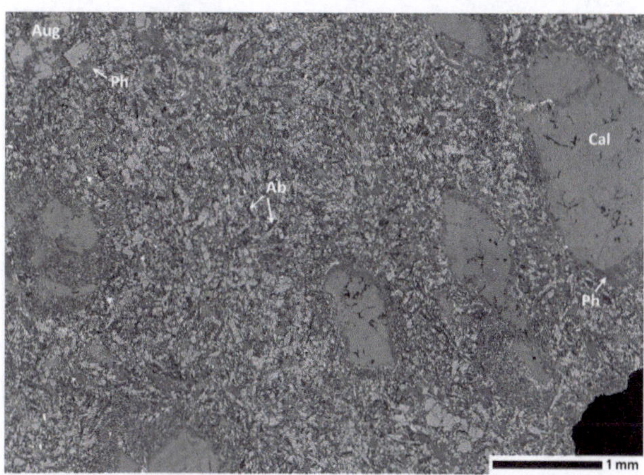

Plate V-11 Low magnification backscattered electron image of augite-phyric, amygdaloidal metabasalt with calcite-filled vesicles (330/79)

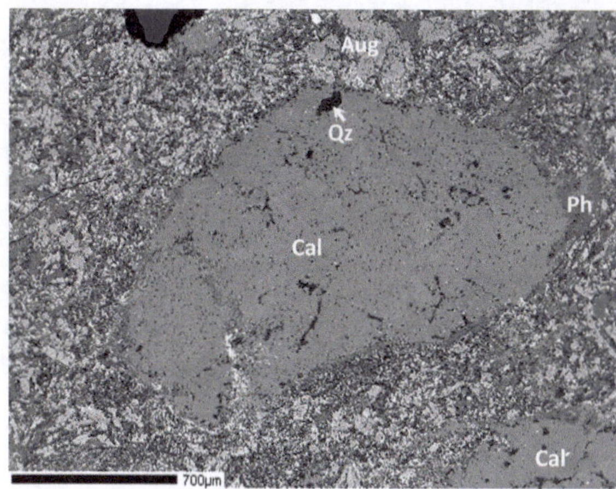

Plate V-12 Magnified backscattered electron image of calcite-filled vesicles lined with phengite in augite-phyric metabasalt 330/79. The vesicles also contain minor quartz, and the groundmass contains fine-grained disseminated Fe-oxide (330/79)

Plate V-13 Porphyritic basalt showing phenocrysts of subhedral clinopyroxene marginally altered to green or brown chlorite and plagioclase microlites in a fine-grained groundmass (330/79-1)

Plate V-14 Porphyritic basalt 330/79 between crossed polars showing sector zoning in clinopyroxene phenocryst (330/79-2)

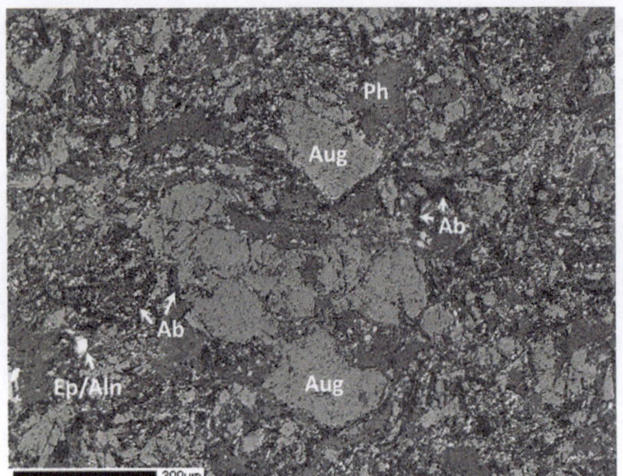

Plate V-15 Backscattered electron image of augite phenocrysts surrounded by a groundmass of augite + albite + phengite + epidote/allanite + fine-grained Fe-oxide in metabasalt 330/79

Plate V-16 Backscattered electron image of a phengite microphenocryst surrounded by augite + albite + epidote/allanite groundmass in metabasalt 330/79

Plate V-17 Porphyritic basalt showing a euhedral clinopyroxene phenocryst with marginal alteration to chlorite set in a fine-grained groundmass of randomly orientated plagioclase microlites and recrystallised glass. The rock contains accessory allanite and titanite, and secondary high-K illite (EPMA) (585/79/2)

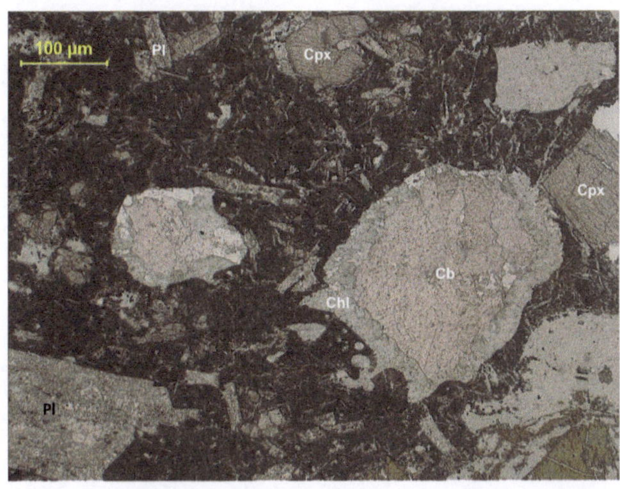

Plate V-18 Porphyritic amygdaloidal basalt 585/79-2 containing phenocrysts of plagioclase (*left corner*) and clinopyroxene with chlorite alteration rims, and vesicles filled with carbonate (*calcite*) and walled by green chlorite (585/79-4)

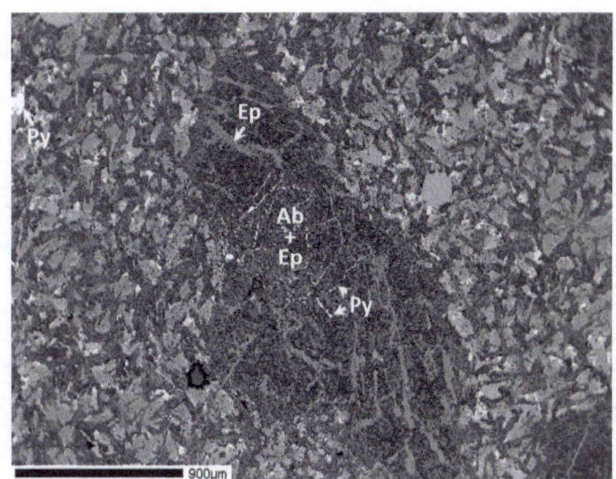

Plate V-20 Magnified backscattered electron image of a glomerocrystic aggregate of albite + epidote/zoisite pseudomorphs after plagioclase megacrysts with veins of epidote/zoisite and pyrite in metabasalt 371/79. The matrix is composed of an intergrowth of augite + epidote/albite + titanite (371/79)

18.4 Basalt with Plagioclase Megacryst

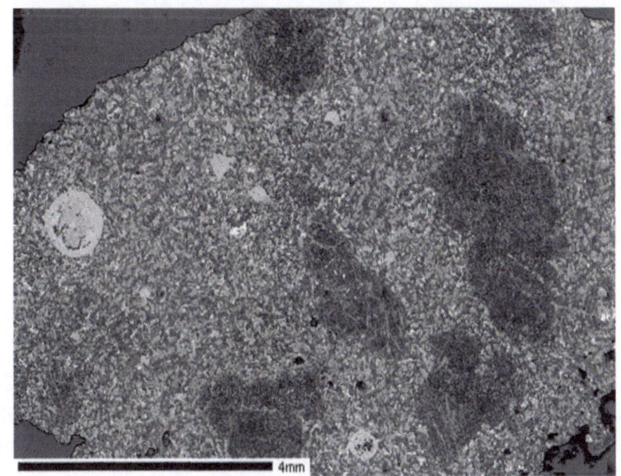

Plate V-19 Low magnification backscattered electron image of greenschist facies metabasalt with pseudomorphs after plagioclase megacrysts occurring in glomerocrystic aggregates and chlorite ocelli (371/79)

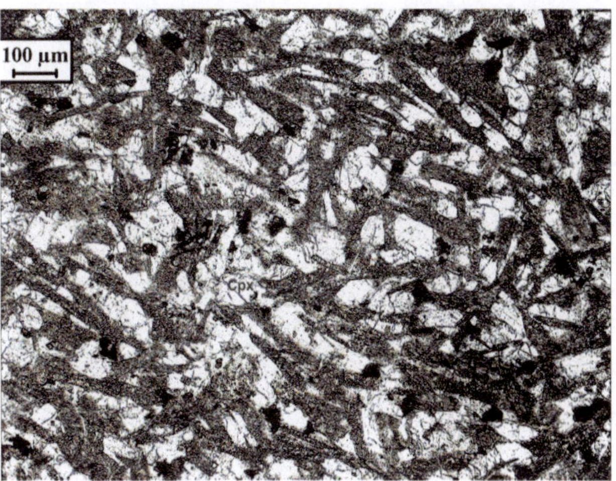

Plate V-21 Groundmass of basalt 371/79 showing intersertal texture with short prismatic clinopyroxene, skeletal plagioclase and opaque minerals set in a recrystallised groundmass with greenschist facies minerals (371/79-2)

Plate V-22 Short prisms of resorbed clinopyroxene and opaque in recrystallised groundmass of green chlorite in basalt (371/79-1a)

Plate V-25 Basalt containing small grains of clinopyroxene, elliptical yellow chlorite and a large plagioclase megacryst (*top right*) altered to albite and epidote (refer to V-20) (371/79-3)

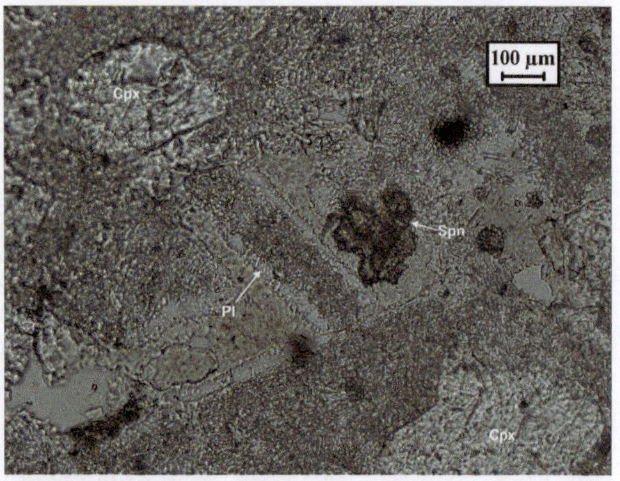

Plate V-23 Phenocrysts of clinopyroxene and zoned plagioclase, and a groundmass containing sphene in basalt (371/79-4d)

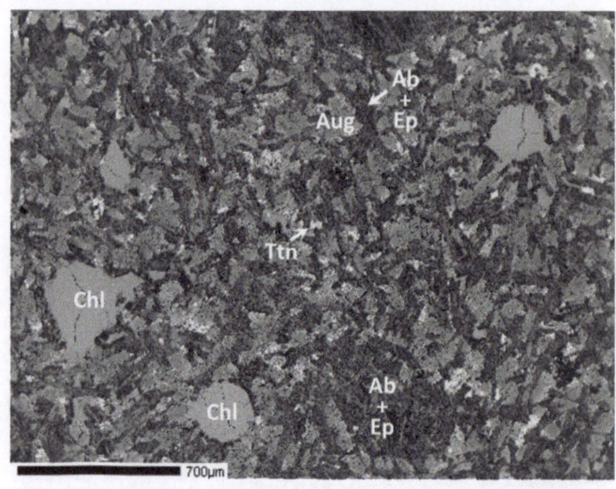

Plate V-26 Backscattered electron image of chlorite-filled vesicles in metabasalt with albite + epidote/zoisite pseudomorph after plagioclase megacryst surrounded by augite + epidote/albite + titanite intergrowth (371/79)

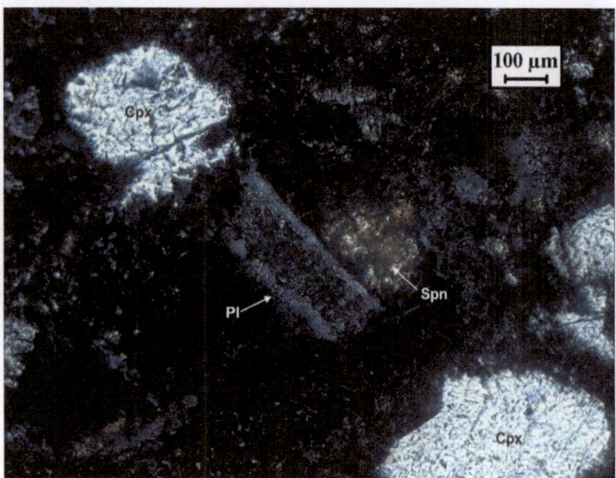

Plate V-24 Groundmass of basalt 371/79 between crossed polars (371/79-3)

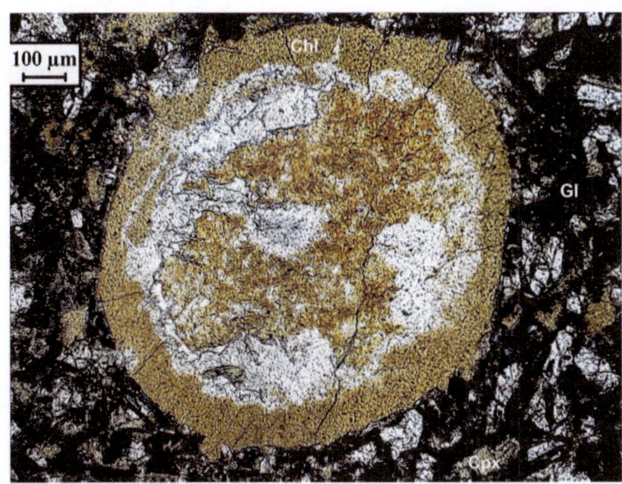

Plate V-27 Magnified view of ocellus composed of Fe-rich chlorite (*orange*) and quartz (*white*) between crossed polars in basalt. The pyroxene grains are oriented in different directions around the ocellus (371/79-1a)

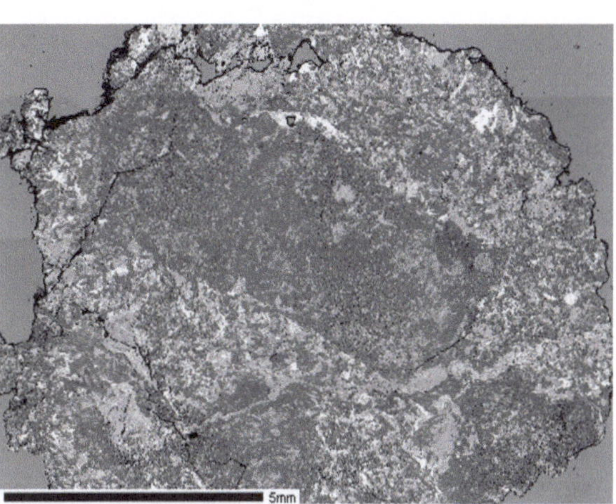

Plate V-29 Low magnification backscattered electron image of greenschist facies metabasalt with altered plagioclase megacrysts and chlorite-epidote veins (90/80)

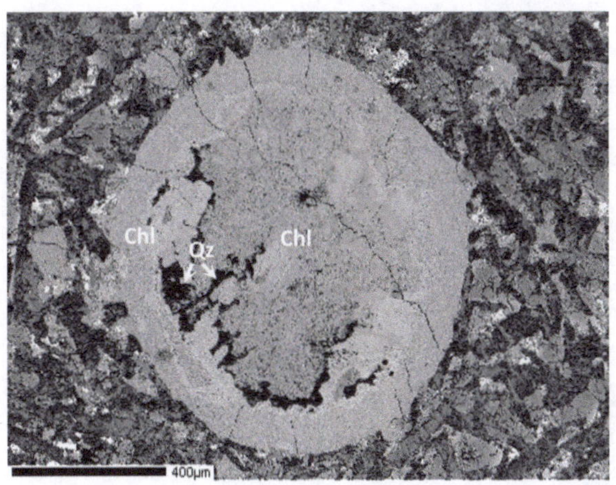

Plate V-28 Magnified backscattered electron image of ocellus in metabasalt containing Fe-rich chlorite and quartz (371/79)

Plate V-30 Magnified view of margin of a plagioclase megacryst (*white*) surrounded by a fine-grained groundmass of plagioclase, mafic minerals and Fe-oxide in basalt (90/80-1)

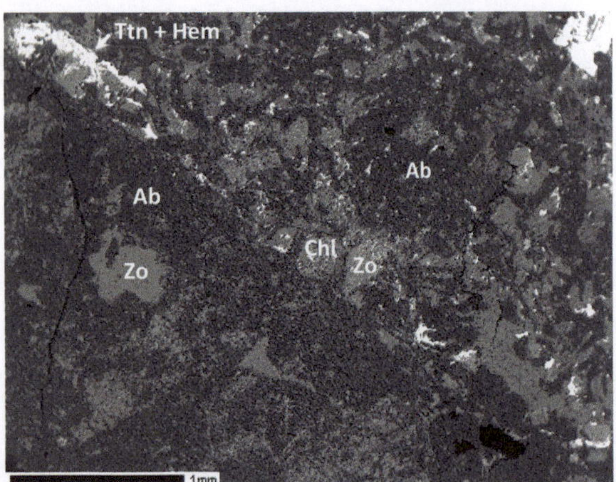

Plate V-31 Backscattered electron image of the margin of the altered plagioclase megacryst (*lower left*) in metabasalt. Plagioclase is altered to pure albite and zoisite. The basalt matrix contains albite, zoisite/epidote, chlorite, titanite and hematite. Accessory apatite is also present (not shown) (90/80)

Plate V-33 Intergranular texture in basalt 90/80-1. The interstices between plagioclase laths are filled by crystals or filaments of magnetite associated with chlorite. Magnetite is altered to red hematite (90/80-3)

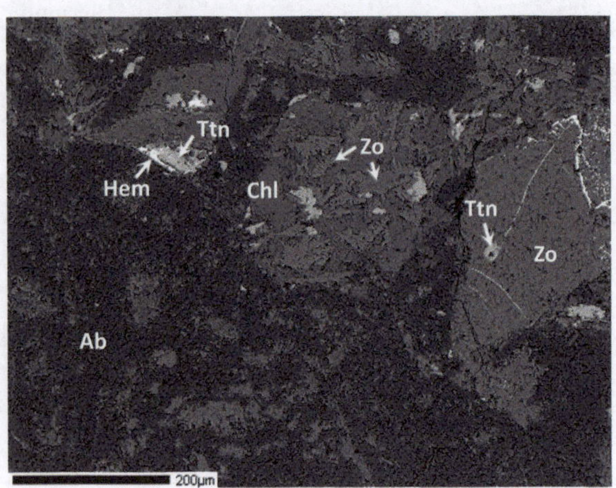

Plate V-32 Magnified backscattered electron image of zoisite + chlorite intergrowth at the margin of the altered plagioclase megacryst in metabasalt. Titanite inclusions are present within both zoisite and chlorite, and are rimmed by hematite (90/80)

Plate V-34 Alteration of primary mafic constituents in porphyritic basalt 90/80-1 to a mixture of albite, chlorite, zoisite/epidote, titanite and hematite (90/80-2)

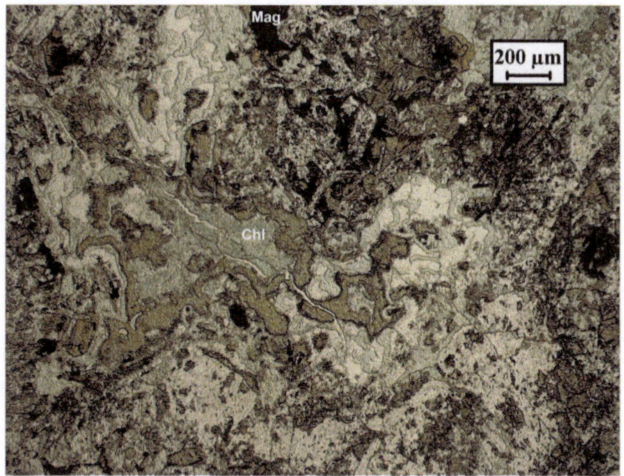

Plate V-35 Altered patches/veins filled with green and brown chlorite in porphyritic basalt 90/80-1 (90/80-5)

Plate V-36 Backscattered electron image of chlorite vein lined with epidote/zoisite in metabasalt 90/80. The matrix consists of albite laths and chlorite + epidote/zoisite intergrowths with titanite + hematite inclusions (90/80)

18.5 Andesite

Plate V-37 Porphyritic andesite between crossed polars showing a large plagioclase phenocryst with twinning. The groundmass plagio clase grains are randomly oriented and display an intersertal texture. The phenocryst contains glass and marginal alterations (62/80-2)

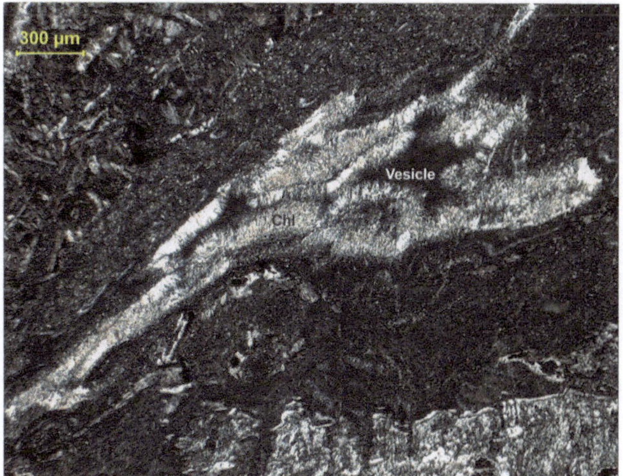

Plate V-38 A fracture in the andesite 62/80-2 filled with a mixture of chlorite and sericite around a glassy core (62/80-4)

18.6 Trachybasalt

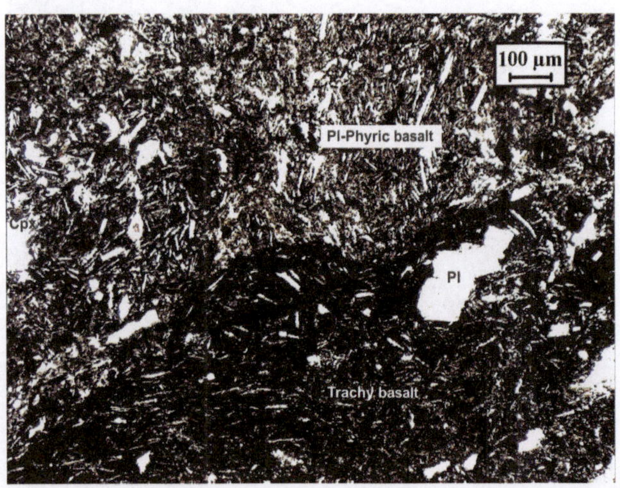

Plate V-39 Chilled contact between plagioclase-phyric basalt (*top*) and trachybasalt (*bottom*). Note the different orientations of feldspar crystals in the two basalt layers. The trachybasalt contains large phenocrysts of plagioclase (584/79-1)

Plate V-40 Xenolith of trachybasalt in plagioclase-phyric basalt. The contact of the two basalts is marked by a sinuous quartz vein (584/79-5)

Plate V-41 Phenocrysts of plagioclase in trachybasalt showing alteration to a mixture of chlorite and epidote. Microlites of feldspars show a well-defined flow texture (434/79-4)

Plate V-42 Brecciated trachybasalt (*dark blocks*) separated by veins of white albite (434/79-1)

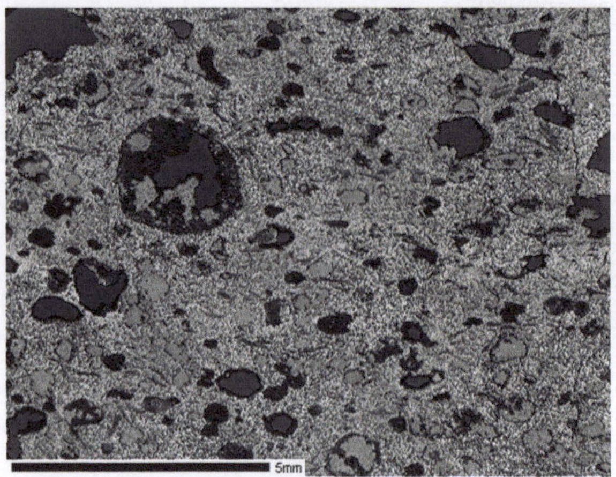

Plate V-43 Low magnification backscattered electron image of metamorphosed amygdaloidal trachybasalt with chlorite ocelli. (The *darker vesicles* are voids) (580/79)

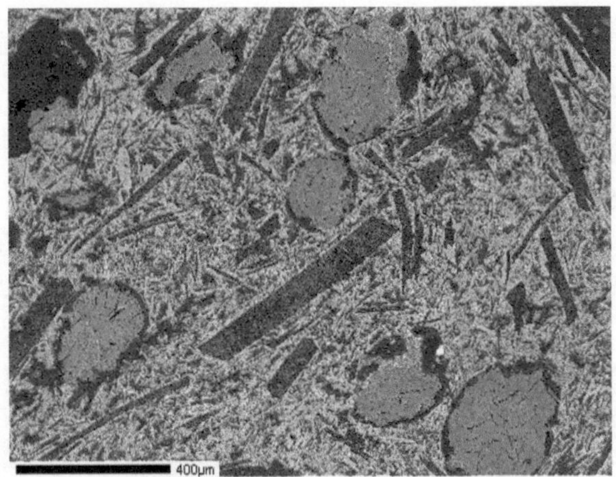

Plate V-44 Backscattered electron image of ocelli of chlorite with albite rims and albite laths in a fine-grained groundmass in amygdaloidal trachybasalt 580/79

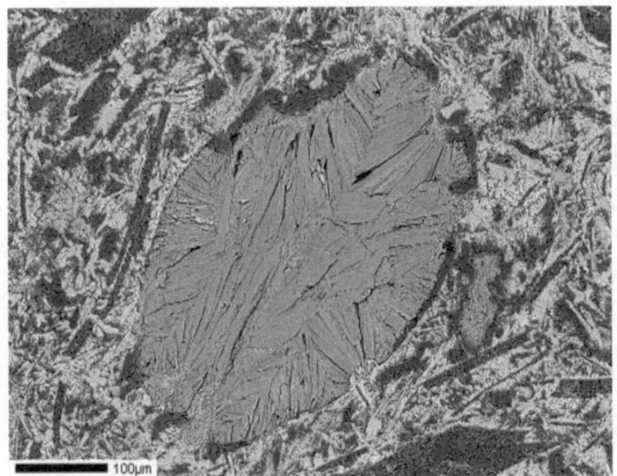

Plate V-45 Backscattered electron image of ocellus in trachybasalt 580/79 showing spherulitic growth of chlorite and a rim of albite

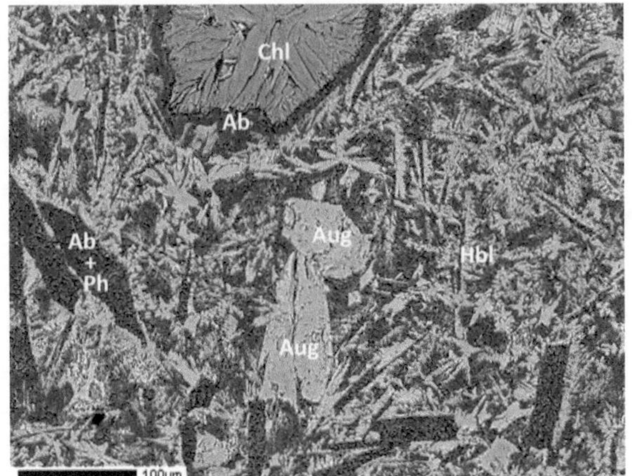

Plate V-46 Backscattered electron image of augite phenocrysts, chlorite ocellus, and albite + phengite pseudomorphs surrounded by a groundmass of dendritic edenitic hornblende + albite in trachybasalt 580/79

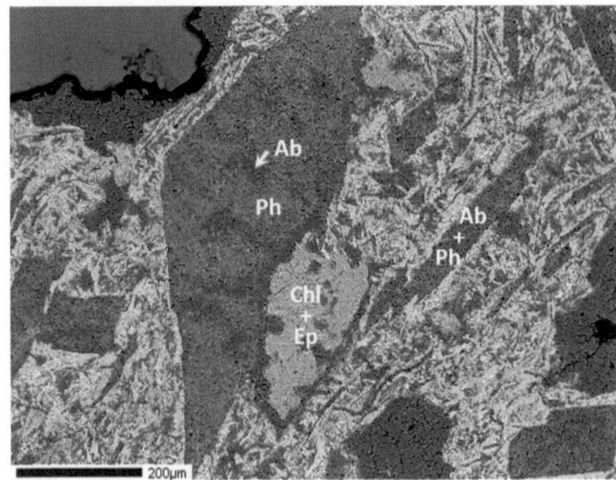

Plate V-47 Backscattered electron image of trachybasalt with albite + phengite and chlorite + epidote pseudomorphs in a groundmass of edenitic hornblende + albite. The chlorite + epidote pseudomorph is rimmed by albite (580/79)

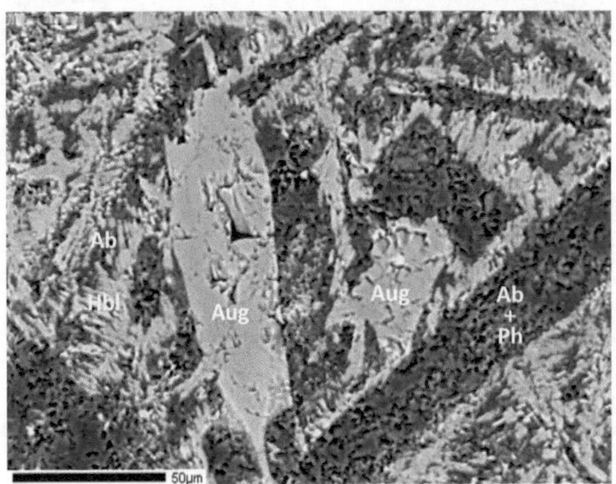

Plate V-48 Backscattered electron image of augite phenocrysts and albite + phengite pseudomorphs surrounded by a groundmass of dendritic edenitic hornblende + albite in trachybasalt 580/79

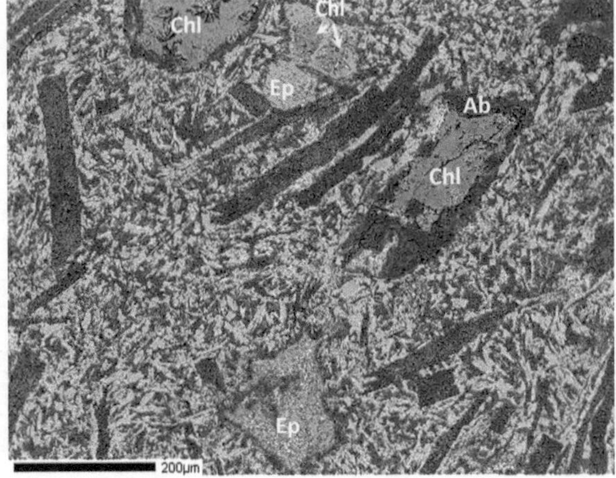

Plate V-49 Backscattered electron image of epidote phenocrysts (some enclosing chlorite), and chlorite-filled round and irregular vesicles walled by albite in trachybasalt 580/79

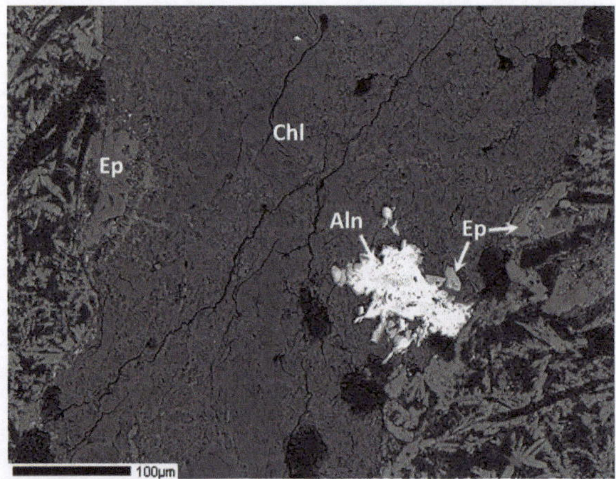

Plate V-50 Magnified Backscattered electron image of vesicle in trachybasalt 580/79 with chlorite filling and epidote and allanite at margin (580/79)

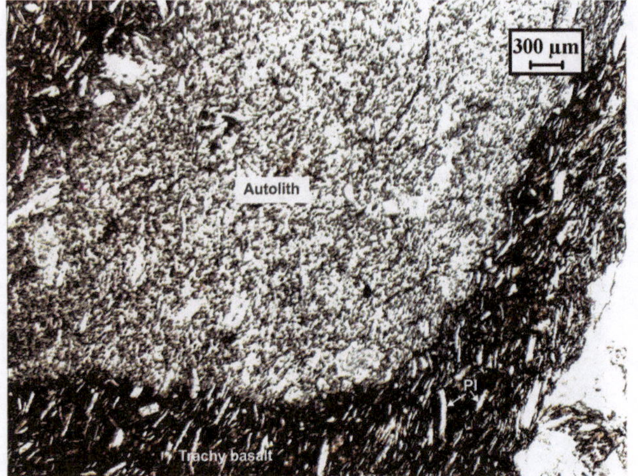

Plate V-51 An autolith of feldspar-rich fragment in trachybasalt. Chilled margin is present at the bottom edge of the autolith (584/79-1)

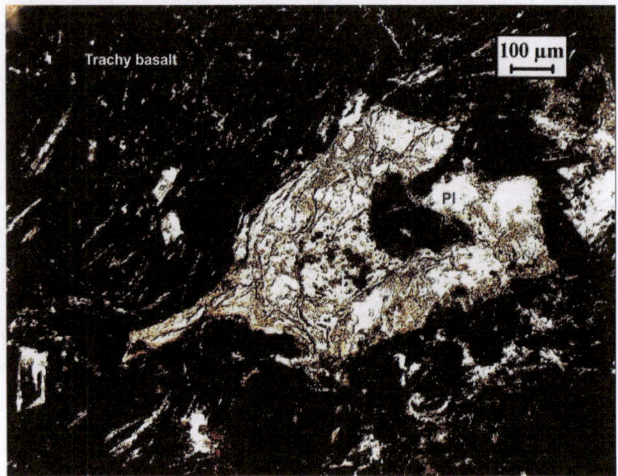

Plate V-52 Altered plagioclase phenocryst in trachybasalt. Fe-oxide is altered to red Fe-hydroxide (555/79-3)

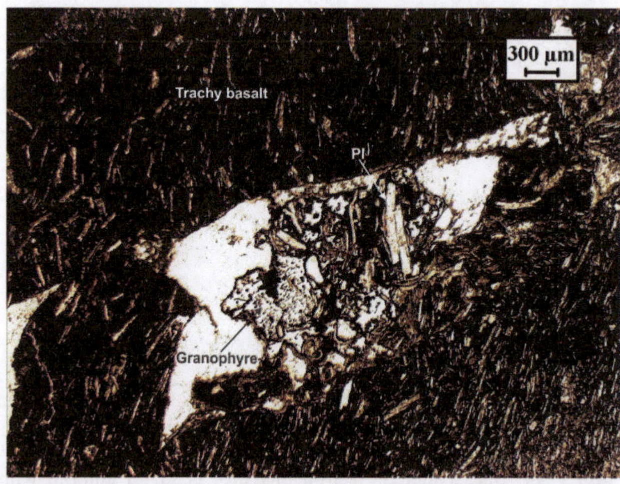

Plate V-53 Trachybasalt containing a lensoid xenolith of argillite partially assimilated by the host basalt. The assimilation reaction resulted in the formation of plagioclase laths (*right*) and granophyre (*left*, *pitted surface*) (584/79-3)

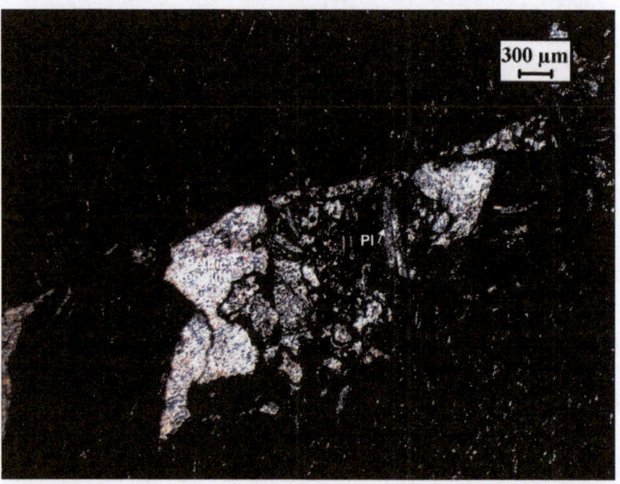

Plate V-54 Xenolith in trachybasalt between crossed polars showing birefringence caused by mica (584/79-3)

18.7 Crater Facies Lava

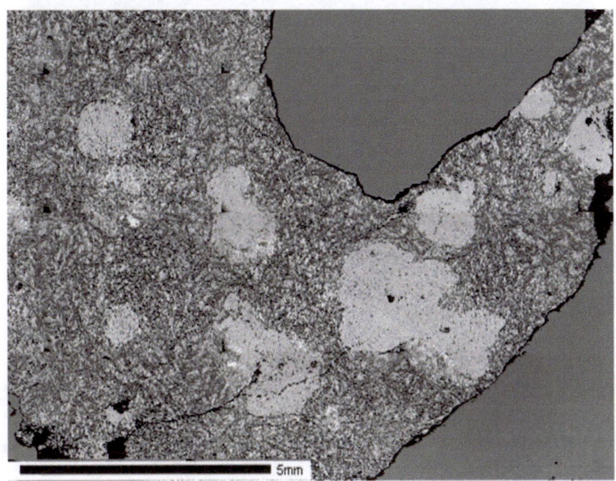

Plate V-55 Low magnification backscattered electron image of greenschist facies amygdaloidal metabasalt with calcite-filled vesicles (the large grey regions are parts of the glass slide on which the thin-section is mounted) (567/78)

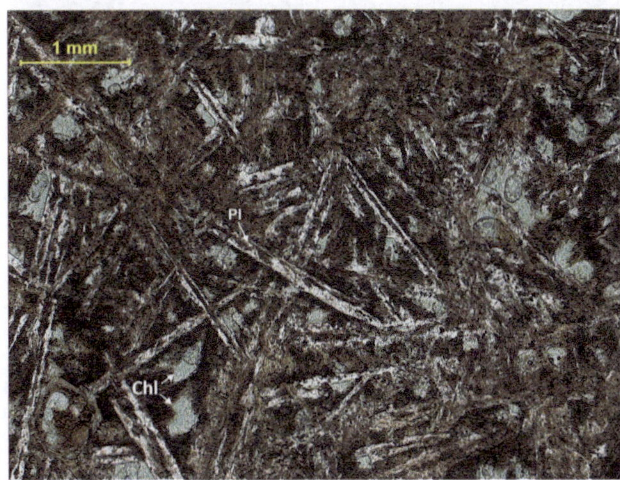

Plate V-56 Radiating laths of plagioclase and chlorite pseudomorphs after pyroxene in basalt showing intersertal texture (567/78-1)

Plate V-57 Amygdaloidal basalt with carbonate-filled vesicles and acicular needles of plagioclase and mafic minerals (epidote) within the vesicle. The host basalt shows an intergranular texture (567/78-3)

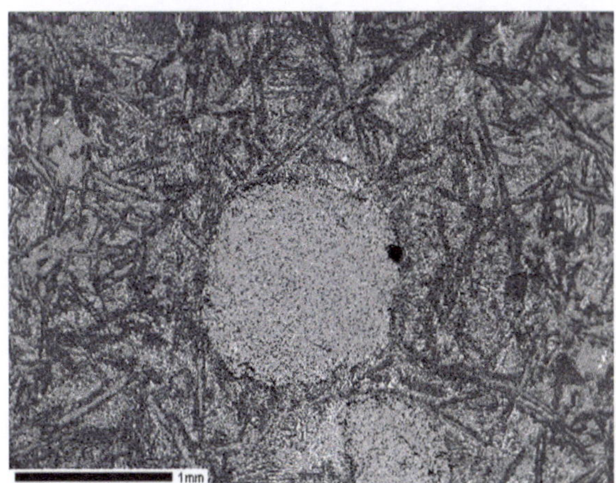

Plate V-58 Magnified backscattered electron image of vesicular calcite surrounded by needle-shaped albite laths and fine-grained intergrowths in metabasalt with intergranular texture (567/78)

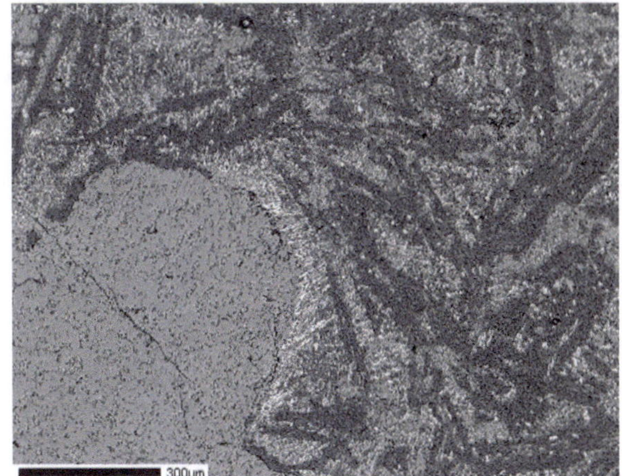

Plate V-59 Backscattered electron image of metabasalt showing late crystallisation of fine-grained aegirine-augite + titanite intergrowth at the contact of a calcite-filled vesicle (567/78)

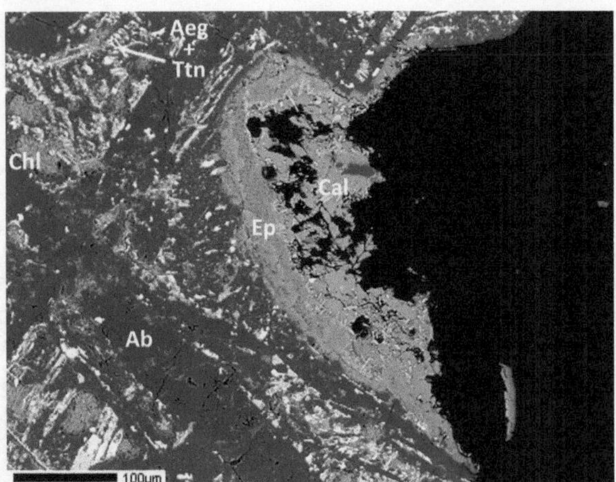

Plate V-60 Backscattered electron image of a calcite-filled vesicle (at the edge of the sample) with a wall of epidote in metabasalt 567/78. The matrix consists of albite laths interspersed with chlorite, and fine-grained intergrowth of aegirine-augite and titanite (567/78)

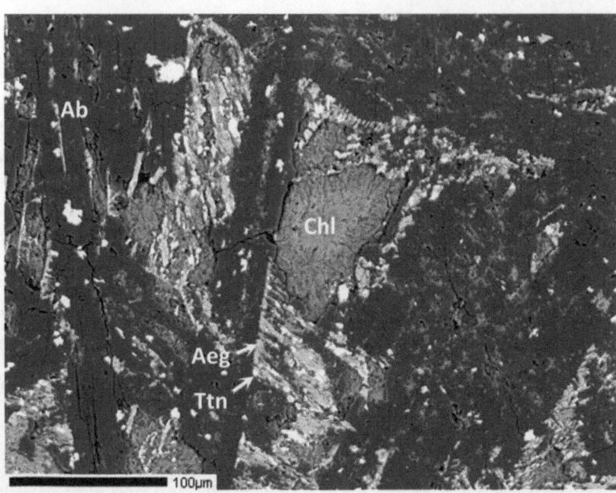

Plate V-62 Backscattered electron image of metabasalt 567/78 with compositionally zoned chlorite pseudomorph after igneous clinopyroxene surrounded by aegirine-augite + titanite intergrowth and albite laths (567/78)

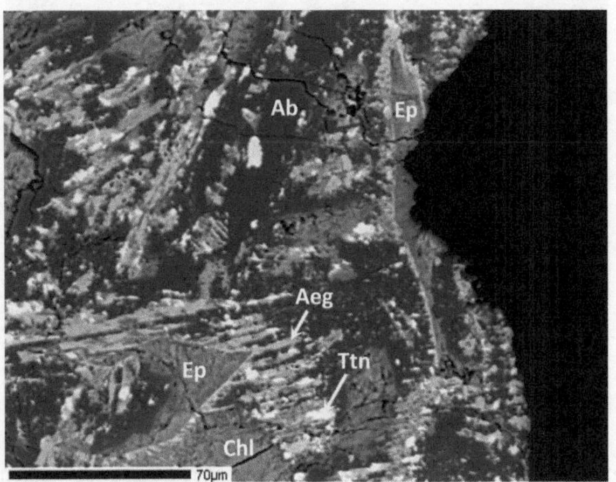

Plate V-61 Backscattered electron image of metabasalt 567/78 (edge of sample) showing euhedral epidote with aegirine-augite + titanite beard surrounded by albite laths and chlorite pseudomorphs after igneous clinopyroxene (567/78)

Plate V-63 Plagioclase-phyric basalt with of plagioclase phenocrysts associated with chlorite and sericite. The groundmass shows an intersertal texture (334/79-2)

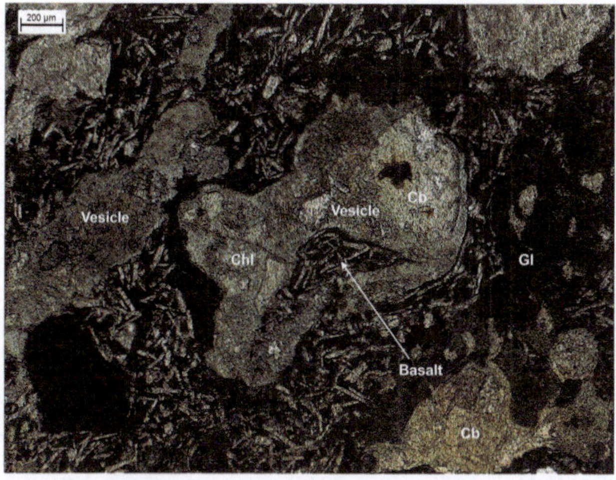

Plate V-64 Plagioclase-phyric basalt showing vesicles filled with a mixture of epidote, chlorite, carbonate and hematite (334/79-1)

Plate V-65 Silicified and albitised andesite with a groundmass of clinopyroxene, plagioclase (albite microlites), opaque and secondary epidote (33/79-1)

Plate V-66 Amygdaloidal basalt containing vesicles filled with devitrified glass and carbonate (calcite) (B69)

18.8 Spilite

Plate V-67 Spilite containing skeletal albite, granular epidote, green chlorite and opaque (180/79-2)

Plate V-68 Folded albite vein in spilite (180/79-1)

Plate V-69 Spilite between crossed polars predominantly consisting of epidote and opaques along with green chlorite (146/80-1)

Plate V-70 Variolitic texture in spilite. Divergent fibres of albite in a glassy groundmass indicates quenched texture

19 Basaltic Hyaloclastite

Plate Hy-1 Folded carbonate and quartz veins (*white*) in a glassy groundmass of fragmented hyaloclastite dissected by micro-shear. The carbonates are intermixed with chlorite in the *lower left corner* (19/79-1)

Plate Hy-2 Hyaloclastite 19/79-1 between crossed polars showing disintegration of lava flow at the interface with water in sub-marine condition. Quartz vein shows folding (19/79-2)

N. C. Ghose et al., *A Petrographic Atlas of Ophiolite*,
DOI: 10.1007/978-81-322-1569-1_19, © Springer India 2014

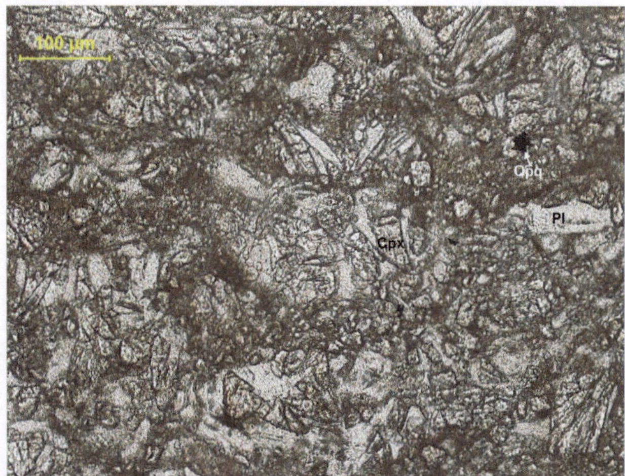

Plate Hy-3 Magnified view of hyaloclastite showing fractured and disjointed grains of clinopyroxene and plagioclase in mafic lava (19/79-3)

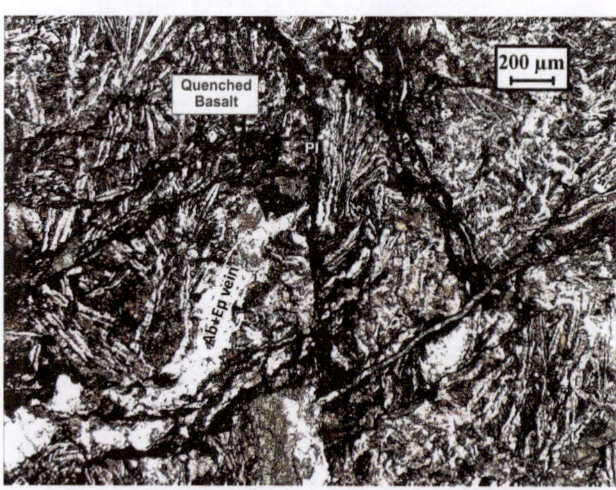

Plate Hy-5 Angular fragments in hyaloclastite indicating disintegration of basaltic lava in sub-marine condition. A leucocratic epidote-albite vein (*white*) cuts across the quenched basalt (27/79-10)

Plate Hy-4 Magnified view of hyaloclastite 19/79-3 between crossed polars (19/79-4)

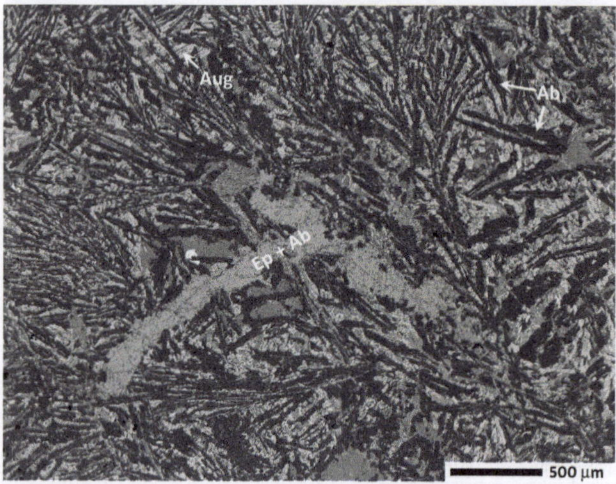

Plate Hy-6 Backscattered electron image of hyaloclastite 27/79, a greenschist facies metabasalt with intergranular texture and a vein of epidote+albite (27/79)

Plate Hy-7 Quenched intergranular texture in mafic lava (hyaloclastite 27/79-10) showing long needles of radiating plagioclase and the interstices are filled with *granular brown clinopyroxene* and chlorite (27-79-8)

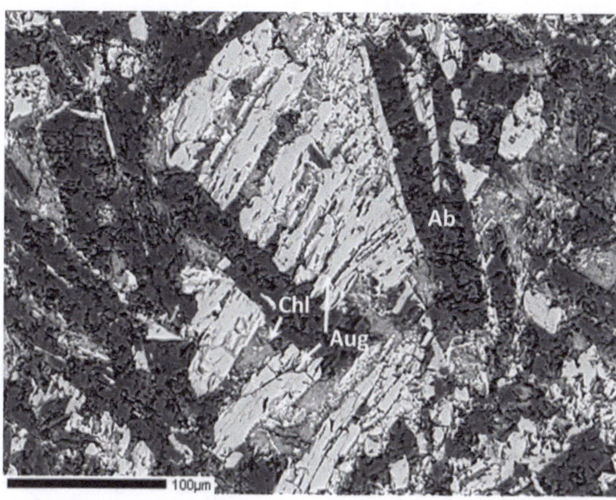

Plate Hy-9 Backscattered electron image of partially chloritised, serrated augite among albite laths in hyaloclastite metabasalt 27/79

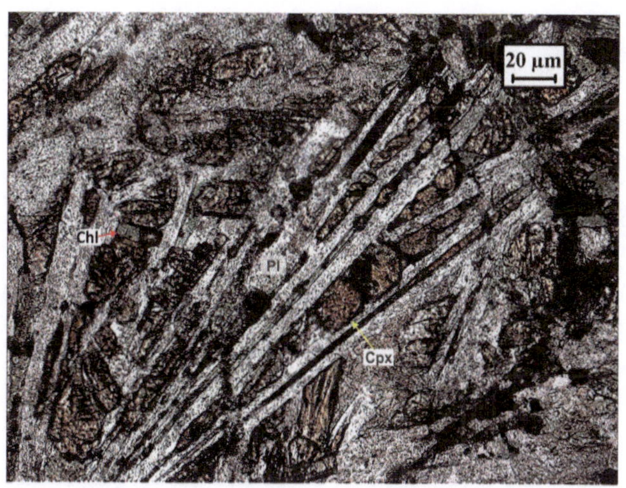

Plate Hy-8 Magnified view of mafic lava fragments containing laths/needles of plagioclase and granular clinopyroxene altered to chlorite in hyaloclastite (27/79-2)

Plate Hy-10 Backscattered electron image of tabular primary augite flanked by calcite in hyaloclastite metabasalt 27/79

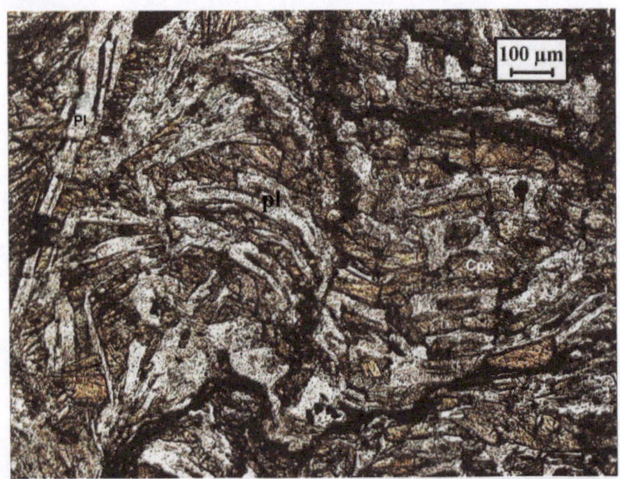

Plate Hy-11 *Curved plagioclase* and *elongated brown clinopyroxene* of mafic lava in fractured fragments of hyaloclastite showing intergranular texture (37/79-2)

Plate Hy-13 Magnified view of dog-tooth augite with serrated margins in glassy groundmass of hyaloclastite (37/79-1). *Brown granular clinopyroxene* altered to chlorite occupies the interstitial spaces between the plagioclase grains (37/79-2a)

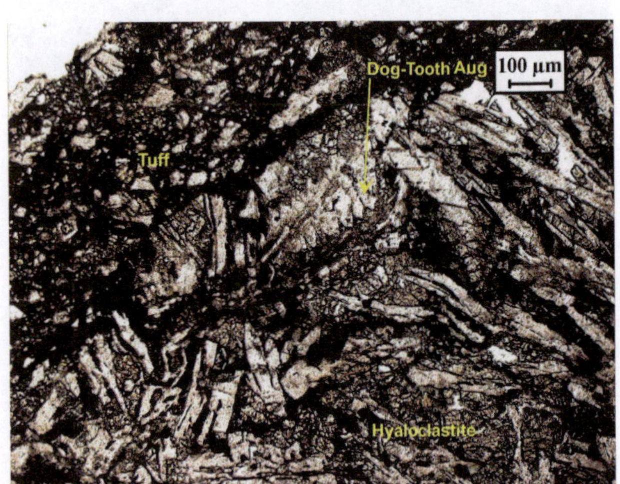

Plate Hy-12 Contact between tuff and mafic lava of hyaloclastite fragments containing curved plagioclase and dog-tooth augite. Pyroxenes are present in the angular interstices between plagioclase grains (37/79-1)

Plate Hy-14 Contact between fine- and very fine-grained basalt separated by chilled glassy margin (26/79-3)

Plate Hy-15 Magnified view of the contact in basalt 26/79-3 showing laths of randomly oriented plagioclase in a glassy groundmass (26/79-2)

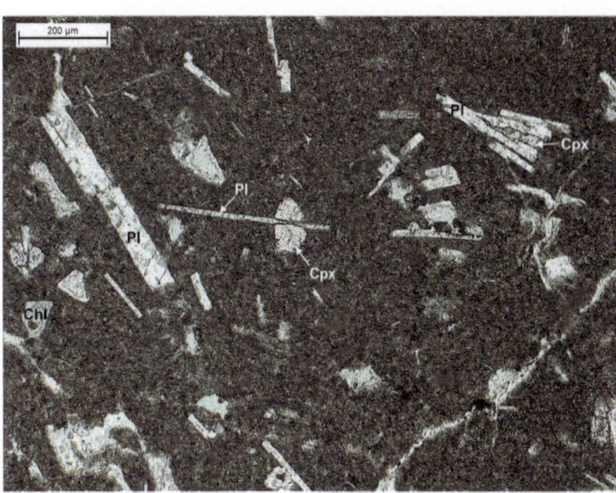

Plate Hy-17 Magnified view of plagioclase and clinopyroxene in plagioclase-phyric basalt G/15-1 showing intergranular and inter-penetration textures. The groundmass contains glass, plagioclase microlites and secondary chlorite (G/15-2)

Plate Hy-16 Multiple flows with undissolved mesostasis in glassy lava (G/15/1)

Plate Hy-18 Hyaloclastite 37/79-1 fragments with clasts of *yellowish brown carbonate* surrounded by curved plagioclase and intergranular pyroxene. The carbonate is rimmed by *brown palagonite glass* (37/79-3)

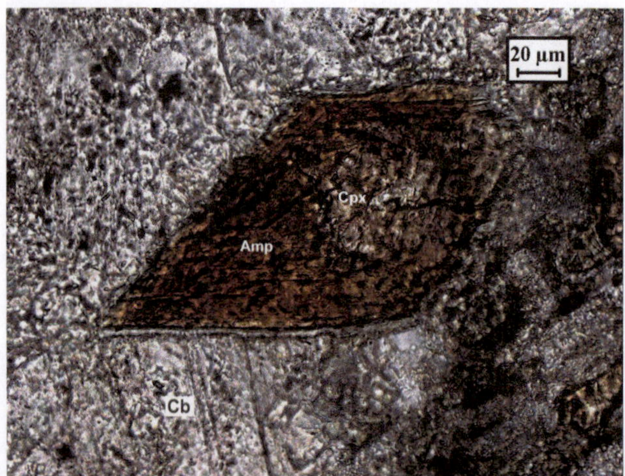

Plate Hy-19 Magnified view of carbonate fragment in hyaloclastite 37/79-3 between crossed polars showing trapped inclusion of *brown*, rhombic pleochroic amphibole with a core of light *brown clinopyroxene*. The amphibole is rimmed by chlorite (37/79-9)

Plate Hy-21 Magnified view of the epidote-albite vein in basaltic hyaloclastite 27/79-7 between crossed polars showing intergranular texture (27/79-8)

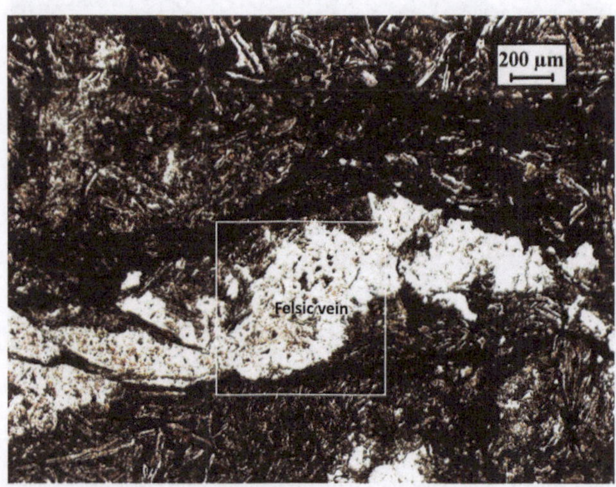

Plate Hy-20 Epidote-albite vein in hyaloclastite 27/79-10 (27/79-7)

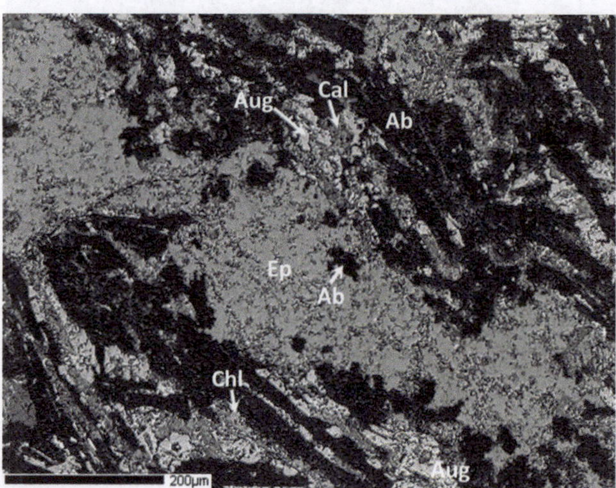

Plate Hy-22 Backscattered electron image of epidote-albite vein hyaloclastite metabasalt 27/79 comprising albite laths interspersed with tabular augite, calcite and chlorite. Minor titanite and apatite (not shown) are also present in the sample (27/79)

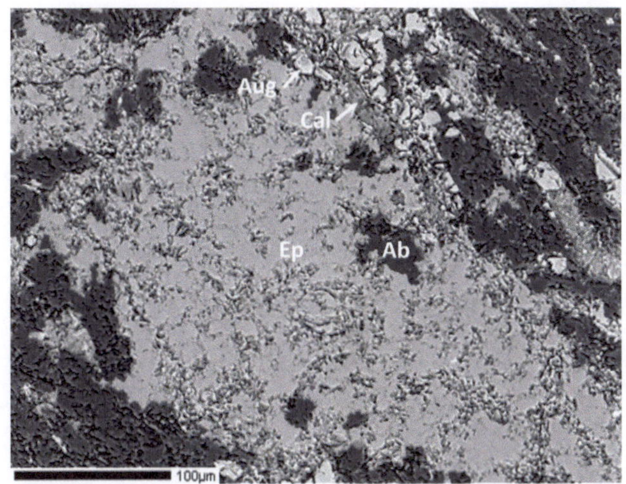

Plate Hy-23 High magnification backscattered electron image of the epidote-albite vein in metabasalt 27/79. Augite+calcite intergrowth borders the vein at its upper margin (27/79)

Plate Hy-25 Magnified view of albite in felsic vein between crossed polars containing needles and laths of skeletal albite, epidote and iron hydroxides (*red*) (27/79-9a)

Plate Hy-24 The epidote-albite vein in 27/79 between crossed polars showing linear orientation of epidote, albite and quartz as a result of late deformation and recrystallisation (27/79-9)

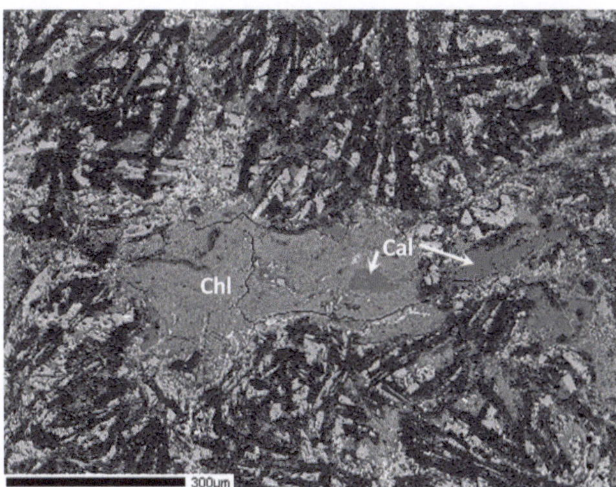

Plate Hy-26 Backscattered electron image of chlorite vein with calcite in metabasalt 27/79

20 Volcaniclastics

20.1 Vitric Tuff (Ash and Pumice)

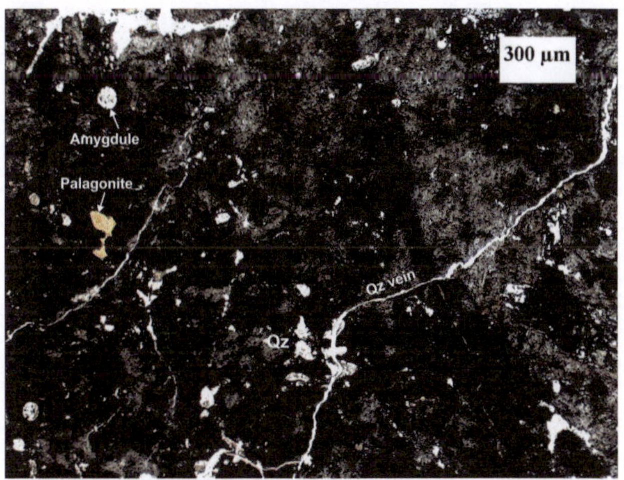

Plate Py-1 Fine tuff (ash) between crossed polars containing fine particles of quartz, orange glass and amygdule. Late silica vein runs through quartz particles (86/80-1)

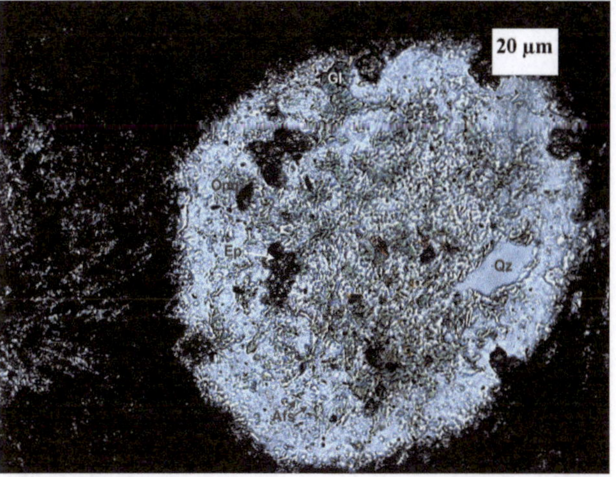

Plate Py-2 Enlarged view of an amygdale in fine ash 86/80-1 between crossed polars containing microlites of feldspar, quartz, epidote, green glass, opaque and ferromagnesian minerals (86/80-4)

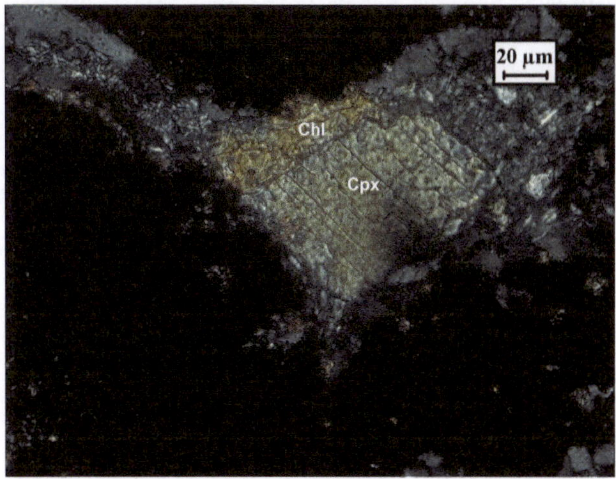

Plate Py-3 Magnified view of a clinopyroxene microcryst between crossed polars attached to a particle in fine ash 86/80-1. The pyroxene shows marginal alteration to pale brown chlorite (86/80-2)

N. C. Ghose et al., *A Petrographic Atlas of Ophiolite*,
DOI: 10.1007/978-81-322-1569-1_20, © Springer India 2014

Plate Py-4 Inter-layered glass (*dark colour*) and chert (*white*) indicating deformation. The glass is variably altered to chlorite (ml-197-1)

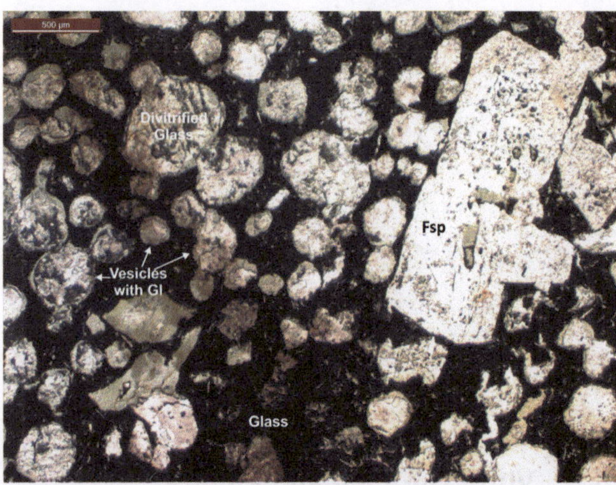

Plate Py-6 Vitric tuff contains twinned feldspar and vesicles filled with devitrified glass. The feldspar grain and the vesicles show alteration (414/79-2)

Plate Py-5 Vitric tuff between crossed polars showing devitrified glass and microlites of feldspar intruded by late veins (119/80-1)

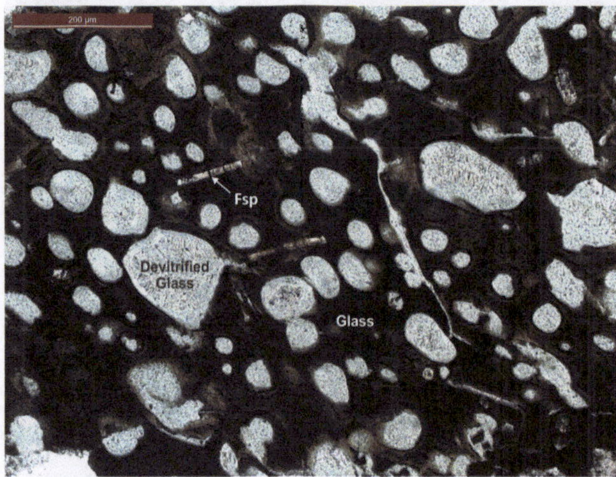

Plate Py-7 Vesicles in pumice filled with devitrified glass and rimmed by glass shards and chlorite. The glass contains microlites of feldspar (428/79-2)

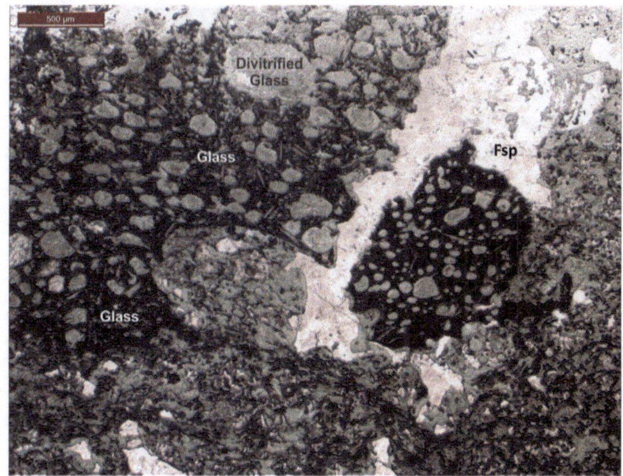

Plate Py-8 Quartzofeldspathic vein in pumice showing hyaline vesicular texture filled with devitrified glass and microlites of feldspar matrix (428/79-1)

Plate Pyr-10 Magnified view of welded crystal-vitric tuff with highly corroded feldspar and rimmed by glass (17/79-4)

20.2 Crystal-Vitric Tuff (Welded Tuff)

Plate Py-9 Welded crystal-vitric tuff containing feldspars rimmed by glass and shards. Channels and grooves inside a cuspate glass shard probably represent escape routes for volatiles upon consolidation (17/79-1)

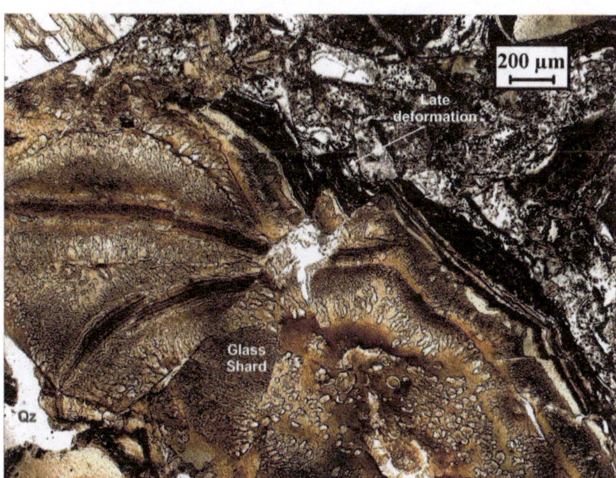

Plate Py-11 Magnified view of a cuspate glass shard in crystal-vitric tuff containing channels, grooves, and fine vesicles filled with secondary silica. Deformation has resulted in the development of microfolds and faults (17-79-3)

Plate Py-12 Magnified view of glass shard with conical edges in crystal-vitric tuff showing fine vesicles and channels filled with secondary silica (17/80-3)

Plate Py-14 Magnified view of titanite and opaques in crystal-lithic tuff in (Py-13) between crossed polars (439-4)

20.3 Crystal-Lithic Tuff

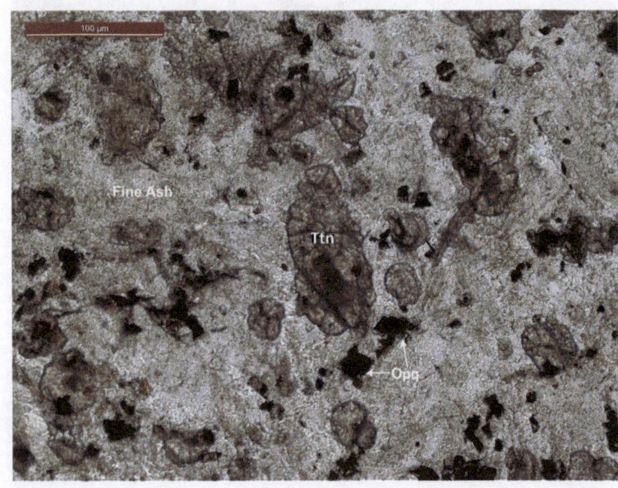

Plate Py-13 Granular aggregates of grey titanite and magnetite (*opaque*) embedded in ash of crystal-lithic (trachybasalt) tuff (refer to Py-17). Ilmenite occurs as inclusion in titanite (439-1a)

Plate Py-15 Micrographic texture in rhyolitic tuff between crossed polars (124/79-1)

Plate Py-16 Fragments of trachybasalt enveloped by feldspar vein in crystal-lithic tuff between crossed polars (426/79-1)

Plate Py-18 Late hydrothermal alteration of pyroxene and plagioclase resulting in the formation of secondary magnetite, chlorite, and epidote in crystal-lithic tuff (439-6)

Plate Py-17 Clasts of trachybasalt containing oblong-shaped particles surrounded by plagioclase crystals in crystal-lithic tuff (439-1)

20.4 Volcanic and Tectonic Breccia

Plate Py-19 Volcanic tuff breccia between crossed polars containing angular rock fragments (124-79-7)

Plate Py-20 Pseudo-tachylite vein in tectonic breccia between felsic volcanics (*grey*) and a brecciated rock (*white*) (124-79-7)

20.5 Ignimbrite

Plate Py-21 Feebly foliated ignimbrite consisting of angular quartz, feldspar and opaque in a groundmass of sericite and chlorite. Laminated grains of opaque in the basal part indicating flow (500/79-1)

Plate Py-22 Ignimbrite showing devitrified glass shards with diffused outlines, poorly sorted and angular quartz, feldspar and opaque grains embedded in chloritic groundmass. Random orientation of minerals indicates crystallisation in a plastic flow by rheomorphism (LKM3/89-5)

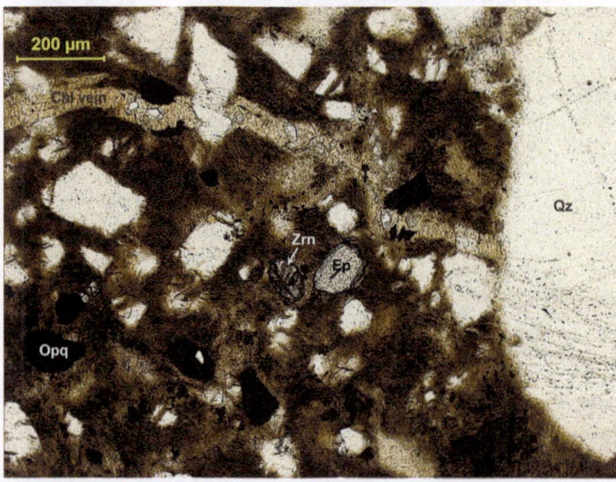

Plate Py-23 Ignimbrite containing angular quartz, opaque, epidote (high relief), and zircon (with *opaque* rim) in a groundmass of *brown* glass and shards. Quartz grains of varied dimension showing resorption. A chlorite-epidote vein cuts through the rock. A large sub-rounded grain of quartz is seen on *right* (LKM4/89-3)

Plate Py-24 Ignimbrite with angular and corroded quartz, skeletal and disseminated opaque, and zircon set in a groundmass containing green chlorite admixed with fine brown glass (LKM5/89-1)

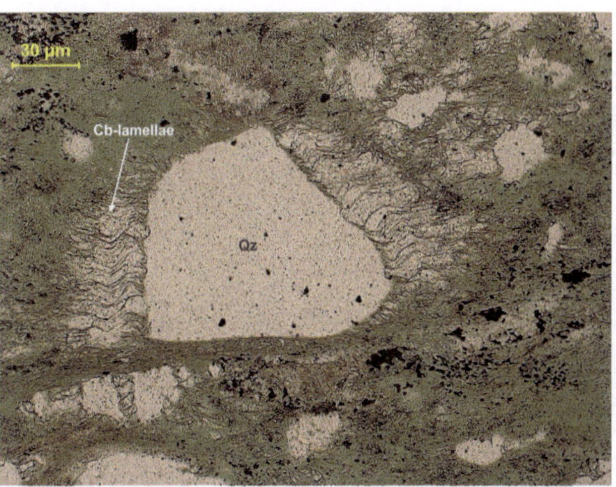

Plate Py-26 Magnified view of symmetrical pressure fringes showing development of calcite fibres perpendicular to the face of quartz in ignimbrite (LKM3/89-1)

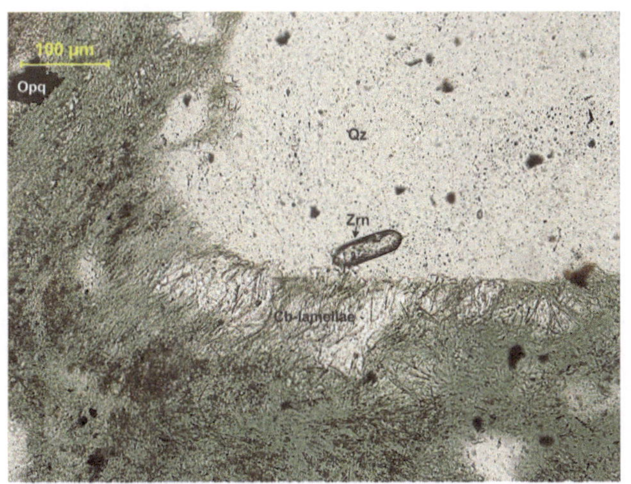

Plate Py-25 Magnified view of inclusion of zircon in sub-rounded quartz showing simple pressure fringes with growth of calcite fibres (carbonate-Cb) perpendicular to quartz in ignimbrite. The groundmass is made of chlorite (LKM3/89-6)

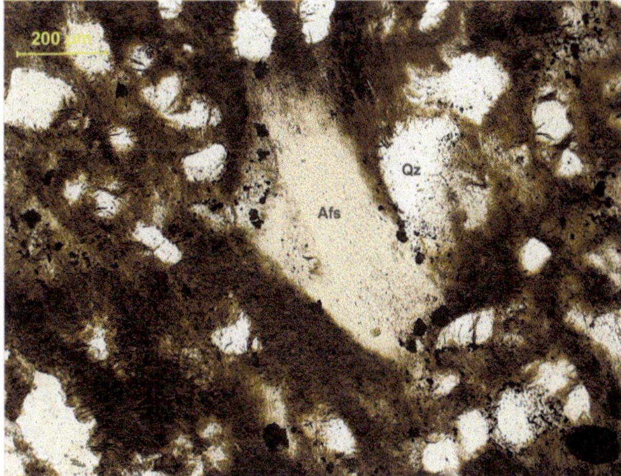

Plate Py-27 Angular and resorbed quartz and perthitic alkali feldspar surrounded by disseminated grains of opaque in brown glassy groundmass with admixture of shards in ignimbrite (LKM4/89-5)

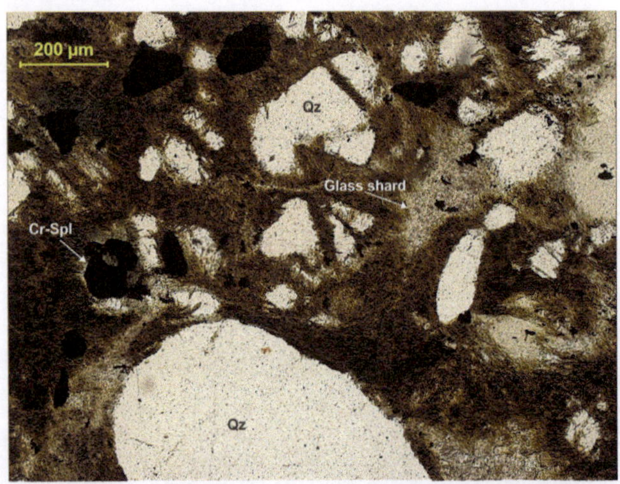

Plate Py-28 Ignimbrite containing resorbed grains of bimodal quartz and opaques with green halo. The groundmass is composed of brown glass, glass shard and chlorite (LKM4/89-2)

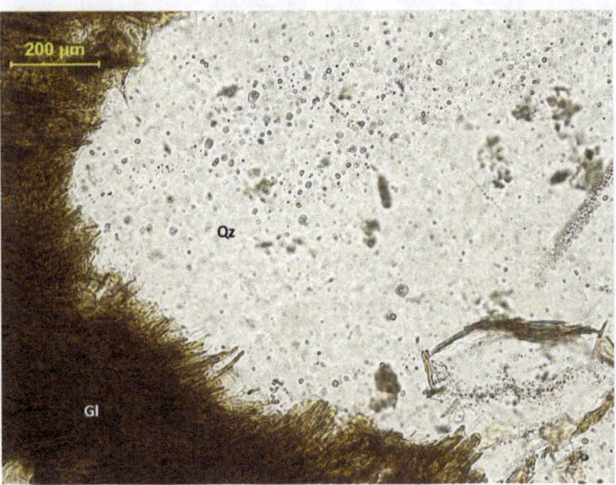

Plate Py-29 A close view of fibrous brown glass in contact with a large resorbed porphyroclast of quartz in ignimbrite (LKM4/89-6)

Plate Py-30 Ignimbrite showing broken and angular quartz fragments with interstitial carbonates surrounded by a groundmass of brown glass and devitrified glass shards (LKM4/89-4)

21 Very-Low Grade Metamorphics

21.1 Chlorite-Natrolite-Albite Association

Plate Vlgm-1 Very-low grade, zeolite facies metamorphite between crossed polars containing spherulitic natrolite, twinned albite, epidote, *green* chlorite and opaque (K-79-7)

Plate Vlgm-2 Low temperature recrystallisation of pyroxene to *green* chlorite and opaque (Fe-oxide) in zeolite facies rock (K-79-5)

Plate Vlgm-3 Chlorite-natrolite-albite rock K-79-7 between crossed polars showing radiating crystals of natrolite and twinned albite (K-79-3)

Plate Vlgm-4 Intergranular texture in epidote-chlorite-natrolite-albite rock between crossed polars (K-79-1)

N. C. Ghose et al., *A Petrographic Atlas of Ophiolite*,
DOI: 10.1007/978-81-322-1569-1_21, © Springer India 2014

Plate Vlgm-5 Porphyroblast of sassuritised albite in very-low grade zeolite facies metamorphite between crossed polars containing epidote (*orange*) and chlorite (*green*) (K-79-4)

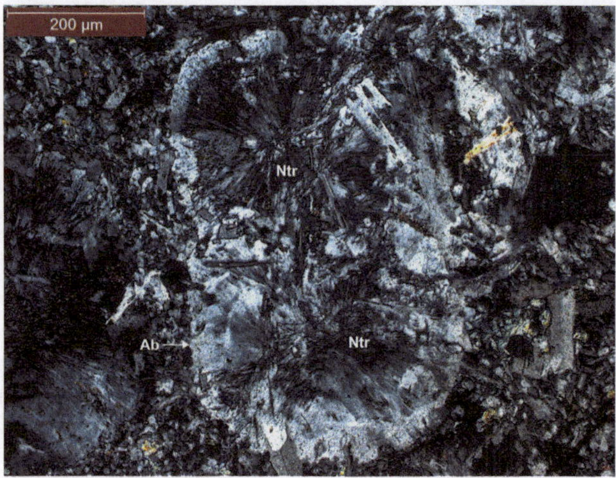

Plate Vlgm-6 Transformation of albite to radial acicular natrolite between crossed polars retaining the original morphology. Relict twinning of albite may be seen at the edges (K-79-6)

21.2 Chlorite-Prehnite Schist

Plate Vlgm-7 Sheaf-like aggregates of prehnite and fine flakes of chlorite (dark) in chlorite-prehnite schist containing clinozoisite and opaque as accessory minerals (390-79-1)

Plate Vlgm-8 Contact between radiating prehnite blades and chlorite between crossed polars showing interfingering relationship (390-79-2)

Plate VIgm-9 'Owl's eye' texture in chlorite-prehnite schist between crossed polars showing radiating blades of birefringent prehnite around a core of chlorite (*dark*) (390-79-5)

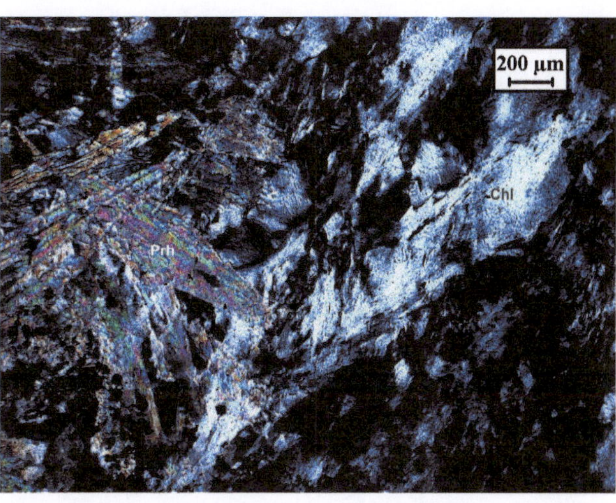

Plate VIgm-10 Magnified view of chlorite-prehnite schist 390-79-5 between crossed polars (390-79-2)

22 Greenschist

Plate GS-1 Greenschist consisting of actinolite, chlorite, albite, epidote and quartz. The amphiboles impart lineation (23/80-1)

Plate GS-2 Finely foliated greenschist containing rows of *brown* epidote + *green* chlorite, and *bluish green* actinolite between quartzofeldspathic (albite) bands (43/80-1)

Plate GS-3 Three phases of deformations in greenschist traversed by quartz veins and secondary ferruginous fillings (43/80-3)

Plate GS-4 Finely foliated greenschist 43/80-3 between crossed polars (43/80-4)

N. C. Ghose et al., *A Petrographic Atlas of Ophiolite*,
DOI: 10.1007/978-81-322-1569-1_22, © Springer India 2014

Plate GS-5 Coarse actinolite schist dominated by long prisms of bluish *green* actinolite and albite. Quartz, epidote, opaque and sphene are present as accessory phases (22/80-1)

Plate GS-7 Coarse meta-pyroxenite (140/80-2) containing *brown* fibrous aggregates of secondary actinolite (Act_2) between the grain boundaries of *greyish brown* actinolite (Act_1). Minor amount of sphene is uniformly distributed (140/80-3)

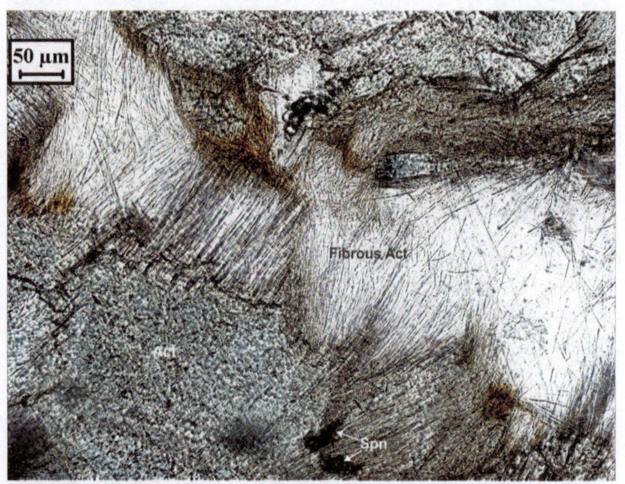

Plate GS-6 Magnified view of fibrous actinolite in greenschist 22/80-1. Iron released through retrograde reaction is recombined to form iron oxide and sphene at grain boundaries (22-80-3)

Plate GS-8 Magnified view of *brown* secondary actinolite (Act_2) recrystallised at grain boundaries of primary iron-rich clinopyroxene/actinolite (Act_1) (140/80-4)

Plate GS-9 Metabasalt showing granoblastic texture and containing actinolite, chlorite, albite, opaque and epidote (y-38-1)

Plate GS-10 Metabasalt y-38-1 between crossed polars (y-38-2)

23 Glaucophane Schist

23.1 Epidote-Chlorite-Phengite-Glaucophane-Omphacite Schist

Plate Gln-1 Low magnification backscattered electron image of *blueschist* showing *top left-bottom right* foliation and porphyroblasts of omphacite, glaucophane, epidote, phengite and chlorite. Minor calcite occurs in pockets and minor albite occurs only in retrograde, late-stage veins. Accessory rutile is partially altered to titanite (212/80)

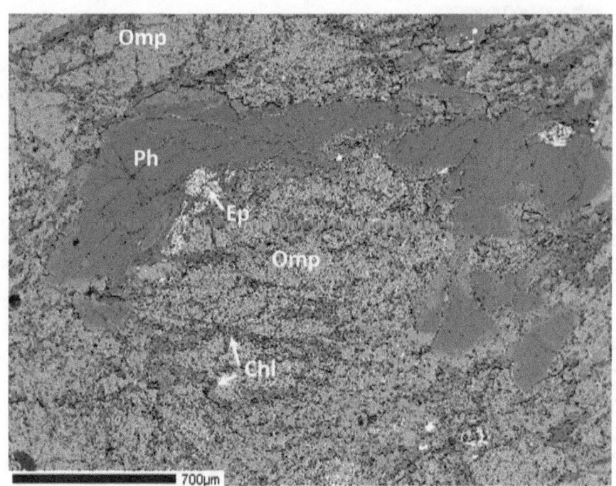

Plate Gln-2 Backscattered electron image showing intergrowth of omphacite with chlorite alterations, epidote and phengite in *blueschist* (212/80)

Plate Gln-3 Backscattered electron image of *blueschist* showing porphyroblasts of epidote, chlorite, omphacite and glaucophane in equilibrium. Rutile occurs as an accessory mineral. Tiny needles of titanite inclusions are present within chlorite (212/80)

N. C. Ghose et al., *A Petrographic Atlas of Ophiolite*,
DOI: 10.1007/978-81-322-1569-1_23, © Springer India 2014

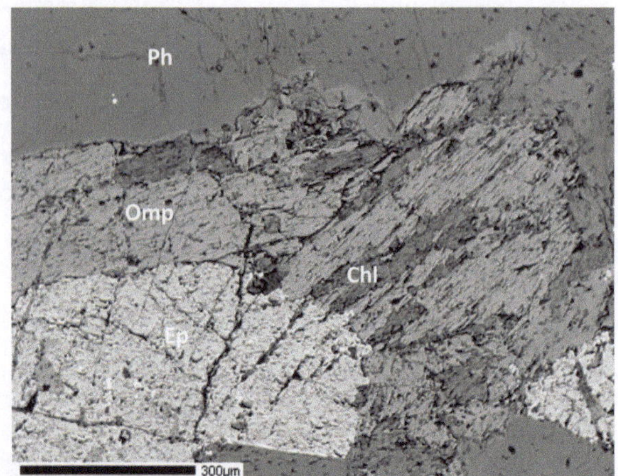

Plate Gln-4 Backscattered electron image showing omphacite+-chlorite intergrowth between epidote and phengite in *blueschist* (212/80)

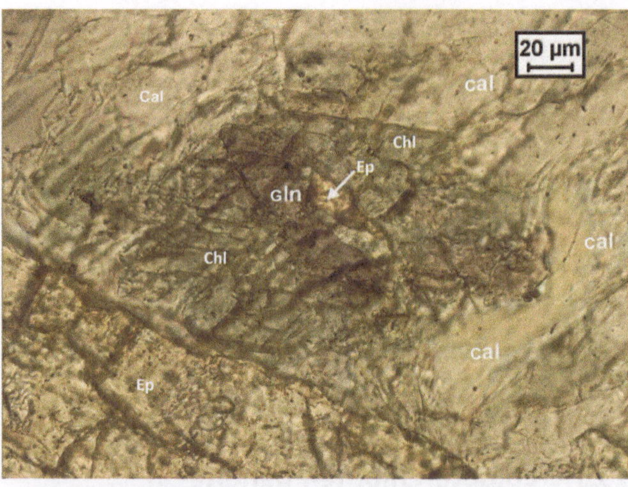

Plate Gln-7 Glaucophane-epidote intergrowth flanked by chlorite and surrounded by calcite and epidote porphyroblast in Ep-Chl-Ph-Gln-Omp schist (212/80-9)

Plate Gln-5 Intergrowth of glaucophane, chlorite and zoisite at the margin of prismatic calcite in Ep-Chl-Ph-Gln-Omp schist (212/80-3)

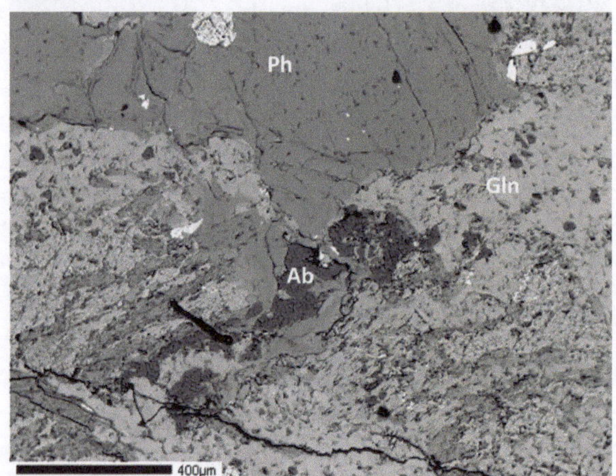

Plate Gln-8 Magnified backscattered electron image of late-stage albite-phengite veins in *blueschist*. The veins are surrounded by porphyroblasts of glaucophane and phengite (212/80)

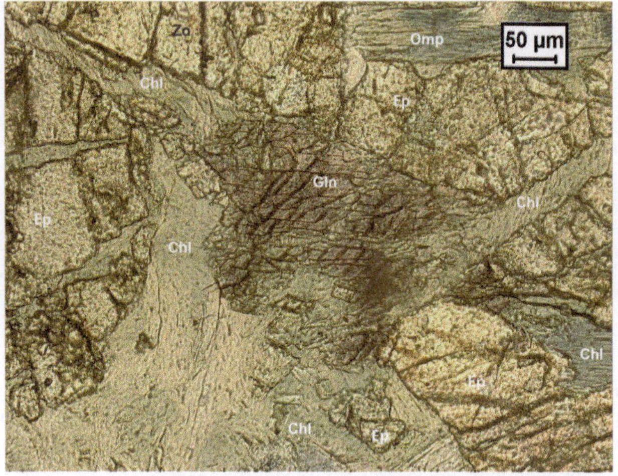

Plate Gln-6 Late-stage chlorite veins cutting through glaucophane-epidote intergrowth in Ep-Chl-Ph-Gln-Omp schist (212/80-11)

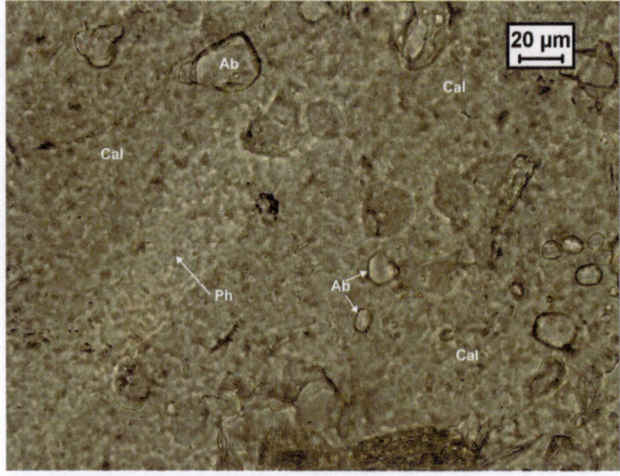

Plate Gln-9 Small crystals of albite in a phengite-calcite matrix in Ep-Chl-Ph-Gln-Omp schist (212/80-15)

Plate Gln-10 Small crystals of albite in a carbonate matrix cross-cut by phengite flakes in Ep-Chl-Ph-Gln-Omp schist between crossed polars (212/80-12)

Plate Gln-12 Glaucophane needles and prismatic rutile in chlorite-epidote-phengite-barroisite-glaucophane schist (ML-69-1)

23.2 Chlorite-Albite-Epidote-Actinolite-Glaucophane Gneiss/Schist

Plate Gln-11 Rhombic idioblastic crystals of late secondary blue amphibole (Amp_2) across the main foliation in quartzofeldspathic gneiss containing glaucophane (Amp_1), phengite, chlorite and epidote (X-11-1)

Plate Gln-13 Quartz vein showing pinch-and-swell structure in chlorite-albite-epidote-green amphibole-glaucophane schist dominated by epidote (N7-1)

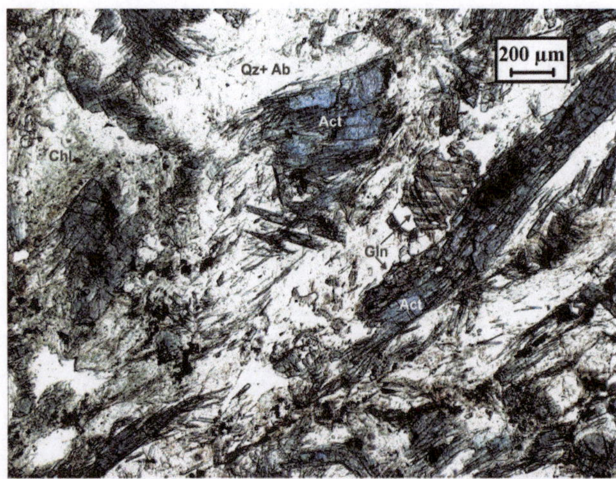

Plate Gln-14 Intergrowth of glaucophane (*purple*) and *bluish green* actinolite in epidote-phengite-quartz-rutile-calcite-titanite-actinolite-glaucophane meta-arenite (2-70)

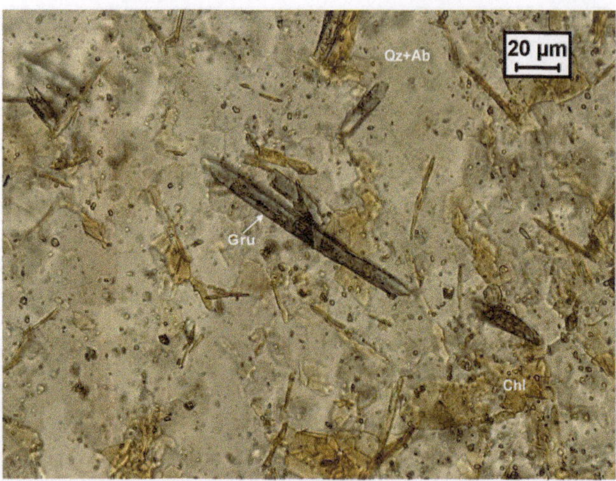

Plate Gln-16 High magnification image of grunerite in metachert 855-1 between crossed polars containing quartz, albite, *brown* chlorite and opaque (855-3)

23.3 Blue Amphibole-Bearing Ferruginous Metachert and Arenite

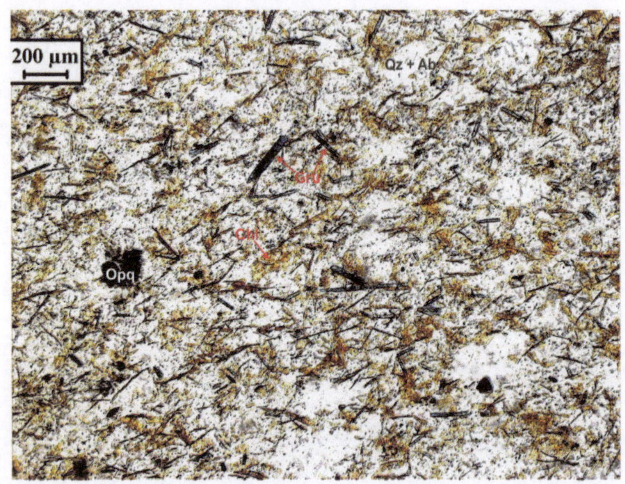

Plate Gln-15 Fine-grained metachert containing brown chlorite, opaque, needles of blue amphibole (Mn-bearing grunerite) and quartz (855-1)

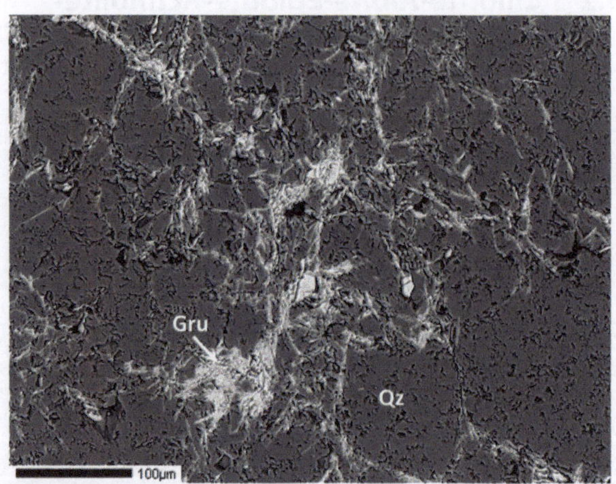

Plate Gln-17 Backscattered electron image showing Mn-bearing grunerite (and minor ferro-anthophyllite) amphibole needles along grain boundaries and as inclusions within quartz in metachert 855

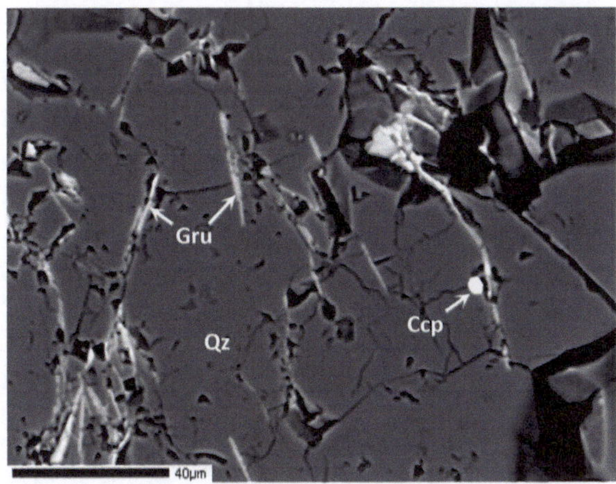

Plate Gln-18 Backscattered electron image showing accessory chalcopyrite in metachert with grunerite needles along quartz grain boundaries. Accessory galena and apatite are also present (not shown) and show similar textural relations (855)

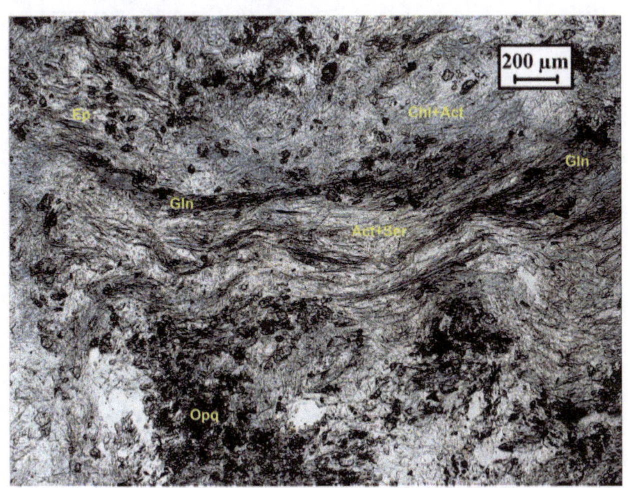

Plate Gln-19 Broad warp in ferruginous meta-arenite containing epidote, sericite, chlorite, actinolite, quartz and glaucophane (M-23A-4)

Plate Gln-20 Glaucophane-rich band between crossed polars containing epidote, opaque and chlorite in fine-grained meta-arenite (M23A)

23.4 Epidote-Garnet-Glaucophane Schist

Plate Gln-21 Two generations of glaucophane in Zo-Ep-Grt-Act-Gln schist: **a** short prismatic crystals along the foliation (Gln_1), and **b** broad and stout, platy crystals of late origin (Gln_2) across the foliation. Late crystallised granular *red garnets* are randomly distributed across the foliation (620/79-4)

Plate Gln-22 Idioblastic garnet grains of different shapes in Zo-Ep-Grt-Act-Gln schist of above. A large porphyroblast of epidote is in contact with the glaucophane and idioblastic garnet on the *left* (620/79-6)

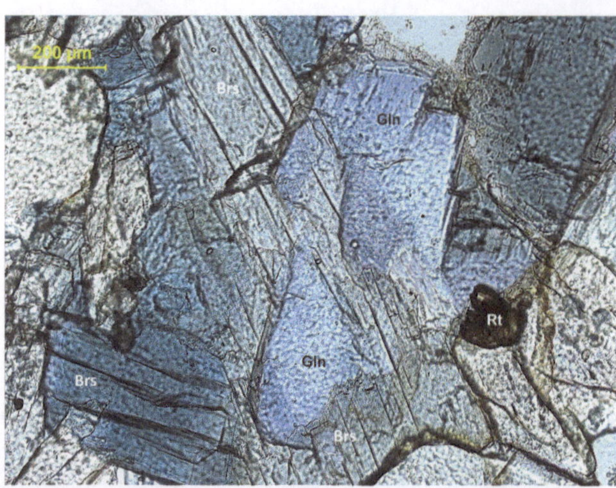

Plate Gln-24 Interpenetration twinning between prismatic barroisite and glaucophane in epidote-garnet-glaucophane-barroisite schist 139/79 (139/79-4)

Plate Gln-23 Epidote-garnet-glaucophane-barroisite schist showing a granoblastic texture with minor garnet, phengite, rutile and late albite (139/79-1)

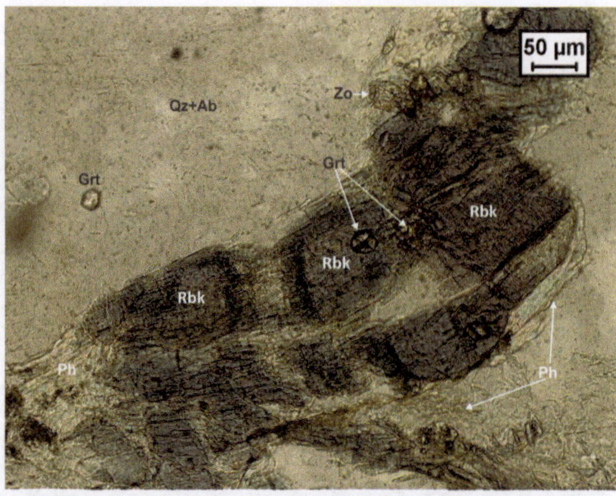

Plate Gln-25 Boudins of zoned sodic amphibole (riebeckite) separated by phengite parallel to foliation in *blueschist*. The matrix is composed of quartz and albite/K-feldspar interspersed with idioblastic garnet and zoisite (Z-74-1)

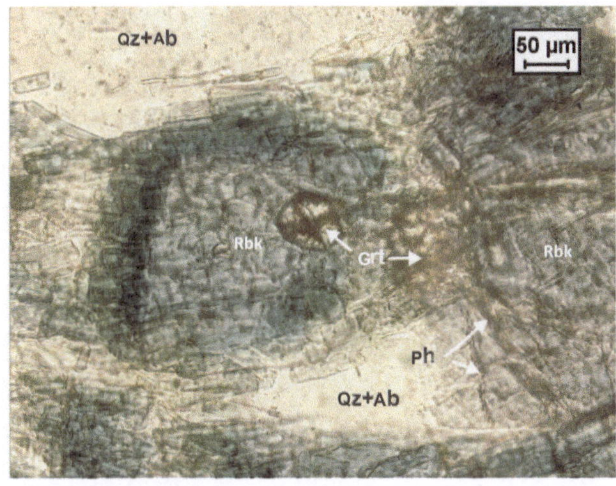

Plate Gln-26 Magnified view of riebeckite boudins swathed by phengitic mica in *blueschist* Z-74. Idioblastic garnet occurs both within and outside the boudins (refer to Gln-27, 28, 29) (Z-74-2)

Plate Gln-28 Backscattered electron image of *blueschist* showing an intergrowth of compositionally zoned sodic amphibole (*core* magnesio-riebeckite, rim: winchite), phengite, quartz and idioblastic almandine-spessartine garnet of a second generation (Grt_2) (Z74)

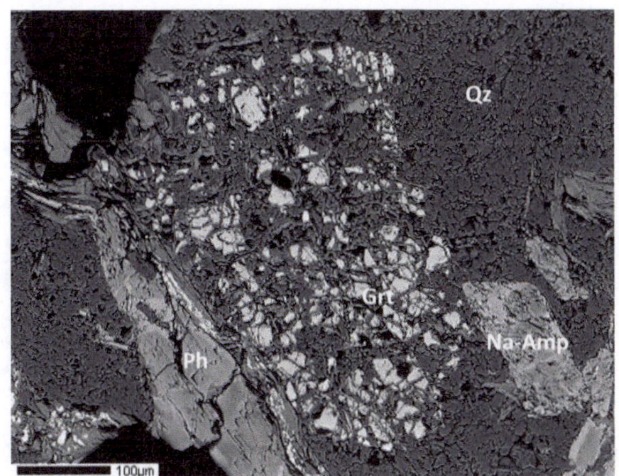

Plate Gln-27 Backscattered electron image showing a highly resorbed almandine-spessartine garnet porphyroblast of an early generation (Grt_1) surrounded by porphyroblasts of compositionally zoned sodic amphibole, quartz and phengite (Z74)

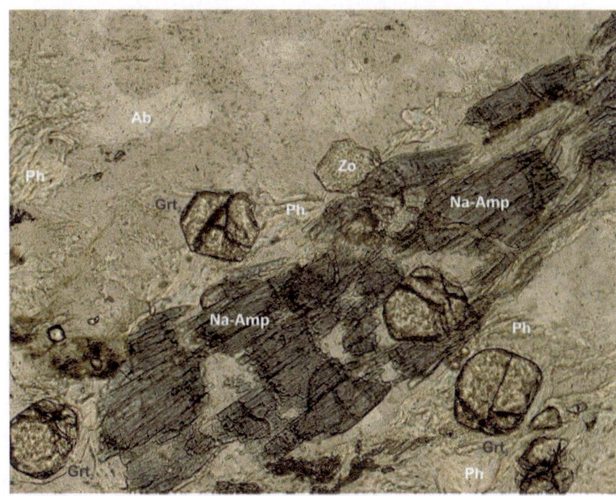

Plate Gln-29 Development of the second generation idioblastic garnet (Grt_2) across the foliation in *blueschist* Z-74 (Z-74-3)

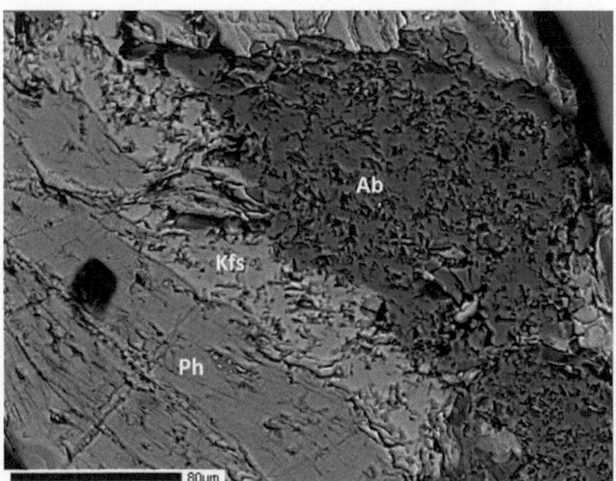

Plate Gln-30 Backscattered electron image showing albite associated with K-feldspar bordering phengite in *blueschist* Z74

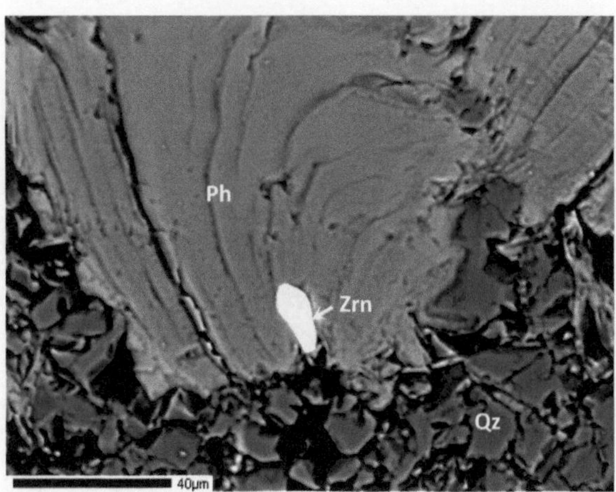

Plate Gln-31 Backscattered electron image of a zircon inclusion in phengite in *blueschist* Z74

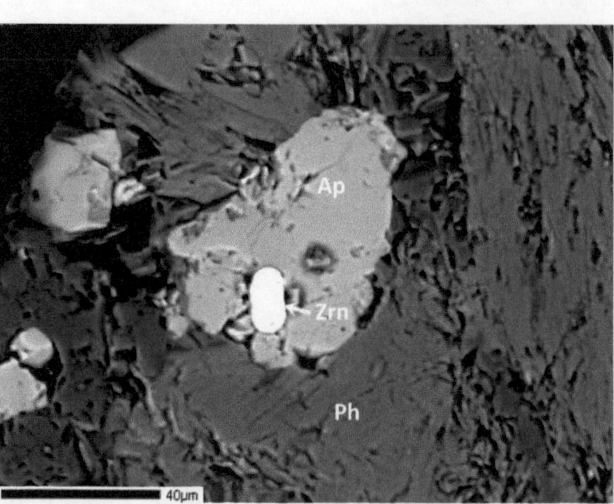

Plate Gln-32 Backscattered electron image showing an inclusion of apatite + zircon in phengite in *blueschist* Z74

23.5 Mylonitic Glaucophane Schist

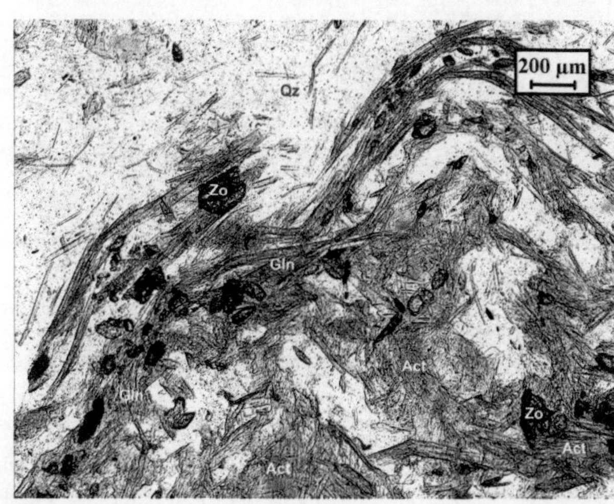

Plate Gln-33 Small warp in albite-phengite-actinolite-glaucophane schist with minor zoisite, magnetite and sphene in meta-arenite (ML-225-4)

Plate Gln-34 S–C mylonite in zoisite-actinolite-glaucophane schist ML-225 between crossed polars showing development of 'mica-fish' indicating shear along flattened quartz (ML-225-2)

Plate Gln-36 'Mica fish' in meta-arenite 78/80 between crossed polars (78/80-3)

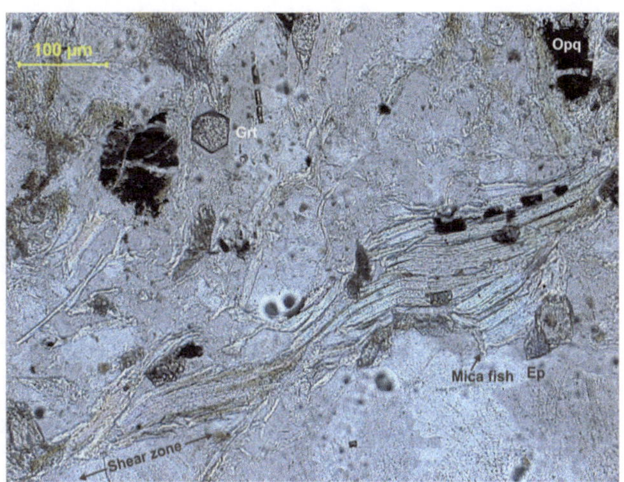

Plate Gln-35 'Mica-fish' in garnet-epidote-opaque-actinolite-glaucophane meta-arenite indicating ductile deformation (78/80-1)

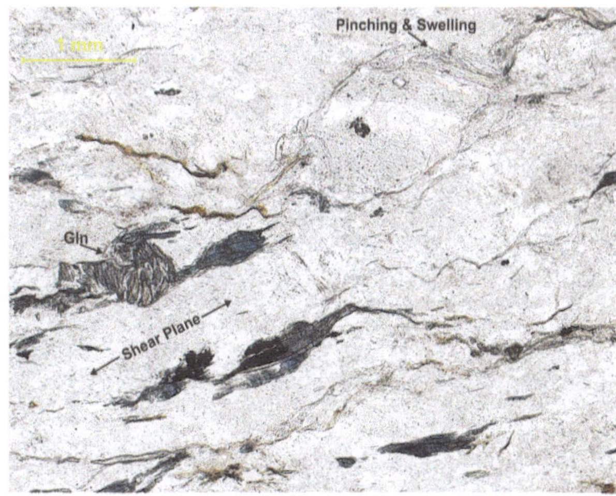

Plate Gln-37 Refolded glaucophane and quartz with pinch-and-swell structure in garnet-glaucophane mylonitic gneiss. One fold axis runs NW–SE and the other is nearly EW, passing through the knee of glaucophane (622/79-1)

Plate Gln-38 Magnified view of refolded glaucophane in garnet-bearing mylonitic gneiss 622/79 (622/79-1)

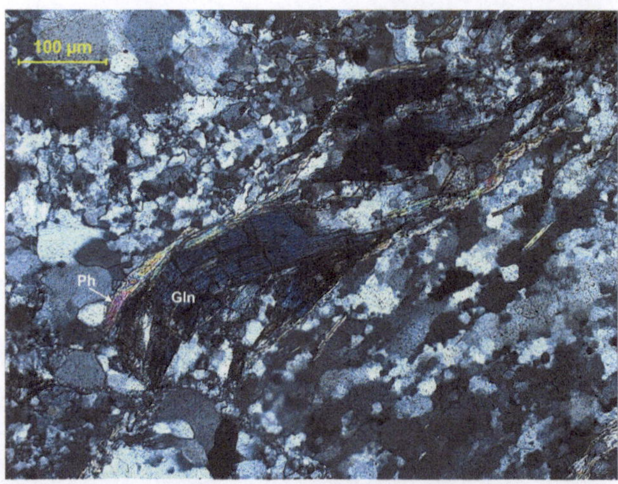

Plate Gln-40 Magnified view of prismatic glaucophane swathed by phengite in shear zone of mylonitic garnet-glaucophane gneiss 622/79-2 indicating ductile deformation (622/79)

23.6 Ultra-Cataclastic Glaucophane Schist

Plate Gln-39 Two generations of foliation planes (F_1, F_2) in mylonitic garnet-glaucophane gneiss. The younger one cross-cuts the shear plane at an acute angle (622/79-2)

Plate Gln-41 Brecciated glaucophane schist consisting of phengite-rutile-zoisite-epidote-actinolite-glaucophane and traversed by secondary quartz and carbonate veins (N-146-6)

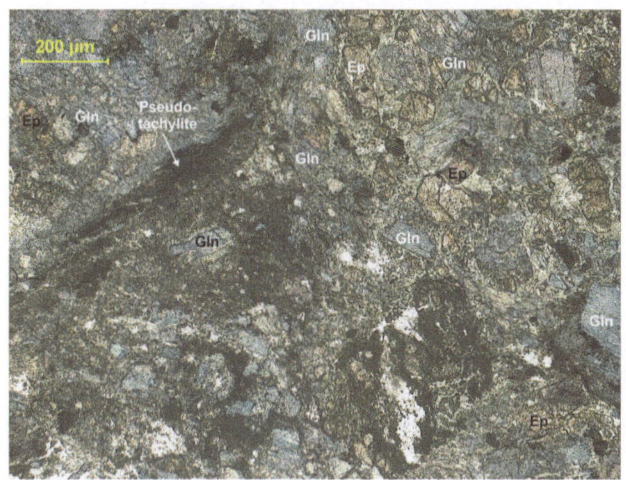

Plate Gln-42 Brecciated glaucophane schist N-146-6 with dark branching veins of pseudotachylite with fragmental material indicating melt formation at high temperature (N-146-1a)

Plate Gln-44 S–C mylonite in brecciated glaucophane schist N-146-6. The glaucophane and epidote crystals are tapered at both ends similar to phengite indicating their involvement in shearing (N-146-3)

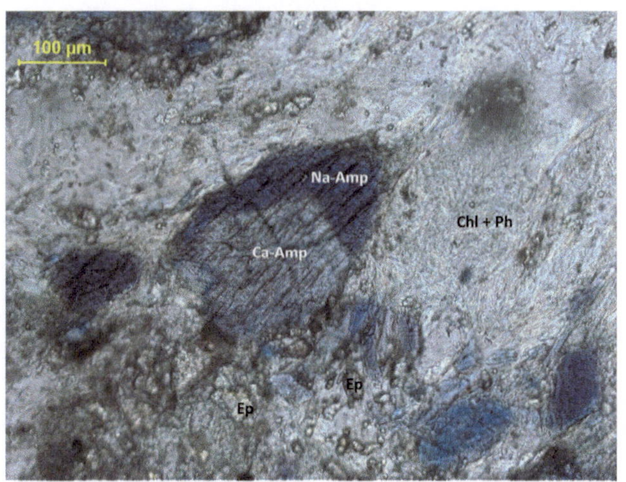

Plate Gln-43 Porphyroblast of amphibole in brecciated glaucophane schist N-146-6 showing transformation between *greenish blue* and *purple* amphiboles. The groundmass is composed of chlorite, phengite and epidote (N-146-1)

Plate Gln-45 Two generations of glaucophane in actinolite-glaucophane schist: **a** prismatic grains along foliation plane (Gln_1) and **b** post-tectonic rhombic crystals with basal cleavage as (Gln_2) (76/80-5)

Plate Gln-46 Actinolite-glaucophane schist 76/80-5 between crossed polars showing micro-shear defined by actinolite, and brecciation and growth of new crystals with triple-junction contacts (76-80-3)

Plate Gln-48 Mortar texture in glaucophane schist 128-III-1a produced by cataclasis, a process of mechanical crushing of large matrix crystals and recrystallisation of *green* chlorite at grain boundaries. *Note* the crushed grains forming a rim around epidote. Random orientation of granular epidote indicates that these are post-tectonic porphyroblasts (128-III-9)

Plate Gln-47 Ultra-cataclastite glaucophane schist showing pseudo-tachylite vein with fragmental material produced by compression and shearing. *Brown* granular grains are epidote (128-III-1a)

Plate Gln-49 Formation of 'glaucophane-fish' in ultra-cataclastic glaucophane schist. A quartzofeldspathic vein runs parallel to the shear plane. Thrusting and buckling under strain might be co-eval (128 III-2)

Plate Gln-50 Two generations of glaucophane in cataclastic glaucophane schist: **a** deep *blue* prismatic grains along foliation and **b** post-tectonic, stout and platy crystals developed across foliation. Lenticular crystals of *dark brown* epidote are aligned parallel to the foliation (128III-4)

24 Eclogite

Plate Ec-1 Eclogite with granoblastic texture and an assemblage containing barroisite, glaucophane, chlorite, epidote, rutile, quartz and secondary albite (77/80-1)

Plate Ec-2 Backscattered electron image of weakly foliated eclogite with porphyroblasts of omphacite, barroisite/glaucophane amphiboles, epidote and chlorite. Minor phengite, rutile/titanite and quartz and late albite (in vein) are also present (77/80)

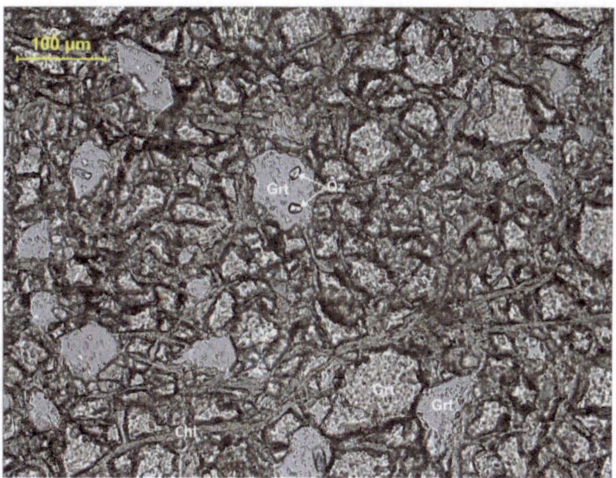

Plate Ec-3 A portion of a large, highly resorbed garnet porphyroblast in eclogite showing secondary chlorite veins cutting through islets (*pieces*) of the original garnet containing quartz inclusions (77/80-7)

N. C. Ghose et al., *A Petrographic Atlas of Ophiolite*,
DOI: 10.1007/978-81-322-1569-1_24, © Springer India 2014

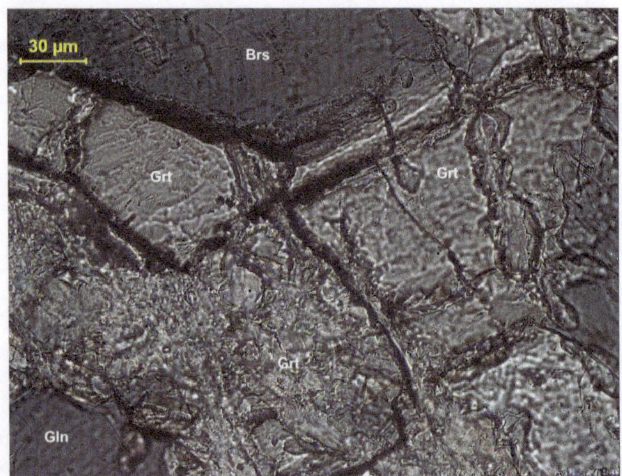

Plate Ec-4 High magnification image of resorbed garnet porphyroblast containing inclusions of barroisite and glaucophane (77/80-3)

Plate Ec-6 Backscattered electron image showing an omphacite porphyroblast with epidote inclusions in eclogite (77/80)

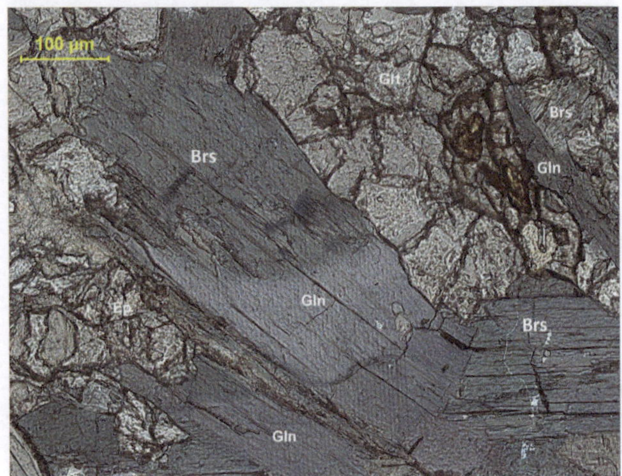

Plate Ec-5 Eclogite showing prismatic grains of barroisite/glaucophane intergrowing with epidote at the edge of a garnet porphyroblast that also contains inclusions of barroisite/glaucophane (77/80-4)

Plate Ec-7 Interpenetration texture between glaucophane and barroisite in association with epidote in eclogite (77/80-2)

Plate Ec-8 Transformation of blue-green sodic-calcic amphibole barroisite to purplish sodic amphibole glaucophane in eclogite also containing epidote and rutile (77/80-2)

Plate Ec-10 Backscattered electron image showing epidote porphyroblast with secondary veins of phengite and quartz in eclogite (77/80)

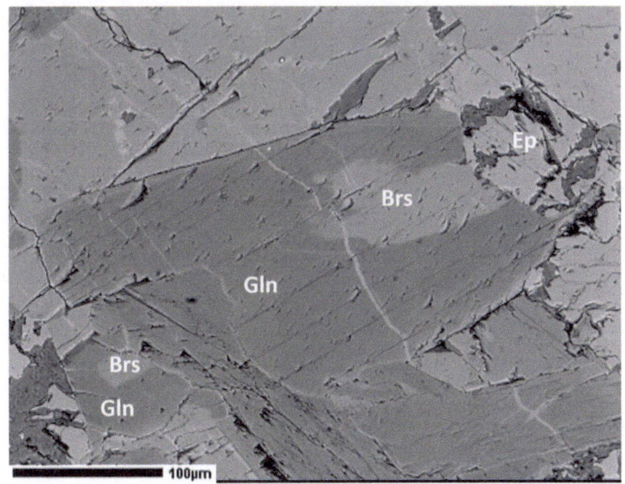

Plate Ec-9 Backscattered electron image showing amphibole grains with barroisite core and thick glaucophane rim in eclogite (77/80)

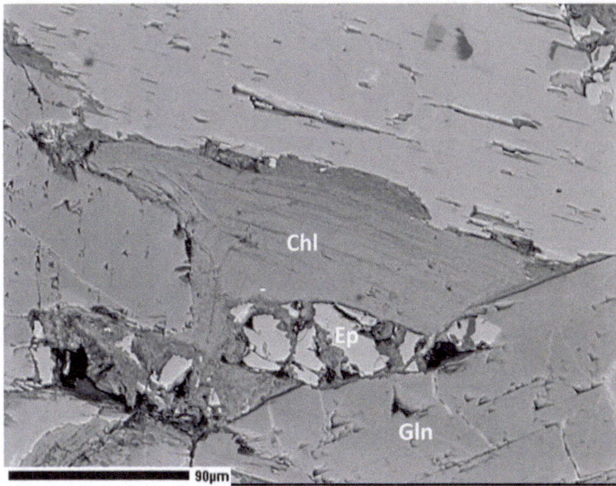

Plate Ec-11 Backscattered electron image showing early formed chlorite along foliation in eclogite (77/80)

Plate Ec-12 Spherulitic chlorite in association with epidote in eclogite (77/80-3)

Plate Ec-13 Large flakes of phengitic mica in association with barroisite/glaucophane and epidote in eclogite (77/80-4)

Plate Ec-14 A cataclastic zone in eclogite showing multiple veins of albite cutting through brecciated clusters of epidote, glaucophane and barroisite (77/80-5)

25 Oceanic Sediments

25.1 Tuffaceous Sediments

Plate OS_T-1 Pressure solution fracture in tuffaceous sediment. Slippage and plastic flow developed by ino-silicates under stress such that the strain is limited to a secondary fabric (507/79-1)

Plate OS_T-2 Randomly oriented very fine needles of amphibole (*green actinolite*) in fine ash bed leading to disjunctive foliations. Both the large and fine grains lack any preferred orientation (507/79-2)

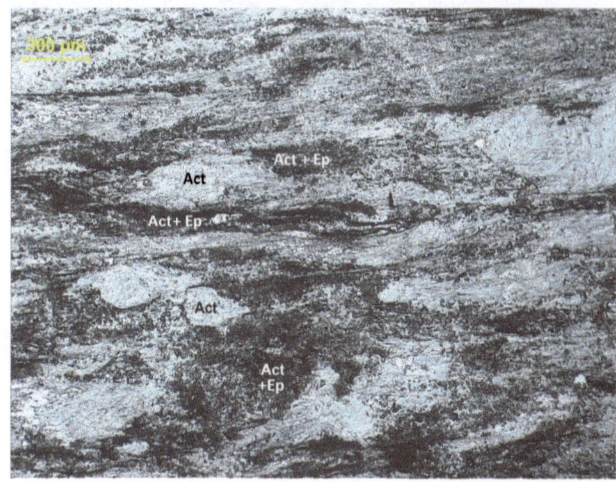

Plate OS_T-3 Very fine-grained epidote-actinolite-albite schist derived from tuffaceous source of basaltic composition under green-schist facies in the oceanic crust (< 500 °C at 3–4 kms depths). Dark patches are aggregates of acicular actinolite and granular epidote (35-3)

N. C. Ghose et al., *A Petrographic Atlas of Ophiolite*,
DOI: 10.1007/978-81-322-1569-1_25, © Springer India 2014

Plate OS_T-4 Magnified view of above rock (OS_T-3) showing actinolite needles at the centre and granular epidote below. The *grey* layer at the bottom is essentially composed of a mixture of actinolite and epidote. Rectangular ferroactinolite is pleochroic from *pale greenish yellow* to *green* (35-2)

Plate OS_T-6 A close view of albite and veins of carbonate in mafic tuff containing actinolite, tremolite and opaque. The carbonate-albite veins represent advanced alteration stage (N183-5)

Plate OS_T-5 Very fine-grained mafic tuff between crossed polars consisting of randomly oriented fine needles of actinolite and tremolite, albite and opaque. This is an example of deformation during crenulation cleavage formation (N183-4)

25.2 Argillaceous Sediments

Plate OS_Arg-1 Phyllite showing small-scale folding with angular and corroded quartz of volcanic origin. The bedding circumvents the quartz grain. Foliation has developed in the phyllite due to recrystallisation of clay minerals (567/78-1)

Plate OS_Arg-2 Convolute bedding in phyllite caused by minor disturbance during sedimentation or due to escape of fluids resulting in crumpling of beds (567/78-2)

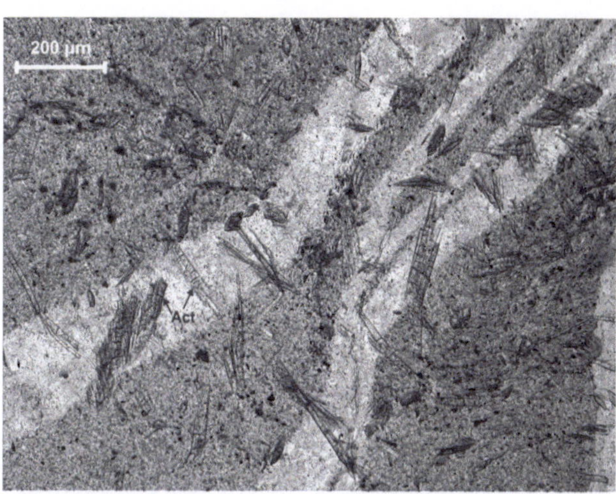

Plate OS_Arg-4 Late crystallisation of actinolite in siliceous argillite developed under greenschist facies (56-1). Note the formation of amphibole across quartz vein which is due to thermal relaxation, probably accompanied by uplift and erosion (56-2)

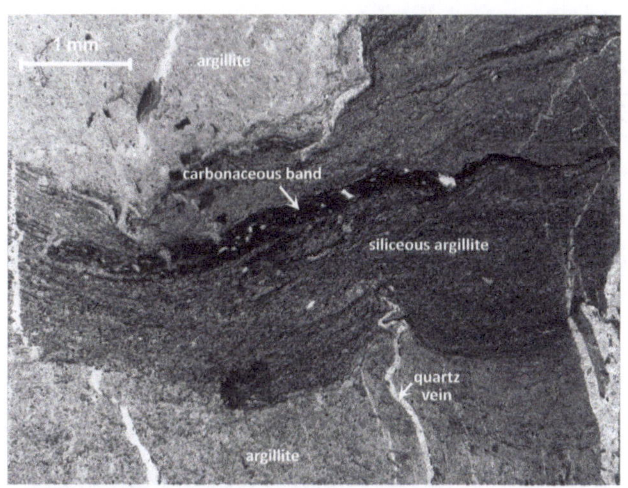

Plate OS_Arg-3 Load cast structure developed by protuberance of silicified (*dark grey*) argillite containing a band of carbonaceous material formed under hypoxic environment into soft (*light grey*) laminated argillite/clay band. Disruption of siliceous veins indicates turbidity flow (56-1)

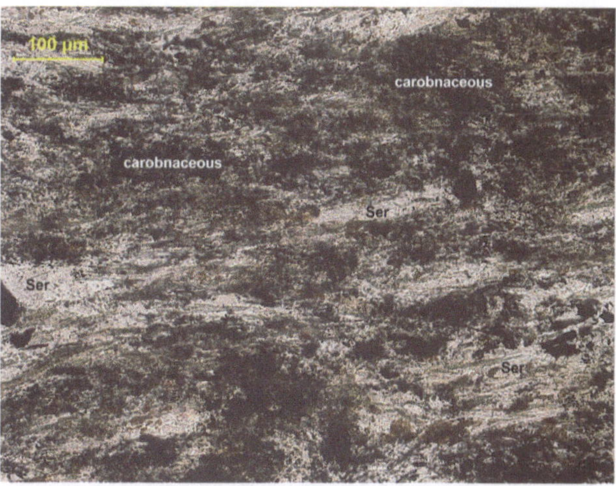

Plate OS_Arg-5 Impersistent layers of quartz-muscovite (*light coloured*) and carbonaceous material-sericite (*dark coloured*) in carbonaceous phyllite indicating hypoxic condition of deposition during sedimentation (473/79-1)

Plate OS_Arg-6 Fine-grained quartz-biotite-muscovite schist consisting of fine flakes of mica, and flattened, corroded and angular quartz of volcanic ancestry. Chlorite, epidote, opaque and zircon constitute minor components. Minor faulting at the *top left* has caused dislocation of the foliation (109-1)

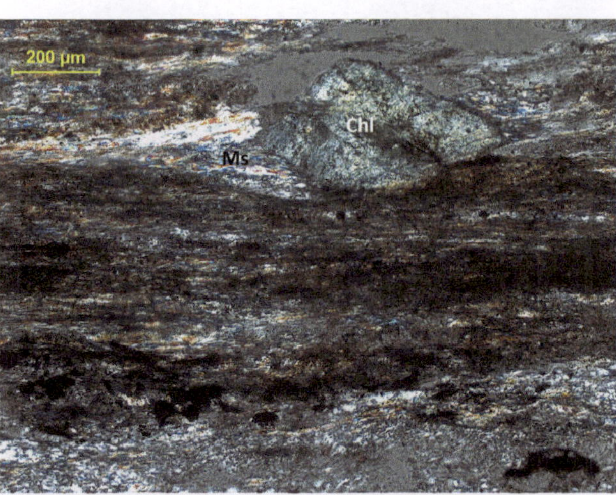

Plate OS_Arg-8 Alternation of metamorphosed argillite and siliceous rock (*dark*) between crossed polar. The *top* and *bottom* layers of argillite are dominated by mica indicating very high settling time for the deposited sediments. A porphyroblast of chlorite is present at the *top* (592-2)

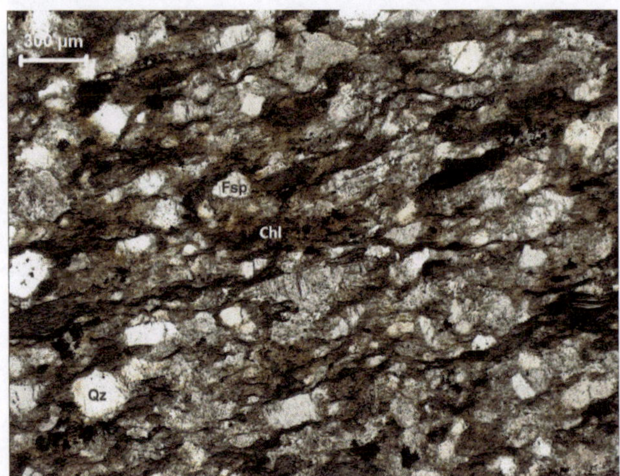

Plate OS_Arg-7 Fine-grained feldspathic schist (meta-arkose) containing feldspar, sericite, opaque, chlorite and quartz (565/95-1)

25.3 Arenaceous Sediments

Plate OS_Sst-1 Deformed sandstone between crossed polars showing development of tectonic foliation. The rock consists of angular to sub-angular quartz showing anomalous extinction and minor sub-rounded potash feldspar and plagioclase (458/79-1)

Plate OS_Sst-2 Feldspathic wacke between crossed polars showing tectonic foliation with development of a weak S–C fabric in the central part. Quartz shows anomalous extinction. Perthitic feldspar dominates over plagioclase (446/79-1)

Plate OS_Sst-4 Vaguely foliated feldspathic sandstone between crossed polars showing granulation along grain boundaries of randomly oriented quartzofeldspathic minerals. The rock contains quartz, microcline and plagioclase, showing a quartz dominated residual framework. Potash feldspar dominates over plagioclase (440/79-1)

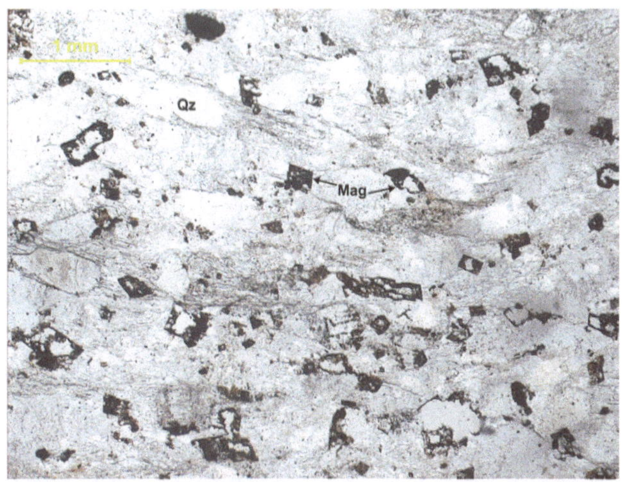

Plate OS_Sst-3 Foliated feldspathic wacke consisting of quartz, feldspar and concentration of heavies include skeletal magnetite and apatite, indicative of proximity to source rock and quick deposition (408/79-1)

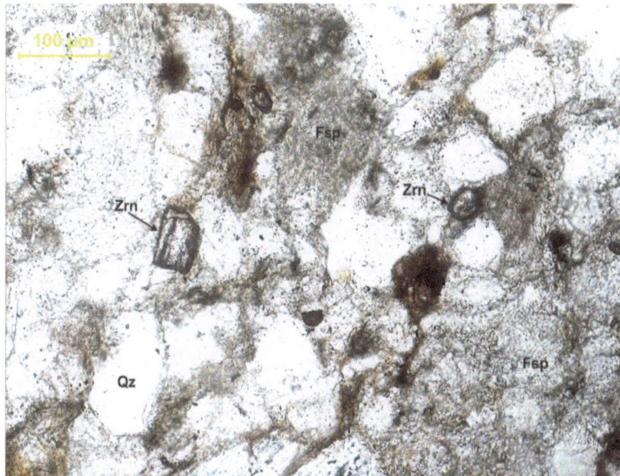

Plate OS_Sst-5 Feldspathic arenite with flattened and angular quartz, microcline (~ 15 vol %), opaque and zircon (prismatic and granular) in a fine-grained matrix of sericite and chlorite (183/80-2)

Plate OS_Sst-6 Feldspathic wacke 183/80-2 between crossed polars showing quartz and feldspar in an iron-rich argillaceous matrix. A flake of muscovite may be seen at the lower left corner. Bimodal and dispersed grains indicative of rapid free fall deposition and quick burial (183/80-1)

Plate OS_Sst-8 Quartz-magnetite massive rock containing randomly oriented euhedral crystals of magnetite, minor sericite and residual clay matter (682-3)

Plate OS_Sst-7 Chert wacke with fragments of radiolarian chert and mafic volcanics (*dark*) in a ferruginous matrix. Angular feldspar and argillaceous (*brown*) fragments occur between quartz grains. Recycling of grains very strongly developed; indicates quick deposition (n-137-1)

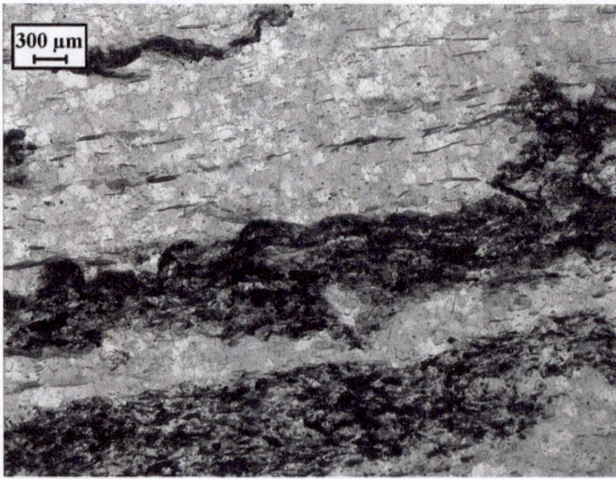

Plate OS_Sst-9 Banded quartzite with mafic-rich bands comprising epidote, chlorite and opaque. Bedding planes show a regularity in distribution, indicating a quiet sedimentation in deep water condition (n140-1)

Plate OS_Sst-10 Magnified view of mineralised quartz-sericite schist 682-2 containing tourmaline, *honey yellow* coloured, translucent granular sphalerite, cubic crystal of *brownish yellow* pyrite and magnetite (682-1)

Plate OS_Lst-2 Siliceous limestone under crossed polar containing quartz, twinned plagioclase and epidote (N145-1)

25.4 Carbonates

Plate OS_Lst-1 Limestone under crossed polar, with medium to coarse crystals of calcite along with serpentine and muscovite. The grey band at the base is composed of cryptocrystalline-carbonates (K68-2)

Plate OS_Lst-3 Magnified view of impure dolomitic limestone (118/79) under crossed polar associated with phengite and opaque (118/79-2)

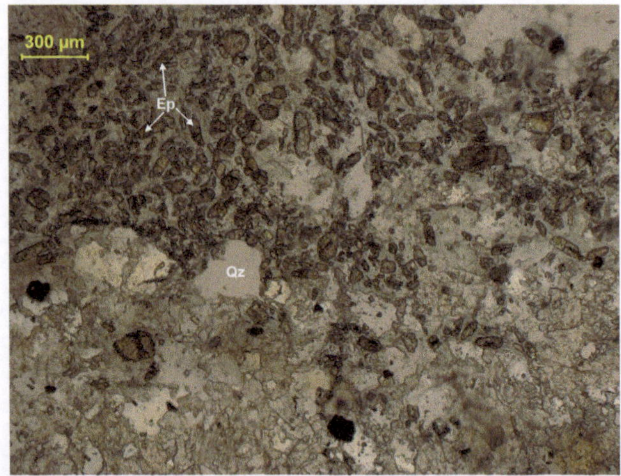

Plate OS_Lst-4 Recrystallised calcareous shale or marl consisting of epidote and white mica (parent rock, refer Lst-5), carbonate and opaque in the lower-half (n145-1)

Plate OS_Lst-6 Recrystallised calcareous shale (n145-1) containing parallel alignment of thin flakes of graphite with smeared edges, in association with opaques and epidote in a sericite matrix (n145-2)

Plate OS_Lst-5 Magnified view of groundmass of recrystallised calcareous shale (n145-1) with weak crenulations, containing sericite and minor epidote (n145-5)

25.5 Radiolarian Chert

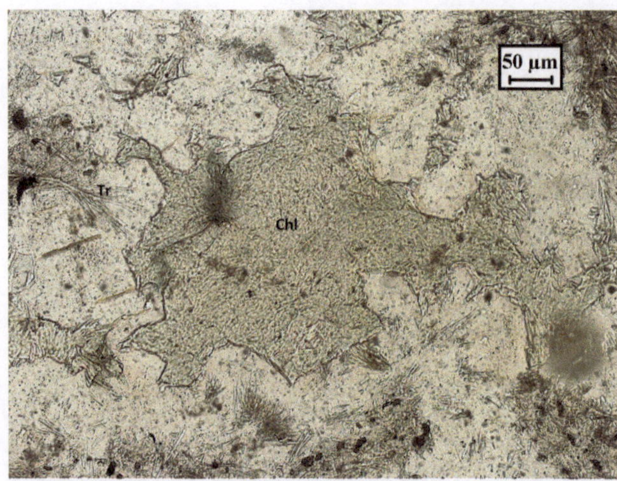

Plate OS_Ch-1 Tuffaceous chert containing quartz, chlorite, acicular radiating tremolite and opaque (5/80-3)

Plate OS_Ch-2 Weathered chert 5/80-3 between crossed polars showing twinned albite, chlorite and rutile. Alteration along grain boundaries is noticed (5/80-1)

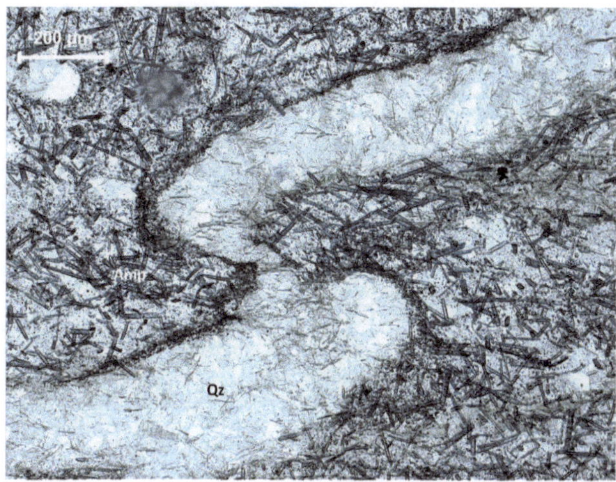

Plate OS_Ch-4 Flattened chert band severed by tectonism and containing fine needles of blue amphibole with tensional patterns (618-1)

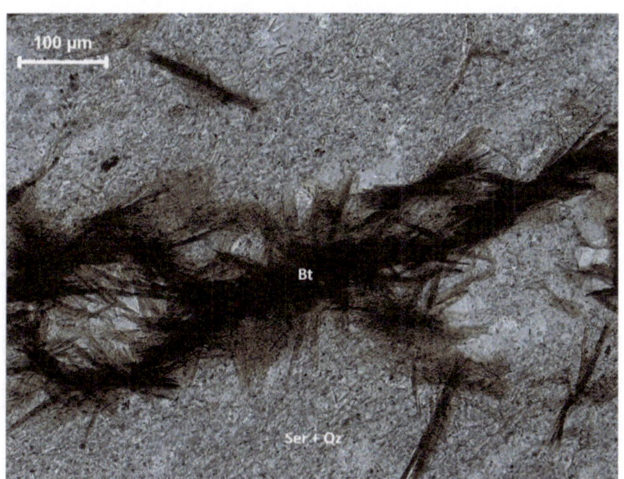

Plate OS_Ch-3 Acicular biotite and sericite in chert showing deformation. Biotite is developed late (Z-55-2)

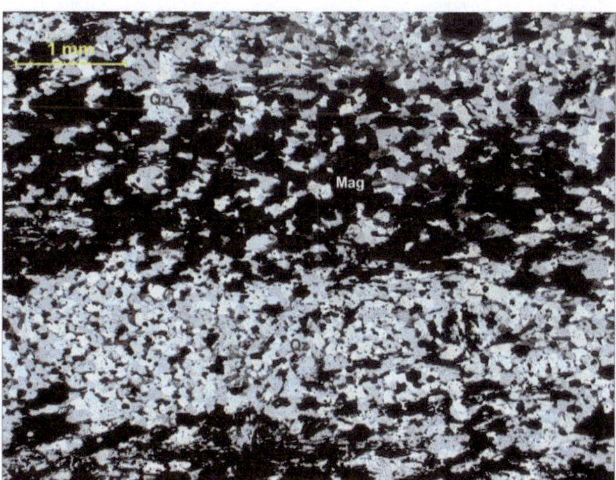

Plate OS_Ch-5 Banded ferruginous chert comprising alternate layers of dark iron oxide/fine sericite and *greyish white* chert intermixed with fine flakes of muscovite (N-24-1)

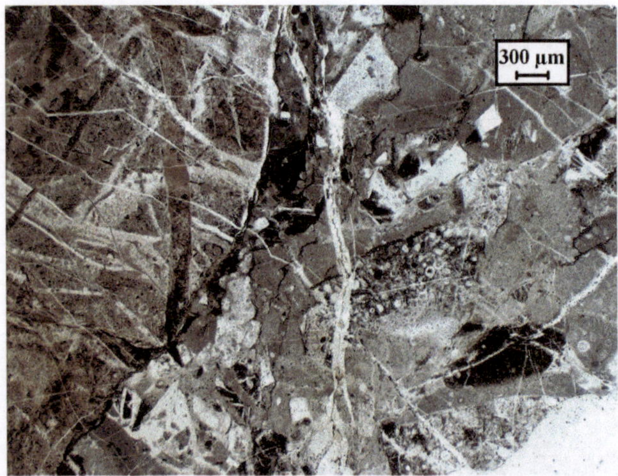

Plate OS_Ch-6 Brecciated radiolarian chert showing network of veins along fractures. Two sets of quartz veins are discernible (44/79-3)

Plate OS_Ch-8 Series of normal and reverse faults in radiolarian chert. The *greenish* fragments at the centre are iron-rich fine pelagic sediment (9/79-1)

Plate OS_Ch-7 Radiolarian chert traversed by recrystallised quartz veins showing minor deformation between crossed polars (N-110-1)

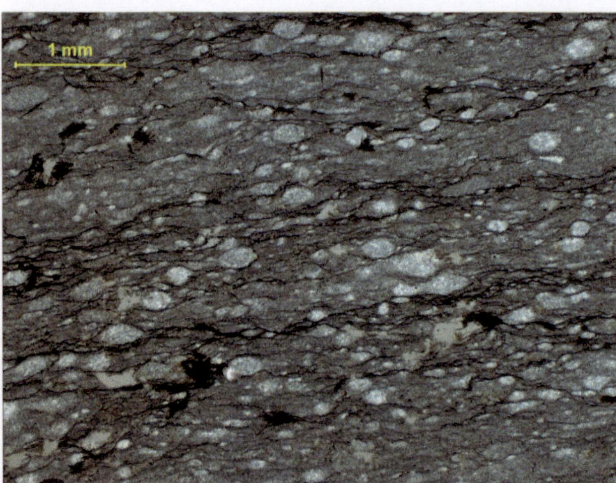

Plate OS_Ch-9 Tuffaceous chert showing radiolaria and fine biotite flakes along foliation. Crude crenulation is developed (Z55-1)

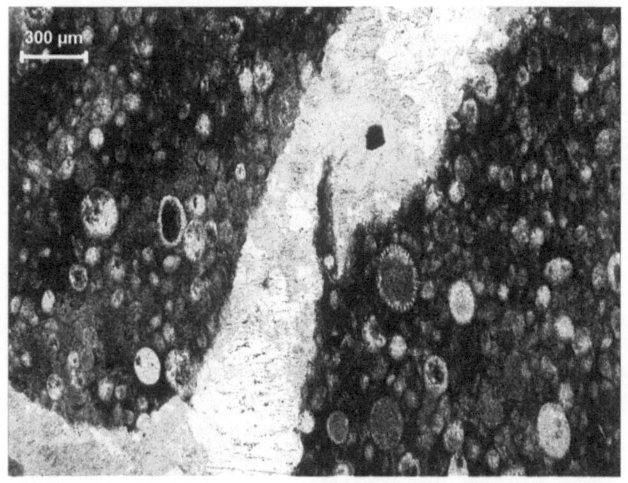

Plate OS_Ch-10 Folded band of recrystallised chert containing several species of radiolaria (N-19-1)

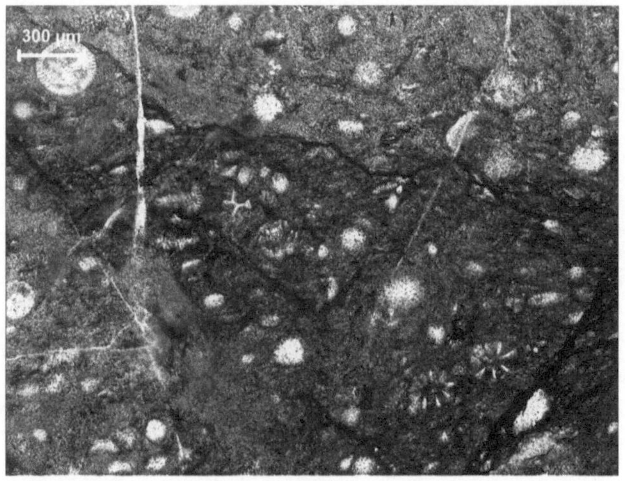

Plate OS_Ch-11 Tectonised radiolarian chert containing several species of radiolaria. Recrystallised quartz veins of post-tectonic origin cut across the rock (N105-1)

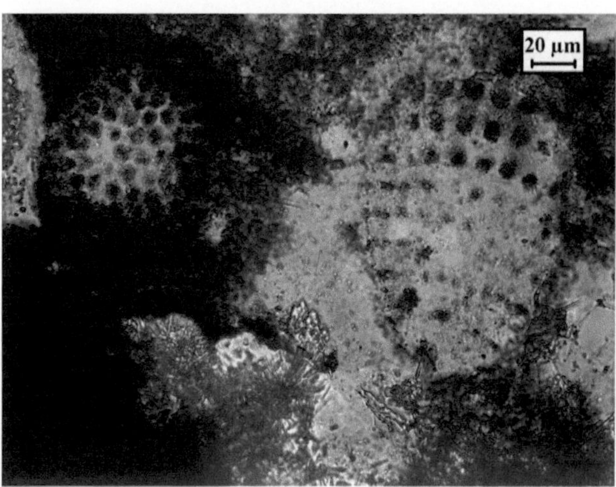

Plate OS_Ch-12 A colony of radiolaria in red chert (139/79-3)

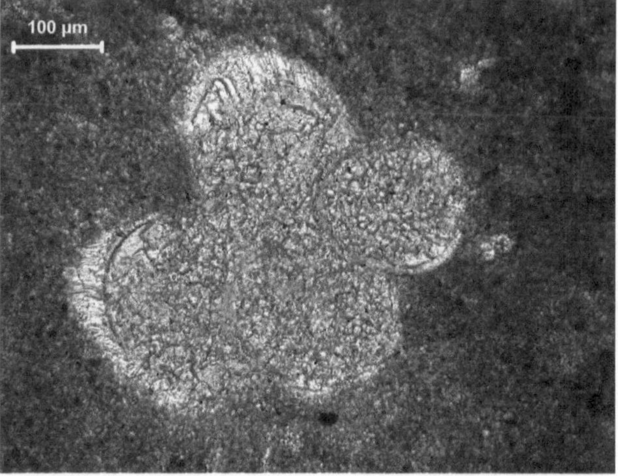

Plate OS_Ch-13 Conglomeration of radiolaria in chert (m4-1)

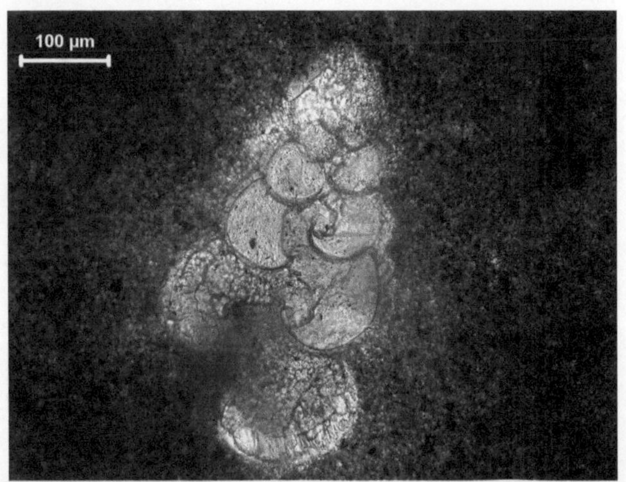

Plate OS_Ch-14 Conglomeration of radiolaria in chert (m4-2)

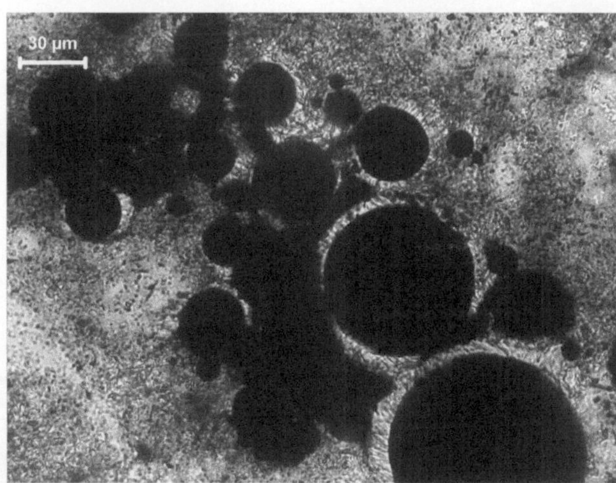

Plate OS_Ch-16 Conglomeration of dark circular radiolaria in sericite-rich chert (m4-5)

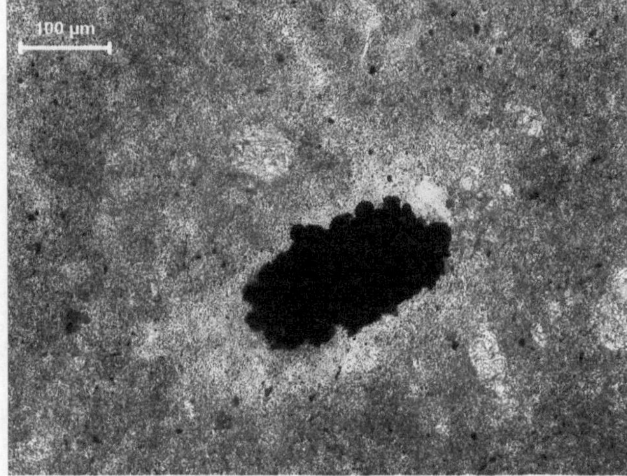

Plate OS_Ch-15 Dark rectangular serrated radiolarian in chert. (m4-4)

26 Cover Sediments (Jopi Formation)

26.1 Polymictic Tuff Breccia: Part 1 (sample C21/79)

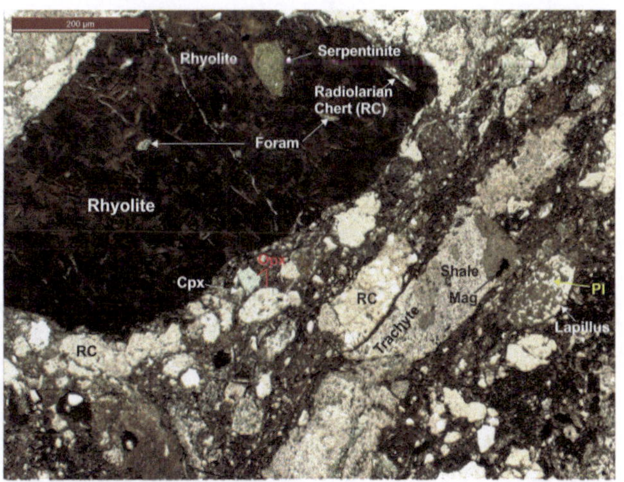

Plate Ptb-1 Fragment of *brown lithic clast* (rhyolite) in polymictic tuff breccia containing fine clasts of **a** serpentinite, **b** radiolarian chert and **c** foraminifera. The *brown lithic clast* is bordered by a girdle of rounded ortho- and clino-pyroxenes (**d**) and chert grains succeeded by a short band of radiolarian chert, a long band of trachyte (**e**) with particles of shale/argillite and **f** lapillus (C21/79-4)

Plate Ptb-2 Magnified view of *brown rhyolite* C21/79-4 between crossed polars showing crystals of radiating and acicular needles of alkali feldspar and flattened quartz in slightly devitrified glass. A large phenocryst of feldspar may be seen at the *top left corner* (C21/79-8)

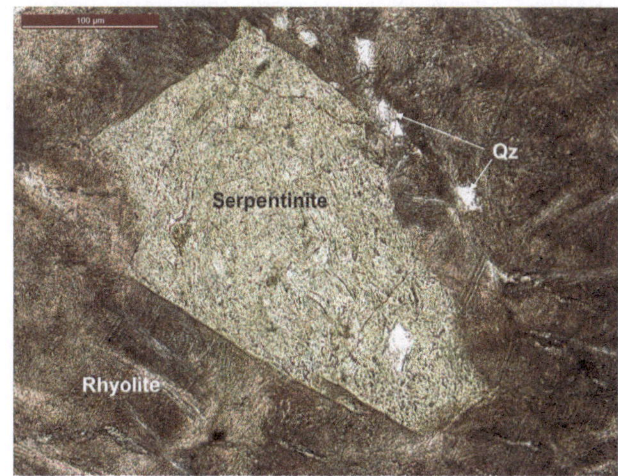

Plate Ptb-3 Clast of serpentinite with sharp edges (**a**, refers to Ptb-1) in rhyolite C21/79-4 between crossed polars (C2/79-9)

N. C. Ghose et al., *A Petrographic Atlas of Ophiolite*,
DOI: 10.1007/978-81-322-1569-1_26, © Springer India 2014

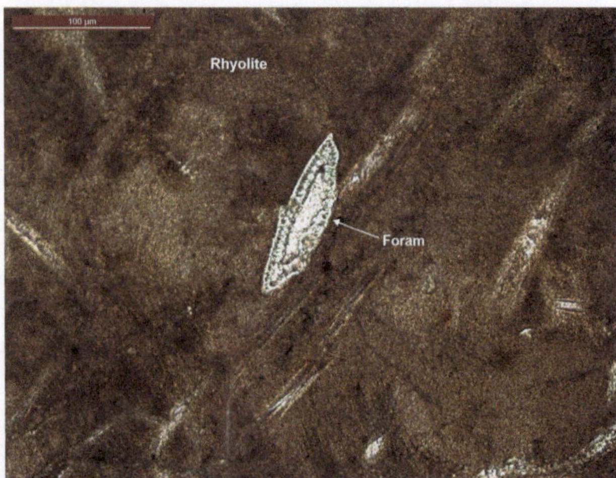

Plate Ptb-4 Magnified view of lenticular foraminifera (**c**, refers to Ptb-1) between crossed polars trapped in the rhyolite flow (C21/79-10)

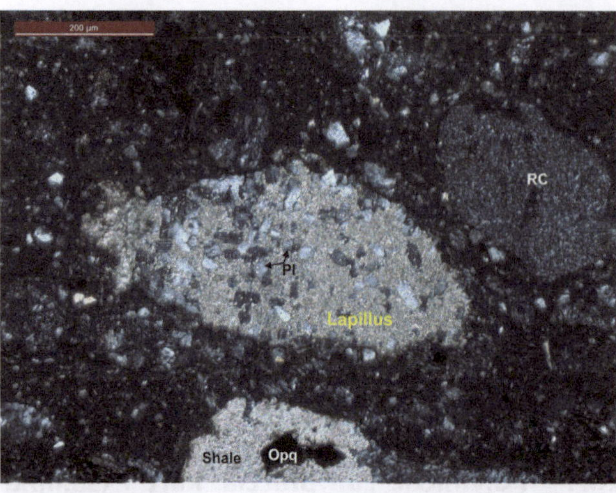

Plate Ptb-6 A volcaniclastic lapillus (**f**, refers to Ptb-1) between crossed polars embellished with plagioclase laths (C21/79-21)

Plate Ptb-5 Magnified view of pleochroic orthopyroxene (neutral to green) and clinopyroxene (**d**, refer to Ptb-1) occurring as girdle around rhyolite C21/79-4. The hypersthene contains inclusions of green and *brown glass* and opaque (C21/79-5)

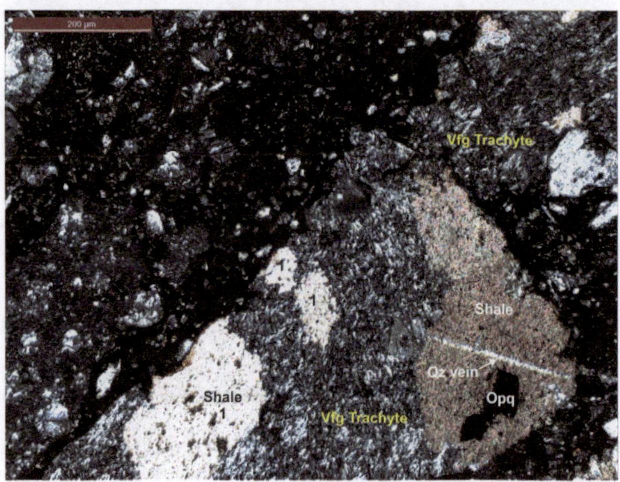

Plate Ptb-7 A band of very fine-grained (Vfg) trachyte (**e**, refers to Ptb-1) in fine ash between crossed polars containing particles of shale/argillite with inclusion of opaque. A fine quartz vein cuts through the shale fragment (C21/79-14)

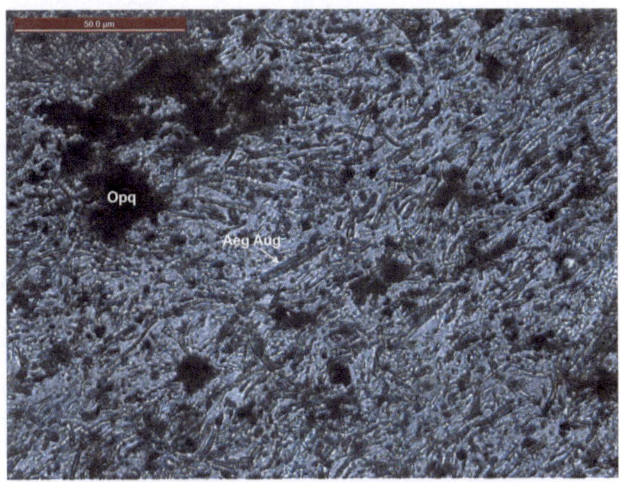

Plate Ptb-8 Magnified view of the very fine-grained trachyte (Vfg in C21/79-14) showing trachytic texture with sub-parallel orientation of feldspars, clinopyroxenes and opaque. The clinopyroxenes are rimmed by pleochroic green aegirine (C21/79-16)

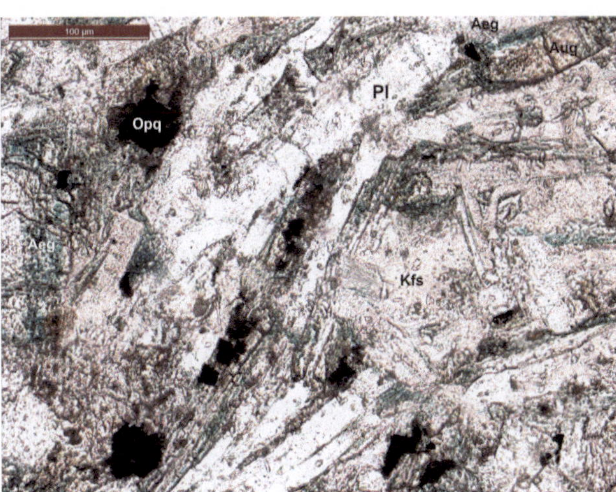

Plate Ptb-10 Fine-grained trachyte (Fg) with pilotaxitic texture between crossed polars consisting of plagioclase laths, pink alkali feldspar, augite rimmed by green pleochroic aegirine (*pale yellowish green* to *bluish green*) and opaque (C21/79-21)

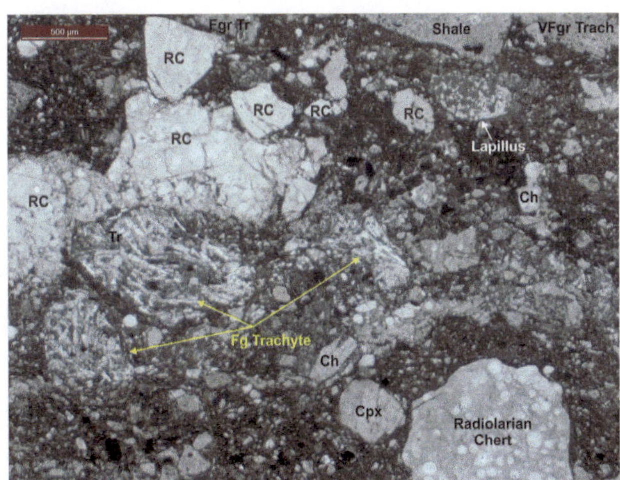

Plate Ptb-9 Clasts/fragments of fine-grained trachyte (Fg) sandwiched between radiolarian chert as tuff breccia in fine ash. A lapillus containing laths of plagioclase (C21/79-21) is seen at the *top right corner*. A grain of clinopyroxene occurs close to radiolarian chert in the *lower corner* (C21/79-14a)

Plate Ptb-11 Fragment of porphyritic basalt containing phenocryst of plagioclase. The basalt is surrounded by chert fragments (C21/79-6)

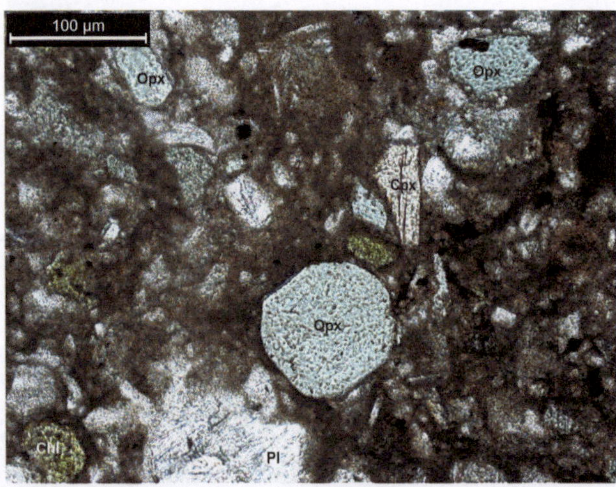

Plate Ptb-12 Sub-alkaline basalt fragment in tuff breccia containing discrete phenocrysts of plagioclase and pleochroic orthopyroxene and prismatic grain of clinopyroxene (C21/79-1)

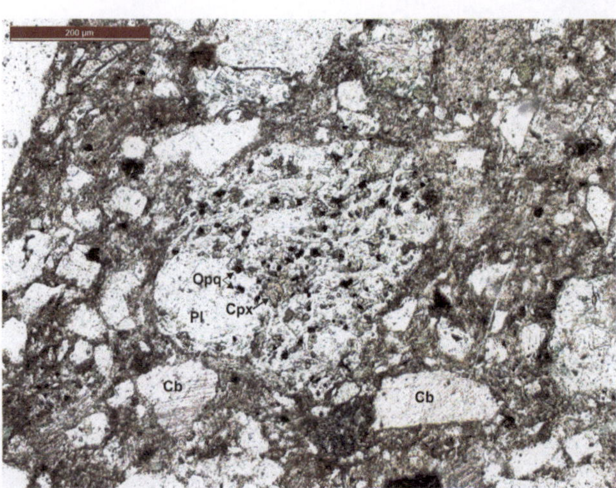

Plate Ptb-14 Clast of gabbroic anorthosite in tuff breccia consisting dominantly of plagioclase (~80 by vol.), subhedral grains of *brown clinopyroxene*, minor orthopyroxene and opaque (~3 vol %). The anorthosite is surrounded by clasts of radiolarian chert and carbonates (calcite) (C21/79-18)

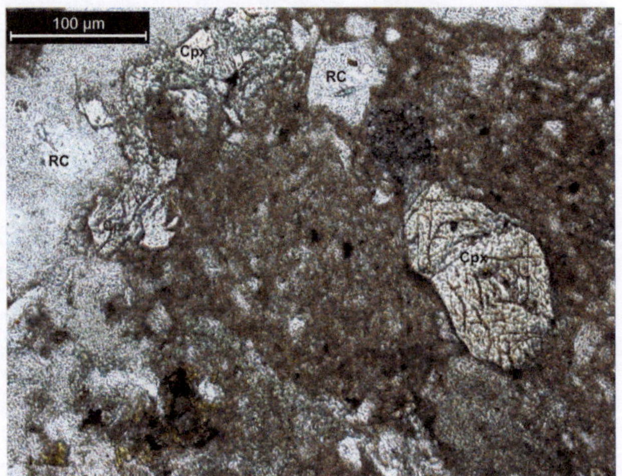

Plate Ptb-13 Two types of clinopyroxene fragments—fractured and anhedral in tuffaceous matrix surrounded by fragments of chert. The former shows secondary alteration to chlorite whereas the latter shows prominent cleavage, cross fractures and inclusions of opaque (C21/79-3)

26.2 Polymictic Tuff Breccia: Part 2 (Other Samples)

Plate Ptb-15 Tuff/epiclastic breccia between crossed polars containing sub-angular, poorly sorted grains of quartz and feldspar with >5 % matrix (immature sandstone/feldspathic wacke). Plagioclase dominates over quartz. Both monocrystalline and polycrystalline quartz are present. Some plagioclase grains show twinning, but most feldspar grains are untwinned (134/79-2)

Plate Ptb-16 Brecciated tuff/epiclast between crossed polars consisting of clasts and particles of varied dimensions. Clasts of rectangular chert on the *left*, recrystallised quartzite in the *top corner* and iron stained polycrystalline quartz on the *right* are set in a matrix of recrystallised glass, oxidised opaque minerals and quartz (95/80 1)

Plate Ptb-18 Matrix of tuff/epiclastic breccia comprising angular and sub-angular grains of quartz (*white*), chert (*greyish white*, angular), argillite (*grey*), sandstone (*fawn*), metabasite (*greenish*), greywacke (*brown* with *white speckle*) and glass (94/80-3)

Plate Ptb-17 Tuff/epiclastc breccia dominantly composed of lithic fragments of metabasic (*top left corner*), argillite (*grey*) and chert. Green chlorite in the centre is associated with opaque minerals. Matrix is largely composed of fragments of metabasics, chert, opaques and non-volcanic grains of bioclasts (*on right*) (Z-63-1)

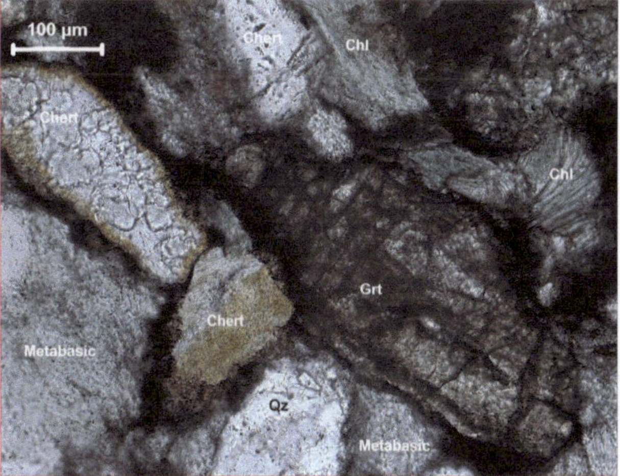

Plate Ptb-19 Magnified view of tuff breccia containing chert, metabasite, *green chlorite*, quartz and elongated porphyroblast of brown isotropic mineral (Grt). The *yellow stain* around globular chert on the *left* is due to limonitic iron oxide/hydroxide (82/80-12)

Plate Ptb-20 Angular lithic fragments of metabasite, chert and serpentinite (1) in polymictic tuff breccia co-existing with discrete minerals, viz., glaucophane, chlorite and quartz, set in a glassy matrix. The iron-rich chlorite is faintly pleochroic (82/80-10)

Plate Ptb-22 Angular lithic fragments in tuff breccia containing quartzite in the centre with *pleochroic blue amphibole* (Gln) associated with opaque minerals. It is surrounded by fragments of serpentine (1) and chert (2) with iron stains (94/80-1)

Plate Ptb-21 Tuff/epiclastic breccia between crossed polars showing sub-angular poorly sorted fragments of chert (1), quartzite (2), metabasiite (3), arkose, pyroxenite and *greywacke*, and cross-cutting quartz vein. The rock is devoid of any matrix indicating little transport (84/80-1)

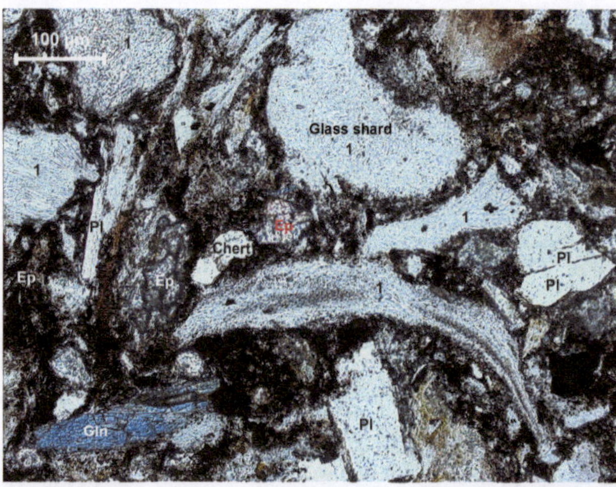

Plate Ptb-23 Tuff breccia between crossed polars showing poorly sorted angular lithic fragments of glass shards (1), chert, *pleochroic blue amphibole* (Gln), plagioclase (Pl) and epidote (Ep) (94/80-4)

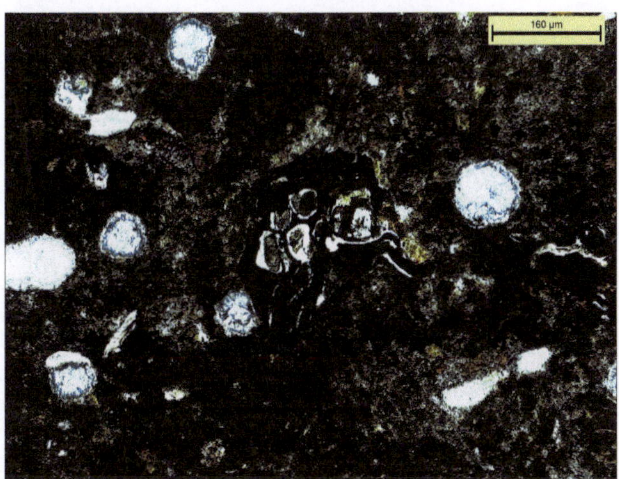

Plate Ptb-24 Magnified view of tuff breccia between crossed polars showing concentric ooids/ooliths with thick nucleus and thin cortices (oolitic coating). Small aggregates of ooids with or without internal structure in the centre are cemented together by fine sparite. Laminar particles of bioclasts are seen at the *right edge*, below the ooid aggregates, and the *lower left corner*. Replacement of calcite by authigenic quartz is observed in *top left corner* (R-45-2)

27 Late Tertiary Granites

Plate Ltg-1 Late felsic intrusive under crossed polars showing post-crystalline deformation with parallel orientation of quartz and feldspar grains. Plagioclase is saussuritised and twinned. K-feldspar, muscovite and epidote are minor constituents of the rock (369/70-2)

Plate Ltg-2 Brecciated quartz porphyry containing quartz (*white* and *grey*, uniaxial) and magnetite (opaque). The rock is devoid of feldspar (57/79-1)

Plate Ltg-3 Brecciated quartz porphyry under crossed polars (57/79-2)

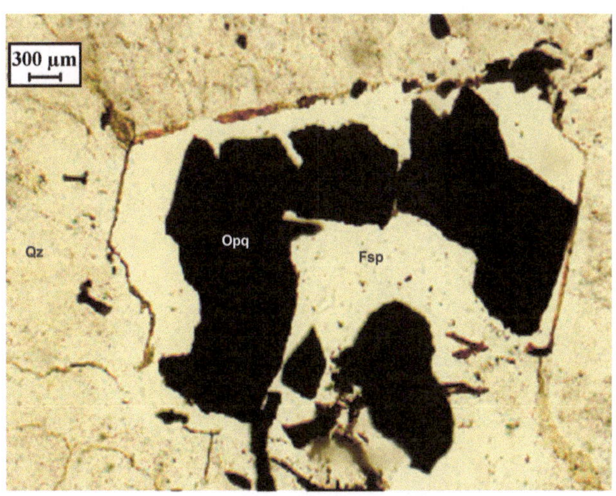

Plate Ltg-4 K-feldspar-hosted opaque magnetite showing atoll structure. The background is due to leaching of iron (59/79-1)

N. C. Ghose et al., *A Petrographic Atlas of Ophiolite*,
DOI: 10.1007/978-81-322-1569-1_27, © Springer India 2014

28 Metamorphics of the Nimi Formation

Plate Nimi-1 Elongated tabular crystals of staurolite and muscovite flakes along foliation in chlorite-biotite-muscovite-garnet-staurolite schist with quartz and feldspar from the Nimi Formation at the eastern tectonic margin of the NHO. Both staurolite and biotite are formed by reactions involving chlorite and muscovite in amphibolite facies (32/80-3)

Plate Nimi-2 Magnified view of chlorite-biotite-muscovite-garnet-staurolite schist. The garnet porphyroblast is developed late and enveloped by chlorite, biotite and quartz (32/80-2)

N. C. Ghose et al., *A Petrographic Atlas of Ophiolite*,
DOI: 10.1007/978-81-322-1569-1_28, © Springer India 2014

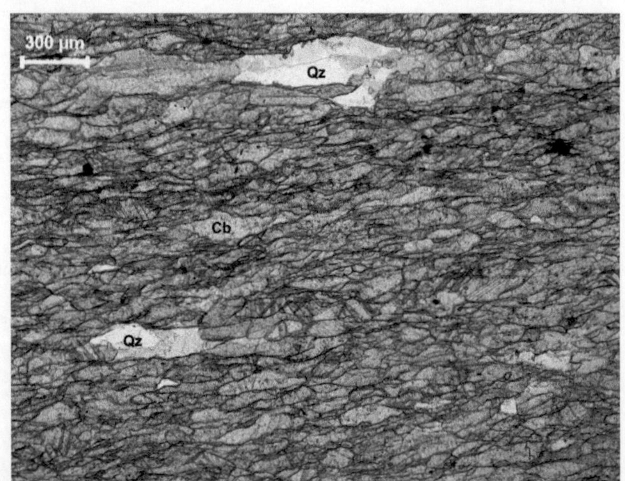

Plate Nimi-3 Asymmetric ductile shear fabric in deformed limestone (57-1)

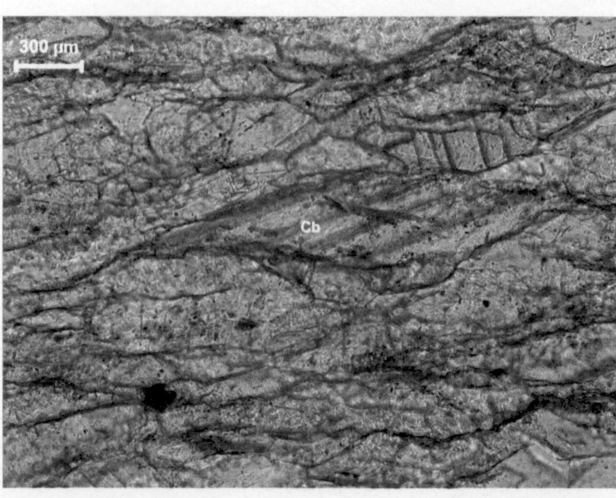

Plate Nimi-4 Magnified view of mylonitic limestone 57-1 showing flattened grains of carbonates (57-2)

Appendices

N. C. Ghose et al., *A Petrographic Atlas of Ophiolite*,
DOI: 10.1007/978-81-322-1569-1, © Springer India 2014

Table 1 Modal abundance (volume %) of minerals in representative ultramafic rocks, gabbroids and plagiogranites from the Naga Hills Ophiolite, India

A. Ultramafics[a]

	Tectonite			Cumulate ultramafic						
Rock	Dunite	Harzburgite	Lherzolite	Dunite	Harzburgite	Lherzolite	Wehrlite	Olivine clinopyroxenite	Clinopyroxenite	Websterite
Olivine	90.8	69.9	77.1	90.8	60.4	62.2	57.4	18.3	4.7	3
Orthopyroxene	5.3	22.2	11.4	3.2	31.9	23.8	2.2	3	1.2	20.8
Clinopyroxene	1.3	1	6.2	…	5.3	9.4	33.3	73.5	92.4	70.8
Opaques	2.6	6.9	5.3	6	2.4	4.6	7.1	5.2	1.7	5.4

B. Gabbroids[b]

Rock	Olivine gabbronorite	Gabbronorite	Leuco-olivine gabbronorite	Leuco-gabbronorite	Gabbro	Meta-gabbronorite
Olivine	20	3	3	…	…	2
Orthopyroxene	7	17	10	…	…	23
Clinopyroxene	21	25	5	29	38	55
Plagioclase	43	55	73	69	60	20
Hornblende	3	…	2			…
Opaques	6	…	7	2	2	…

C. Plagiogranites[b]

Sample No.	326/81	286/81	225/81	324A/81	248/81	311B/61	325/81
Locality	Reguri	Akhen	Molhe	Reguri	Akhen	Reguri	Reguri
Quartz	69	58	48	45	42	40	42
Plagioclase	25	23	45	50	51	48	52
Hornblende	4	8	6	4	6	7	4
Opaques	2	2	1	1	1	1	2

Data Source: [a]Ghose et al. 1986; [b] Venkataramana et al. 1986

Table 2 Chemical analysis of minerals in ultramafic rocks from the Naga Hills Ophiolite, India

Rock	Lherzolite								Cpx-harzburgite								Peridotite	
Sample	KO13				KO41/1				PS30/1				GA8				GM-5	NK-13
Mineral	Ol	Opx	Cpx	Cr-Spl	Ol	Opx	Cpx	Cr-Spl	Ol	Opx	Cpx	Cr-Spl	Ol	Opx	Cpx	Cr-Spl	Cr-Spl	Cr-Spl
Weight per cent																		
SiO_2	40.33	53.67	50.66	0.08	41.54	54.73	52.20	0.06	41.60	54.10	51.75		41.27	55.80	52.94	0.02		0.17
TiO_2		0.10	0.22	0.03	0.02	0.09	0.39	0.10	0.02	0.14	0.40	0.02	0.01	0.04	0.20	0.01	0.03	0.13
Al_2O_3		4.57	5.18	55.84		5.48	6.13	49.11		5.38	5.44	45.25		3.71	3.82	50.92	53.12	46.89
Cr_2O_3	0.06	0.58	1.08	10.44	0.05	0.50	0.77	16.74		0.68	1.11	22.41		0.69	0.95	16.98	15.38	23.47
Fe_2O_3				1.43				1.85				0.84				0.82		
FeO	9.27	5.64	2.67	11.82	9.24	6.45	2.45	11.22	9.30	6.31	2.23	11.61	9.25	5.79	2.51	11.81	12.64	13.10
MnO	0.13	0.07	0.10	0.19	0.15	0.20	0.06	0.06	0.20	0.15	0.13	0.14	0.22	0.10	0.08	0.20	0.06	0.19
MgO	48.36	29.38	15.63	18.43	48.74	31.71	14.53	18.04	50.02	31.47	15.06	17.51	48.33	32.95	16.68	18.12	18.73	15.98
CaO	0.01	4.14	23.48			0.52	21.73		0.03	0.80	22.46		0.01	0.47	23.18		0.04	0.07
Na_2O	0.01	0.06	0.32	0.04	0.03	0.06	1.38	0.04	0.02	0.06	0.81			0.02	0.29			
K_2O		0.03			0.04	0.02	0.04								0.01	0.02		
P_2O_5			0.02		0.06			0.02			0.03	0.02				0.03		
Total	98.17	98.24	99.36	98.30	99.87	99.76	99.68	97.24	101.19	99.09	99.42	97.80	99.09	99.57	100.66	98.93	100.00	100.00
Atomic Proportions																		
O	4	6	6	4	4	6	6	4	4	6	6	4	4	6	6	4	4	4
Si	1.00	1.90	1.86	0.00	1.01	1.89	1.90	0.00	1.00	1.89	1.89		1.02	1.93	1.91	0.00		0.00
Ti		0.00	0.01	0.00	0.00	0.00	0.01	0.00	0.00	0.00	0.01	0.00	0.00	0.00	0.01	0.00	0.00	0.00
Al		0.19	0.22	1.75		0.22	0.26	1.59		0.22	0.23	1.49		0.15	0.16	1.62	1.66	1.51
Cr	0.00	0.02	0.03	0.22	0.00	0.01	0.02	0.36		0.02	0.03	0.49		0.02	0.03	0.36	0.32	0.51
Fe^{3+}				0.03				0.04				0.02				0.02		
Fe^{2+}	0.19	0.17	0.08	0.26	0.19	0.19	0.07	0.26	0.19	0.18	0.07	0.27	0.19	0.17	0.08	0.27	0.28	0.30
Mn	0.00	0.00	0.00	0.00	0.00	0.01	0.00	0.00	0.00	0.00	0.00	0.00	0.01	0.00	0.00	0.00	0.00	0.00
Mg	1.79	1.55	0.86	0.73	1.77	1.64	0.79	0.74	1.80	1.64	0.82	0.73	1.77	1.70	0.90	0.73	0.74	0.65
Ca	0.00	0.16	0.92			0.02	0.85		0.00	0.03	0.88		0.00	0.02	0.90		0.00	0.00
Na	0.00	0.00	0.02	0.00	0.00	0.00	0.10	0.00	0.00	0.00	0.06			0.00	0.02			

(continued)

Table 2 (continued)

Rock	Lherzolite								Cpx-harzburgite								Peridotite	
Sample	KO13				KO41/1				PS30/1				GA8				GM-5	NK-13
Mineral	Ol	Opx	Cpx	Cr-Spl	Ol	Opx	Cpx	Cr-Spl	Ol	Opx	Cpx	Cr-Spl	Ol	Opx	Cpx	Cr-Spl	Cr-Spl	Cr-Spl
K		0.00			0.00	0.00	0.00								0.00	0.00		
P			0.00		0.00			0.00			0.00	0.00				0.00		
Sum	3.00	4.00	4.02	3.00	2.99	3.99	4.00	3.00	3.00	3.99	4.00	3.00	2.98	3.99	4.00	3.00	3.01	2.98
Mg#	90.3	90.3	91.3	73.5	90.4	89.7	91.4	74.1	90.5	89.9	92.3	72.9	90.3	91.0	92.2	73.2	72.5	68.5
Cr#				11.1				18.6				24.9				18.3	16.3	25.1
Fo	90.3				90.4				90.6				90.3					
Fa	9.7				9.6				9.4				9.7					
En		82.7	46.0			88.8	46.1			88.4	46.4			90.2	48.0			
Fs		8.9	4.4			10.2	4.3			9.9	3.9			8.9	4.1			
Wo		8.4	49.6			1.0	49.6			1.6	49.7			0.9	47.9			

Fe^{3+} and Fe^{2+} calculated through charge balancing; Mg# = 100 Mg/(Mg+Fe^{2+}), Cr# = 100 Cr/(Cr+Al)
Mineral abbreviations Cpx clinopyroxene, *Cr* Spl-chrome spinel, *En* enstatite, *Fa* fayalite, *Fo* forsterite, *Fs* ferrosilite, *Ol* olivine, *Opx* orthopyroxene
Data source samples GM-5 and NK-13—Ningthoujam et al. 2012 (SEM-EDS), all other samples—Singh, 2013 (EPMA)

Table 3 Chemical analysis of minerals in high-pressure assemblages from metabasics and meta-arenites of the Naga Hills Ophiolite, India

Rock	Eclogic				Blueschist													Greenschist
Sample	77/80				139/80b			2/70		76/80		212/80		49/70[a]		ML-69		22/80
Mineral	Grt-c	Omp	Gln	Brs-c	Grt-c	Gln	Brs-c	Mrbk	Act	Gln	Tr	Omp	Gln	Grt	Mrbk	Gln	Brs	Mhb
n	6	4	4	4	6	3	3											
Locality	Sataza	Sataza	Sataza	Sataza	Purr	Purr	Purr	Pungro	Pungro	Sataza	Sataza	Purr	Purr	Moya	Moya	Purr	Purr	Sataza
Weight per cent																		
SiO_2	37.41	54.59	55.34	48.22	37.60	55.54	47.34	56.52	54.90	58.18	56.41	54.82	57.11	38.07	55.07	58.26	50.92	50.96
TiO_2	0.09	0.09	0.02	0.27	0.14	0.01	0.25											0.35
Al_2O_3	21.47	8.72	8.95	9.97	21.70	10.15	9.99	5.82	2.82	9.56	1.85	7.22	9.49	21.05	6.18	8.94	8.94	8.07
Cr_2O_3	0.05	0.04	0.01	0.04	0.02	0.01	0.02			0.34	0.31							
Fe_2O_3	0.53	7.65	2.74	3.55	0.01	2.02	2.88	5.56	2.08	1.67	0.87	5.34	2.11		4.58	2.35	1.53	1.53
FeO	27.48	4.50	11.70	13.30	26.71	9.19	10.92	11.16	6.98	7.81	5.27	4.00	9.52	13.85	14.94	8.09	8.94	7.98
MnO	3.42	0.05	0.05	0.08	2.89	0.07	0.20	0.36	0.57		0.36			17.04	0.32			0.30
MgO	2.20	5.70	9.42	9.73	3.12	10.58	11.65	11.22	17.45	12.04	20.26	8.49	10.92	3.06	9.03	12.27	14.79	15.66
CaO	7.63	10.42	1.41	5.87	7.71	1.15	7.78	2.09	10.03	1.22	11.60	14.59	1.44	6.76	1.77	1.56	9.47	10.65
Na_2O	0.04	8.05	6.61	4.81	0.00	6.51	3.79	6.29	1.94	6.79	0.65	5.98	6.77		6.91	6.51	3.25	1.94
K_2O			0.03	0.27		0.01	0.32		0.16								0.36	0.73
Total	100.31	99.80	96.28	96.10	99.90	95.24	95.14	99.02	96.93	97.61	97.58	100.43	97.36	99.83	98.80	97.98	98.20	98.17
Atomic proportions																		
O	12	6	23	23	12	23	23	23	23	23	23	6	23	12	23	23	23	23
Si	2.97	1.99	7.88	7.14	2.98	7.87	7.03	7.91	7.79	7.97	7.84	1.98	7.93	3.02	7.86	7.98	7.21	7.21
Ti	0.01	0.00	0.00	0.03	0.01	0.00	0.03											0.04
Al(iv)	0.03	0.01	0.12	0.86	0.02	0.13	0.97	0.09	0.21	0.03	0.16	0.02	0.07	1.97	0.14	0.02	0.79	0.79
Al(vi)	1.99	0.36	1.39	0.88	2.00	1.56	0.78	0.87	0.26	1.51	0.15	0.29	1.48		0.90	1.42	0.70	0.56
Cr	0.00	0.00	0.00	0.00	0.00	0.00	0.00			0.04	0.03							
Fe^{3+}	0.03	0.21	0.29	0.40	0.00	0.22	0.32	0.59	0.22	0.17	0.09	0.15	0.22		0.49	0.24	0.16	0.16
Fe^{2+}	1.83	0.14	1.39	1.65	1.77	1.09	1.36	1.31	0.83	0.89	0.61	0.12	1.11	0.92	1.78	0.93	1.06	0.94
Mn	0.23	0.00	0.01	0.02	0.19	0.01	0.04	0.04	0.10		0.04			1.15	0.04			0.04
Mg	0.26	0.31	2.00	2.15	0.37	2.23	2.58	2.34	3.69	2.46	4.20	0.46	2.26	0.36	1.92	2.50	3.12	3.30
Ca	0.65	0.41	0.21	0.93	0.65	0.18	1.24	0.31	1.52	0.18	1.73	0.57	0.21	0.57	0.27	0.23	1.44	1.62
Na(B)			1.70	0.96		1.72	0.67	1.54	0.41	1.75	0.15		1.72		1.59	1.68	0.52	0.34
Na(A)			0.12	0.43		0.07	0.42	0.17	0.12	0.06	0.06		0.10		0.33	0.05	0.38	0.19
Na(total)	0.01	0.57	1.82	1.38	0.00	1.79	1.09	1.71	0.53	1.80	0.21	0.42	1.82		1.91	1.73	0.89	0.53

(continued)

Table 3 (continued)

Rock	Eclogic				Blueschist													Greenschist
Sample	77/80				139/80b			2/70		76/80		212/80		49/70[a]		ML-69		22/80
Mineral	Grt-c	Omp	Gln	Brs-c	Grt-c	Gln	Brs-c	Mrbk	Act	Gln	Tr	Omp	Gln	Grt	Mrbk	Gln	Brs	Mhb
n	6	4	4	4	6	3	3											
Locality	Sataza	Sataza	Sataza	Sataza	Purr	Purr	Purr	Pungro	Pungro	Sataza	Sataza	Purr	Purr	Moya	Moya	Purr	Purr	Sataza
K			0.01	0.05		0.00	0.06		0.03								0.07	0.13
Sum	8.00	4.00	15.13	15.48	8.00	15.07	15.49	15.17	15.18	15.06	15.06	4.00	15.10	7.99	15.33	15.05	15.44	15.33
Mg#	12.5	69.3	58.9	56.6	17.2	67.2	65.5	64.2	81.7	73.3	87.3	79.1	67.2	28.2	51.9	73.0	74.7	77.8
Fe^{+3}	1.6	36.6	17.5	31.0		12.1	29.3	40.2	46.4	10.2	38.3	33.3	13.0		35.2	14.6	18.9	22.6
Aug		42.9										57.7						
Jd		36.2										28.2						
Aeg		20.9										14.1						
Alm	61.6				59.3									30.6				
Sps	7.8				6.5									38.2				
Prp	8.8				12.4									12.1				
Grs	21.9				21.9									19.1				

Fe^{3+} , Fe^{2+} and site occupancies in amphibole calculated after Leake et al. 1997, and through charge balancing in other minerals
Aug, Jd and Aeg calculated after Morimoto et al. 1988; Mg# = 100 Mg/(Mg+Fe^{2+}); Fe3# = 100 Fe^{3+} /(Fe^{3+} +Al) [a] *Meta-arenite.c* core, *n* number of analyses. *Mineral abbreviations. Act* actinolite, *Aeg* aegirine, *Alm* almandine, *Aug* augite, *Brs* barroisite, *Gln* glaucophane, *Grs* grossular, *Grt* garnet, *Jd* jadeite, *Mhb* magnesiohornblende, *Mrbk* magnesioriebeckite, *Omp* omphacite, *Prp* pyrope, *Sps* spessartine, *Tr* tremolite. *Data Source* Chatterjee and Ghose 2010; Ghose et al. 2010

About the Authors

Prof. N. C. Ghose was born in Uttarpara near Kolkata on 18 June 1940, and he obtained his M.Sc. in 1961 and his Ph.D. in Geochemistry in 1964 from Banaras Hindu University, Varanasi, India. He joined Patna University in 1965 and retired as a Professor in 2000. As a fellow of the Alexander von Humboldt Foundation, he pursued research in Gottingen (1972–1973) and Karlsruhe (1978–1979) on "Experimental Melting of Granite System". He has determined the Eutectic composition and temperature in the system "Quartz–Orthoclase–Anorthite" at 4 and 7 kbars (1973), which has wide implications in crustal melting. He extended the melting experiment studies to natural rocks ranging from gneisses and granites from the Darjeeling Himalaya, Chotanagpur gneisses to the migmatites of South Norway (Dypvag). Prof. Ghose has worked in the field of Earth and Environmental Sciences and authored 7 books and published 90 original research papers in peer reviewed journals. His contributions in the fields of Proterozoics of Chotanagpur gneisses, Massif anorthosite of Bengal, Rajmahal basalts and the Ophiolites of Naga Hills are widely acclaimed. He has published original research papers and books on "Pollution of Chemical, Bacteriocidal and Radioactive (Radon) Constituents of River Ganga in the Mid-Ganga Basin" (1989), "Groundwater Resources of Patna Urban Conglomerate" (1992), "Iodine Contents of Soil–Water–Plant in the Gangetic Plain of North Bihar" (2003), and "Synthetic Detergents (surfactants) and Organochlorine Pesticides in the Surface water and Groundwater of the Deltaic Region of Bengal" (2009), used in the fields of science, medicine, town planning and social work.

Dr. Nilanjan Chatterjee received his Ph.D. in Geochemistry from the City University of New York in 1989. His high school and undergraduate education was in India. He is currently employed as a Principal Research Scientist in the Department of Earth, Atmospheric and Planetary Sciences at the Massachusetts Institute of Technology (MIT), Cambridge, USA. He has over 22 years of experience in Electron Probe Micro-analysis (EPMA) and manages the EPMA facility at MIT. He teaches courses related to the EPMA technique and its applications. Dr. Chatterjee has co-authored 40 papers in peer-reviewed earth science journals, 30 meeting abstracts and a book on thermodynamic database published by Springer. His research interests include igneous and metamorphic petrology and he is currently working on ultrahigh pressure metamorphism and ophiolite emplacement in the Himalayas, high-grade metamorphism in eastern India, Kerguelen Plateau and Antarctica related to Proterozoic-Paleozoic supercontinental break-up and assembly, and the geochemical evolution of the Deccan flood basalts. He has also collaborated in the modelling of hydrous peridotite melting in the mantle wedge. Dr. Chatterjee has reviewed manuscripts for several leading journals and research grant proposals for the National Science Foundation, and has appeared in a BBC documentary on flood basalts.

Dr. Fareeduddin completed his Ph.D. in Geochemistry from Mysore University and joined the Geological Survey of India in 1983. Currently working as Director in Kolkata as a professional geologist, he has mapped parts of Precambrian terrains of southern and western India and was involved in several mineral exploration projects. He has served the Geological Society of India as its Secretary and Editor of, *Journal of the Geological Society of India.* He was the convener of the 10th International Kimberlite Conference held at Bangalore in 2012. He has published 60 research papers and 2 books, viz., "Recent Advances in Earth System Sciences" and "Diamonds and Their Source Rocks in India". He is currently the Editor for the IUGS journal *Episodes* and Member, International Kimberlite Conference Advisory Committee.

N. C. Ghose et al., *A Petrographic Atlas of Ophiolite*,
DOI: 10.1007/978-81-322-1569-1, © Springer India 2014

Glossary

Abyssal Oceanic environment of depth zone >1000 m

Accidental Pyroclasts Pyroclastic fragments of any composition that are derived from the subvolcanic basement not related to the volcanic eruption

Accretionary prism/wedge A series of steeply inclined, fault-bounded wedges of sediments and volcanic rocks above a descending slab in subduction zone

Aegirine A dark green iron-rich sodic pyroxene, strongly pleochroic, with low extinction angle (2°–6°)

Agglomeratic Rock A pyroclastic rock made of fragments in the size range 64–256 mm called 'bombs' that show generally smooth outlines and may contain up to 25 % by volume of finer fragments of volcanic origin

Alkali basalt Basalts composed of olivine, titaniferous augite, labradorite and iron-ore, with accessory apatite. Carry ultramafic xenoliths. Feldspathoid is a phenocrystic phase in many but not all members of the series

Allochthonous Rocks Not formed in situ. These occur in a tectonic belt that shows kinematic indicators suggesting tectonic transport over great distances from the site of their origin; commonly characterised by thrusts and ductile shear zones

Allotriomorphic A texture composed of minerals that are mostly anhedral or do not have well-developed crystal faces

Alpine orogeny Convergence of micro plates in the Mediterranean and South Asia in Late Cretaceous

Alpine peridotite An assemblage also known as an ophiolite suite reported first time from the Alps. It includes an association of peridotite-serpentnite, gabbro, sheeted mafic dyke complex, spilite and deep-sea sediments—radiolarian chert, shales, minor limestone and umbers (Fe–Mn–rich sediments). It is thought that these rocks originate as a part of the oceanic crust and upper mantle

Amphibolite Medium-to coarse-grained metamorphic rock mainly composed of plagioclase and hornblende. It is derived from either sedimentary or igneous source and formed at moderate temperature

Amygdule Cavities in volcanic rocks filled by late-stage or post-magmatic secondary minerals (adjective: amygdaloidal)

Andaman island arc Constitutes 850-km-long chain of islands between the Bay of Bengal and the Andaman Sea, and composed of Cretaceous and Tertiary sediments and magmatic rocks. It comprises two nearly parallel arcuate belts; the western arc is represented by 300 islands of the Mentwai Nicobar and Andaman groups known for occurrence of ophiolite

Andesite A calc alkaline volcanic rock, usually porphyritic, consisting of plagioclase, hornblende and/or biotite the more basic varieties known as basaltic andesite

Anhedral Mineral grains without crystal faces

Anorthosite A leucocratic plutonic rock consisting essentially of plagioclase often with small amounts of pyroxene or hornblende and accessory Fe–Ti oxides

Aphanitic A crystalline rock in which mineral grains are too small to be seen by unaided eye

Arkose A detrital sedimentary rock that contains >20 % by volume of feldspars

Arenite A detrital sedimentary rock with grain size diameter in the range of 0.06–2.0 mm. *Synonym* of sandstone

Argillite A weakly metamorphosed mudstone or shale, without slaty cleavage or fissility

Ash Pyroclastic fragment with particle size in between 0.06 and 2 mm; forms the pyroclastic rock 'Tuff'

Assmilation Physical and chemical interaction of magma with wall rocks and enclaves resulting in change in

N. C. Ghose et al., *A Petrographic Atlas of Ophiolite*,
DOI: 10.1007/978-81-322-1569-1, © Springer India 2014

chemical composition and mineral assemblage of both magma and the assimilated material

Asthenosphere The zone in the earth that extends in depths approximately from 70 to 250 km in which shear wave velocities decrease markedly and is known as the low-velocity zone. The rigid crust and part of the mantle forming the lithosphere occurs above this zone

Authigenic Growth or deposition of mineral matter during diagenesis of sedimentary rocks and is considered a primary sedimentary feature

Autolith Also known as cognate xenolith, and is genetically related to the magmatic rock they occur in and may represent either cumulates or restites

Back arc basin A basinal area behind an island arc and the continent

Barails A thick group of sandstones and shales belonging to Oligocene. It unconformably overlies the Eocene Disangs flysh sediments

Barren Island An active volcanic island of Quaternary age in the Bay of Bengal that lies on the arc system extending to Sunda chain

Barroisite A member of the sodic-calcic group of monoclinic amphiboles. Barroisite is defined as a group member with Mg > Fe2+ and Al > Fe3+ in the C position; mg# 0.5–1.0, Si = 6.5–7.5 (IMA 1997)

Basalt A volcanic rock consisting essentially of calcic plagioclase and pyroxene. Olivine, opaques and minor foids or minor interstitial quartz may also be present

Basaltic andesite A volcanic rock which has feldspar composition as that of andesite and abundance of ferromagnesian minerals commonly found in basalts

Belt of Schuppen An imbricate zone of upthrust sediments that separate the main Naga-Patkai Mountain from the crystalline Proterozoic Shillong-Mikir hills plateau in the northeastern India. Its fish-scale pattern of faults is called Zone or 'Belt of Schuppen'. Its southern limit is the Disang Thrust

Block Angular and sub-angular fragments (>64 mm) derived from volcanic eruption

Blueschist A fine-grained schistose rock characterised by the presence of blue colour sodic amphibole, viz., glaucophane formed under conditions of high-pressure and low-temperature

Bomb A wholly or partially molten rounded pyroclast with diameter >64 mm

Boudinage A process by which a layer of rock is transformed by stretching or shrinking into elongate segments like 'sausage' (boudin in French) when viewed in cross-section

Breccia Faulted angular fragments of wall rocks set in a fine-grained matrix of crushed material

Brittle deformation Breaking of rock in response to stress is called brittle deformation. In contrast, when rocks bend or behave like flow as plastic or clay, it is called *ductile* deformation

Cataclastic texture Sheared and crushed rock; not used where the rock is clearly mylonitic

Chert A non-clastic (authigenic/chemical) siliceous rock composed of finely crystalline precipitated silica (cryptocrystalline)

Coccolith Individual plates of calcium carbonate formed by coccolithophores (single-celled algae viz., Emiliania huxleyi) which are arranged in a coccosphere. Such microscopic calcareous impressions have different shapes

Coesite A dense phase of silica (Sp. Gr = 2.93) formed at pressures >2.5 GPa. Often found in rocks developed under ultra-high pressure (UHP) metamorphic condition in the subducted lithospheric plate of a collision zone

Cognate Pyroclastic fragment or particle belonging to co-magmatic volcanic rocks, but which have been detached and ejected with other pyroclastic debris during a later eruption

Conglomerate A coarse consolidated detrital rock with rounded larger fragments composed of rounded gravel set in a matrix of fine-grained material. If the fragments are angular, the rock is called Breccia. *Synonym*: rudite

Continental crust A relatively thick, hard and brittle upper layer of the earth (30– 80 km) and composed of materials of low density (av. p = 2.7). It underlies the continents and continental shelves

Continental lithosphere It is 100–250 km thick and consists of 60–80 % upper mantle rocks attached to the base of the continental crust. It is rigid and possibly not affected by active convection

Continental margin The zone of transition from a continental landmass to the adjacent continental shelf of ocean basin

Continental shelf The continental margin between the shoreline and continental slope (up to 200 m depth). It has low gradient ($\sim 0.1°$)

Convergence A zone where tectonic plates collide, characterised by earthquakes, formation of mountains and volcanic activities

Cross bedding Layers within a stratified unit oriented at an angle with the regional stratification

Crossite A variety of glaucophane rich in iron, blue in colour

Crust The thin outer layer of earth above the Mohorovicic discontinuity, thicker in the continental regions (30–50 km) and thins out in oceanic region (10–12 km). It represents <1 % of earth's volume

Cryptic layering A change in bulk composition of a rock along height of the intrusion in a layered magmaic series

Crystal tuff A tuff in which crystal fragments are more abundant than either lithic or vitric or glassy fragments

Cumulate Mineral aggregates (e.g. olivine and pyroxene) with variable amounts of interstitial liquid that crystallised from a slowly cooled magma and sank to the base of magma chamber due to relatively high density

Deformed peridotite Refers to bimodal grains of olivine in peridotites: porphyroclasts (2–10 mm) showing variable strain, the other is recrystallised grains (neoblast <0.3 mm) which show mosaic or granuloblastic shapes

Diagenesis Physical and chemical changes that take place within a sediment after its deposition prior to metamorphism or weathering

Diorite A plutonic rock consisting of sodic plagioclase, commonly hornblende and often with biotite or augite

Disang fault The Disang fault is tectonically a thrust from south and east, but west of Haflong, its major physiographic manifestation is that of a hinge zone with a rising horst to the north, and piling of sediments to the south

Disang flysch A folded tract of N–NE trending flysch to the west of ophiolite, consisting of dominantly argillaceous sediments, together with graphitic slate, siltstone and fine-grained sandstone

Divergent plate boundary A plate margin formed where the lithosphere splits into plates that drift apart from one another. The void created is filled by igneous intrusions and accumulation of mafic lavas, e.g. mid-oceanic ridge (MOR)

Dolerite A medium-grained rock composed essentially of plagioclase, pyroxene and opaque minerals, often with ophitic texture. The term is synonymous with diabase and micro-gabbro

Ductile deformation Deformation of plastic materials without rupture. Refer brittle deformation

Dunite An ultramafic plutonic rock consisting essentially of olivine (>90 vol. %)

Dynamic metamorphism Metamorphic changes caused by load, tensional faulting, thrusting or folding limited to narrow zones. It includes processes involving brecciation, cataclasis, mylonitization, recrystallisation and partial melting in extreme cases

Eclogite A high-pressure metamorphic equivalent of basalt containing garnet and omphacite (clinopyroxene)

Epiclastic Volcanic fragment that is produced by weathering and erosion of volcanic rocks. Pyroclastic fragments become epiclastic when reworked by surface (sedimentary) process

Exsolution Breakdown of a homogeneous solid solution

Flysch Thick sequences of marine sedimentary facies, thinly bedded, graded deposits of rhythmically formed interbedded sandstones and mudstone, poor in fossils

Fold mountains The young Tertiary (<70 Ma) mountain ranges (e.g. Himalayas) that are developed by collision of plates. These may contain older deformed metasedimentary and young volcanic and intrusive rocks

Fore-arc basin A basin lying between a deep-sea trench and its adjoining island arc

Foreland basin A basin in a back-arc area that is floored by continental crust. *Synonym* of Back-arc basin

Gabbro A coarse-grained plutonic rock composed essentially of calcic plagioclase, pyroxene and iron oxide. If olivine/ hornblende or quartz is an essential constituent it is olivine gabbro/hornblende gabbro or quartz gabbro

Gabbroid A group name for rocks of the gabbro-norite groups

Gabbronorite A plutonic rock consisting of calcicplagioclase, clinopyroxene and orthopyroxene

Gedrite Orthoamphibole with mg# 0.5–1.0 and Si = 5.0–7.0 belonging to the Mg–Fe–Mn–Li amphibole group (IMA 1997)

Glaucophane A blue to violet soda amphibole, strongly pleochroic and with very small extinction angle (4°–6°) that develops in metabasic and metasediments (greywacke) of ophiolite

Gondwana The old continental landmass that is thought to have split apart during Mesozoic time into the present-day continents principally of South America, Africa, Arabia, Madagascar, India, Antarctica and Australia. Portions of southwestern Eurasia and southern Europe were also the part of Gondwana during much of the Palaeozoic

Graded bedding Beddings characterised by a progressive decrease in grain size from bottom to the top

Granoblastic texture A mosaic of coarse, anhedral grains, in which prismatic or platy grains are randomly oriented

Greenschist Green coloured, low-grade metamorphic rocks composed predominantly of chlorite, epidote, actinolite and albite

Greywacke A poorly sorted sandstone typical of flysch sequence with abundant matrix, feldspar and/ or rock fragments

Harzburgite An ultramafic plutonic rock composed essentially of olivine and orthopyroxene

Himalayan orogeny The deformation of rocks that took place as a result of collision of India with Eurasia during Early Tertiary

Hyaloclastite Fragmental rocks resembling tuff formed by flowing of magma under water or ice producing shattered, small, granular or angular glassy fragments. These are not formed by explosive activity. Instead shattering takes place because of sudden chilling of hot magma against cold water or ice. The granular fragments in lavas exhibit quenched texture with long, acicular grains of plagioclase

Hydrothermal alteration Volatile-rich phase of magma that carries elements which preferentially enter the fluid and transport them to cooler or upper part of magma chamber

Idioblastic texture A euhedral porphyroblast bounded by its own crystal faces

Ignimbrite An indurated tuff consisting of crystal and rock fragments in a matrix of glass shards which are usually welded together

Indo-Myanmar Ranges (IMR) The westward convex 1300 km long belt along the India-Myanmar border; consists of Upper Cretaceous to Tertiary sedimentary rocks. It continues further south for another 1700 km under water embracing the island arc of Andaman and Nicobar in the Bay of Bengal

Indus-Tsangpo Suture (ITS) It designates a zone of collision between northern margin of the Indian plate and the southern margin of Tibet of Eurasia. It lies in the course of two most important E-W flowing rivers, viz., Indus and Tsangpo (Yarlung-Zangbo) and represents surface expression of boundary between the Indian and Tibetian plates in the northern part of Himalayan Ranges

Intergranular texture The angular interstices between feldspar laths are occupied by ferromagnesian minerals, dominantly pyroxene in mafic lavas

Intersertal texture Interspaces between the feldspars when filled by glass or secondary minerals

Island arc An arcuate chain of islands rising from the deep sea floor and located close to a continent

Jopi Formation Immature, ophiolite derived sedimentary lithosequence, deposited in epi-continental sea, exposed unconformably over the ophiolite as canopy. These are equivalent to Barails (Oligocene) of Assam and surrounding areas

Juvenile Pyroclasts form directly from cooling magma during aerial and subaerial transport

Kohima Synclinorium The Kohima Synclinorium lies southwest of the Naga Hills ophiolite belt. It occupies an area where the colliding plates were bringing the Indian continental mass at an angle towards Myanmar

Lapilli A pyroclast of any shape with a mean diameter between 2 and 64 mm

Lamina A layer of sediment of less than 1 cm thick

Laurasia The continents consist of North America, Europe and most of Asia

Lherzolite An ultramafic plutonic rock composed of olivine with subordinate orthopyroxene and clinopyroxene. Minor components of spinel, garnet and plagioclase are common

Lithic tuff A tuff containing more lithic (rock) fragments than either crystal or vitric (glass) fragments

Lithosphere The lithosphere is the outer shell of earth (globe) and includes the crust and part of the upper mantle. It is characterised by both rigidity and elasticity and lies buoyantly upon the asthenosphere

Mantle It lies below the crust and is composed of denser material called peridotite with lenses of eclogite. It is identified by abrupt increase in seismic velocities

Mantle plume A buoyant mass of hot mantle material that rises to the base of the lithosphere. They cause volcanic activity and structural deformation

Marl A mixture of micrite and clay

Megacryst One large single crystal, e.g. olivine, plagioclase, in igneous rocks representing diverse petrogenetic history

Melange A tectonically chaotic mixture of very large dimension of ophiolite litho-unit (pre-existing rocks) set in turbidite clay matrix

Mesozoic An era of geological time scale (250–65 Ma)

Metamorphic facies A set of mineral assemblages occurring in spatially associated rock types of diverse chemical composition formed under a restricted P-T condition

Metasomatism The process of chemical change involving some fluid during metamorphism. The rocks so produced are known as metasomatic rocks

Micrite Microcrystalline calcite or aragonite mud in a carbonate rock

Mid-Oceanic Ridge Basalts erupted at oceanic ridges known as the oceanic ridge basalts or MORB. Most are tholeiite in composition

Mobile belt Linear to curvilinear crustal provinces or blocks that have undergone several episodes of tectono-magmatism and metamorphim

Molasse A continental, deltaic or marine sedimentary facies consisting of primary sedimentary structure, fossiliferous detrital rocks with minor coal and carbonate deposits. Also represents postorogenic accumulation of sediments

Mortar texture Deformational texture resulting from reduction in grain size in matrix surrounding larger original grains

Mylonitic texture Highly sheared, recrystallised grain, foliated and containing relict crystals. The term *proto-mylonitic* or *ultra-mylonitic* is used to indicate slight or extreme degrees of deformation, respectively

Naga thrust It separates the shelf from basinal trough

Nappe Faulted overturned folds; uprooted sheets of rock pile

Neoblast Constitutes smaller grains of olivine and pyroxene (<1 mm) in sheared peridotite. They are equant, free of strain features and recrytallised grains

Neogene An interval of time denoting upper Tertiary, includes Miocene and Pliocene (23–1.6 Ma)

Nimi Formation A thick pile of folded metasediments composed of calc-psammopelitic sequence exposed along the eastern tectonic contact of Naga Hills ophiolite

Norite A plutonic mafic rock consisting plagioclase and pyroxene; orthopyroxene dominates over clinopyroxene

Obduction Thrusting or welding of oceanic crust/ophiolite upon edge of the continental lithosphere

Oceanic crust A thin (4–9 km) layer of brittle material composed predominantly of rocks that are relatively high in density (av. 2.9)

Oceanic Island Occur on or close to ocean ridges and in ocean basins

Ocellus Phenocryst of an igneous rock. The phenocrysts are aggregates of smaller crystals arranged radially. *Plural*: ocelli

Olistostrome A heterogeneous fragmental rock produced by submarine gravitational sliding or slumping and lacks true bedding. Non-tectonic equivalent of mélange

Omphacite Isomorphs of diopside and jadeite restricted to eclogites. This grass green to pale green aluminous pyroxene is an essential component of eclogite and occasionally in glaucophane schist. It is colourless in thin section

Ophiolite A distinctive assemblage of mafic and ultramafic rocks represented by ultramafic tectonite, cumulate gabbros and ultramafics (peridotite-serpentinite-pyroxenite), sheeted diabase dykes, volcanics and oceanic sediments. It should not be used as a rock name or as a lithologic unit in mapping (Penrose Conference 1972)

Ophiolitic mélange/tectonic mélange An olistostomal and tectonic mixture of ophiolitic material and sediments of oceanic origin produced by compressive forces along the upper part of descending slabs at shallow depth. They indicate important suture lines connected to plate boundaries

Ophitic texture Laths of plagioclase largely enclosed in pyroxene

Orogenic belt Pertaining to deformation of a continental margin to the extent that a mountain range is formed

Overturned fold A fold in which at least one limb has been rotated through an angle $>90^{\circ}$

Paleocene An epoch of the upper Tertiary period (65–53 Ma)

Paleogene Lower Cenozoic era comprising Palaeocene, Eocene and Oligocene (65–23 Ma)

Pangea The largest supercontinent during Phanerozoic, which includes most of the existing continents

Paralic Shallow water (lagoonal or littoral) environments in which marine and continental sediments are deposited

Partial melting The process by which minerals with low melting temperature in muti-component systems melt as a result of increase in temperature and pressure or a decrease in pressure

Pelagic deposit Deep-sea sediment lacking terrigenous material

Peralkaline A chemical term for rocks in which the molecular amounts of $Na_2O + K_2O/Al_2O_3$ exceed 1. It produces alkali pyroxenes and/or alkali amphiboles in the mode

Peridotite A collective name for plutonic ultramafic rocks consisting essentially of olivine with pyroxene (ortho- and/ or clino-) and/or amphibole

Phanerozoic An eon of geological time known for an abundance of life (<570 Ma to present)

Pilotaxitic A subparallel arrangement of plagioclase microlites which are visible to naked eye. The synonymous terms 'trachytic' or 'flow' are to be avoided

Plagiogranite Defines a group of plagioclase dominated felsic rocks as a synonym for trondhjemite, leucocratic tonalite and quartz diorite in ophiolite parlance

Plate A segment of the lithosphere that floats on the underlying asthenosphere. Plates are driven chiefly by ridge-push and slab-pull

Plate tectonics The theory of global dynamics in which the lithosphere is divided into rigid individual plates that move in response to convection in the upper mantle. The margins of the plastes are sites of considerable geological activity

Podiform chromitite These are irregular, discontinuous concentrations of chromite varying from lens to pod-like in shape. These are commonly hosted within dunite of the shallow mantle sequence of oceanic lithosphere emplaced in Alpine-type (allochthonous) orogenic belt

Poikiloblastic A porphyroblasts containing numerous inclusions of one or more minerals trapped during metamorphic growth

Porphyritic A large crystal (phenocryst) is surrounded by fine-grained or glassy groundmass in magmatic rocks

Porphyroblast A large crystal of a mineral grown in metamorphic rock and surrounded by smaller grains of other minerals. Comparable to 'phenocryst' in an igneous rock

Porphyroclast A large strained or broken grain in a groundmass of finer grains

Pumice A light coloured pyroclastic fragment, highly vesicular, silicic glass foam which commonly floats on water

Pyroclasts A general term used for crystal, glass and rock fragments generated by disruption as a result of explosive eruption of a volcano. If predominantly consolidated, it is classified as a pyroclastic rock. Refer tephra

Pyroxenite A collective term used for ultramafic plutonic rocks composed dominantly of one or more pyroxenes (> 90 vol. %), and occasionally olivine, hornblende and phlogopite

Quaternary A period of the Cenozoic era (1.6–0.01 Ma) including the present Holocene epoch

Radiolarian chert A bedded chert along tectonically active plate margins which is found in association with graded turbidites, ophiolites and mélanges

Rhyolite A silicic volcanic rock of granitic composition. It consists of phenocrysts of quartz and alkali feldspar, often with minor plagioclase and biotite, in microcrystalline or glassy groundmass

Rhythmic layering Small-scale gravity layering marked by repeated graded layers of mafic (e.g. pyroxene) and felsic (e.g. feldspar) minerals of a gabbroic layered intrusion

Riebeckite Iron-rich alkali amphibole. Dark blue in thin section, strongly pleochroic and small extinction angle (5°)

Rodingite Formed by calcium metasomatism of a source rock of any composition usually during serpentinisation. The rock is characterised by presence of minerals like grossular, diopside, epidote, etc.

Scoria Vitric components of volcanic ejecta include shards or glassy fragments that are highly vesiculated (pumiceous). These can be classified into scoria (dark coloured) and pumice (light coloured)

Sea-floor spreading The theory that sea floor spreads laterally away from the oceanic ridge as new lithosphere is created along the crest of the ridge by upwelling magma

Seamount Small volcanic structure, morphologically similar to sub-aerial shield volcanoes which never grow above sea level

Serpentinite The ultramafic rocks like dunite and peridotites are susceptible to alteration at low temperature in presence of water, resulting in a green rock called serpentinite. Act as 'the ball-bearings of an orogeny' (Hall 1988)

Shard Glassy microscopic magmatic fragments of different shapes. Commonly of Y and cuspate shapes that characterise explosive silicic magmatic eruptions

Shear A strain generated due to stress to cause sliding relative to each other. Strain analysis is carried out to identify directions in a deformed body

Sparry calcite A pore filling cement in arenite or carbonate rocks that consists of a mosaic of calcite crystals larger than micrite

Spherulitic Spheroidal bodies in a rock composed of an aggregate of fibrous crystals of one or more minerals radiating from a nucleus, with glass/crystals in between

Spilite Altered basaltic lava composed of albite and chlorite. Often shows pillow structure due to eruption in aqueous media

Spinel A group of minerals of trivalent ions Al, Fe^{3+} or Cr belonging to isometric system. They display variable colours, darker varieties are opaque in thin section

Sub-alkaline basalt Chemically designates a type of basalt which does not contain normative nepheline

Subduction The process of one crustal block descending beneath another, e.g., along convergent plate margins, allowing the descending slab to merge into the mantle

Sub-ophitic texture Pyroxene grains when partly enclose plagioclase laths. Refer ophitic texture

Suture A suture marks the boundary where two plates are welded together, and is interpreted as evidence for a closed ocean or back-arc basin. Ophiolites are emplaced along a suture zone together with mélange and deep-sea sediments

Syenite A plutonic rock consisting mainly of alkali feldspar with subordinate sodic plagioclase, biotite, pyroxene, amphibole and occasional fayalite. Minor quartz and nepheline may also be present

Syntaxial bend The Himalayan mountain system takes a sharp turn both at its northwestern and the northeastern ends. These are known as syntaxial bends which originated through northward drifting and rotation of the Indian continental block. The whole pile of rock formations is bent and characterised by a succession of faults and thrust planes

Tachylite A volcanic rock consisting essentially of alkali feldspar

Tectonite The peridotites with increasing strain and recrystallisation develop preferred orientation and define a strong foliation. Such peridotites in ophiolite are known as metamorphic tectonite and are distinguished from the cumulate peridotites associated with the layered gabbro

Tephra All particles exploded from volcanoes, predominantly unconsolidated

Tertiary Fold Mountain Young orogenic belt (<70 Ma) that contains older, deformed, metasedimentary rocks together with young volcanic and magmatic intrusives

Tethys Ocean The ocean that occupied a large part on the east of Pangea between Permian to Mid-Triassic

Texture The size, shape and arrangement of crystals or grains of a rock as seen under a microscope

Tholeiitic basalt A subgroup of the sub-alkaline basalt group. The term is recommended by IUGS instead of tholeiite to denote a suite of rocks displaying characteristic chemical features (e.g. a 'tholeiitic trend'). The British Geological Survey recommends that use of the terms 'tholeiite' and 'tholeiitic' as rock names or as part of rock names should be discontinued

Thrust fault A low-angle fault (45° or less) in which hanging wall has moved upward in relation to the footwall

Trachyte A volcanic rock consisting predominantly alkali feldspar, showing a wide range in composition, from oversaturated quartz trachytes to undersaturated feldspathoidal type

Trachytic texture Microlites of feldspar laths disposed in a sub-parallel manner to the direction of flow

Trachybasalt Basaltic volcanic rocks containing labradorite and alkali feldspar on the alkali basalt-phonolite trend (nepheline-normative)

Tuff A pyroclastic rock (solid volcanic ejectas) in which the particle or fragment size range from 2 mm to 1/16 mm

Turbidite A deposit of a turbidity current

Turbidity current A current flowing as a consequence of the load of sediment it is carrying

Ultrabasic A chemical term commonly used for a rock containing <45 wt.% SiO_2

Ultramafic A term commonly used for rocks essentially composed of >90 % mafic minerals

Upper Cretaceous A stratigraphic system belonging to the Late Mesozoic era (99.8–65.5 Ma). The Late Cretaceous period is known for global crustal upheaval including the eruption of the Deccan flood basalts and mass extinctions of dinosaurs at the Cretaceous-Tertiary (K–T) boundary

Variolitic A fan-like arrangement of divergent fibres of albite; the space in-between is occupied by glass or iron oxide

Vesicular Round, ovoid or irregular cavities formed by expansion of gas in volcanic rocks

Vitrophyric texture Phenonocrysts lie in a matrix of glass in volcanic rock

Volcanic breccia Volcaniclastic rocks composed dominantly of angular volcanic particles (>2 mm) set in a tuff or ash called tuff breccia or ash breccia respectively

Volcaniclastic A general term that includes the entire range of clastic materials composed in part or entirely of volcanic fragments, formed by any particle forming mechanism, transported by any mechanism, deposited in any physiographic environment or mixed with any other volcaniclastic type or with any non-volcanic fragment types in any proportion

Vitric tuff A tuff in which glassy fragments are more abundant than either crystal or lithic fragments

Wacke A texturally immature sandstone

Websterite A variety of pyroxenite consisting dominantly of both orthopyroxene and clinopyroxene with or without modal olivine (<10 % by volume)

Wehrlite An ultramafic plutonic rock essentially composed of olivine (>40 % by volume) and clinopyroxene often with minor brown hornblende

Welded tuff Glass shards becoming plastic at high temperature after settling due to compaction

Xenoblast An anhedral porphyroblasts with irregular outline

Xenocryst A foreign mineral that occurs with sharp boundary unrelated to parent magma and incorporated from crustal/mantle wall rock

Xenolith Inclusion of a foreign/country rock unrelated to a crystallizing magma. The xenolith may acquire the mineralogy of the enclosing magma to a limited extent by mechanical disintegration and/or chemical interaction

Zeolite A group minerals of hydrated aluminosilicates of the alkali and alkaline earth metals formed by late-stage hydrothermal solution in cavities and veins. Characterised by lowest refractive index of all minerals

Subject Index

DOI: 10.1007/978-81-322-1569-1, © Springer India 2014

Locality Index

N. C. Ghose et al., *A Petrographic Atlas of Ophiolite*,
DOI: 10.1007/978-81-322-1569-1, © Springer India 2014